T·H·E L·I·V·I·N·G
COSMOS

CHRIS IMPEY is University Distinguished Professor at the University of Arizona and deputy head of one of the largest astronomy departments in the country. Balancing a love of teaching with his work on quasars and distant galaxies, he has written more than 160 research papers and has had two dozen projects approved with the Hubble Space Telescope. Impey has won ten University of Arizona teaching awards and was selected as Arizona's Professor of the Year by the Carnegie Foundation. He has been vice president of the American Astronomical Society, and in 2002 he was chosen as National Science Foundation Distinguished Teaching Scholar. In 2007 he was a Phi Beta Kappa Visiting Scholar. *The Living Cosmos* was Chris Impey's first book of popular science. He has also written a popular book called *How It Ends* and edited a book of astrobiology interviews called *Talking About Life*, published by Cambridge University Press. He lives in Tucson, Arizona.

The Living Cosmos by Chris Impey

"Lively, clear and up-to-date overview of astronomy, cosmology, biology and evolution, specifically as related to the search for extraterrestrial life . . . [Impey] does an impressive job explaining an avalanche of information, including such recent major discoveries as the first planets found orbiting distant stars. A skillful account of the universe, the nature of life and where in the universe life might occur."
Kirkus Reviews

"There has been a recent flood of books about astrobiology – the study of life in the universe – but this latest effort by astronomer Chris Impey is one of the best. It provides a solid overview of the diverse research involved . . . beautifully written."
New Scientist

"Impey has written a wonderfully readable book about the chances of life existing elsewhere in the universe . . . But *The Living Cosmos* is not about just that. It is an overview of everything you need to know about the fundamentals, including how we got here and where we're probably going. More important, the science – a word that often causes eyes to glaze over – is laid out with uncommon clarity and panache."
Sara Lippincott, *Los Angeles Times*

"Chris Impey, one of the world's most distinguished astronomers, takes an exhaustive and illuminating look at astrobiology . . . Consistently engrossing and provocative, and frequently absolutely mind-blowing in its implications, *The Living Cosmos* is filled with scientific details but it remains accessible to readers without a background in astronomy and science. This book is most highly recommended . . ."
Book Loons Reviews

"Impey has clearly done his research thoroughly, and interviewed a great number of the key scientists whilst writing the book . . . *The Living Cosmos* is not only comprehensive in its treatment of the great breadth of astrobiology research, but is also beautifully written. Each chapter opens with an engaging account, full of imagery, of the up-coming topic. On the whole, this is a sterling attempt at making astrobiology accessible to a general audience and I enjoyed reading it immensely."
Dr Lewis Dartnell, UCL, *The Astrobiology Society of Great Britain*

"Chris Impey provides a broad, accessible context for his thoughtful, engaging and up-to-date take on the quest for extra terrestrial life . . ."
Professor Bruce Jakosky, University of Colorado, *Nature*

"Chris Impey surveys the state of the art in this exciting multidisciplinary field. Impey frames his book around three questions: How many habitable worlds are there? Is biology unique to the Earth? And are there other intelligent civilizations? Complete with a companion website featuring podcasts, video clips, interviews, news stories and original artwork, *The Living Cosmos* provides an eloquent summary of humankind's quest for life elsewhere."
Scientific American Book Club

"This is a book about a science that is changing our view of the universe and about what life really means and where it might exist. Impey provides us with a road map to the future of astrobiology, a map that is meant to lead us into a deeper understanding of life and man's station in the universe."
National Space Society

THE LIVING
COSMOS

Our Search for Life in the Universe

CHRIS IMPEY

CAMBRIDGE UNIVERSITY PRESS
Cambridge, New York, Melbourne, Madrid, Cape Town,
Singapore, São Paulo, Delhi, Tokyo, Mexico City

Cambridge University Press
The Edinburgh Building, Cambridge CB2 8RU, UK

Published in the United States of America by Cambridge University Press, New York

www.cambridge.org
Information on this title: www.cambridge.org/9780521173841

Hardback edition published by Random House 2007
Updated paperback edition published by Cambridge University Press 2011

Printed in the United Kingdom at the University Press, Cambridge

A catalog record for this publication is available from the British Library

Book design by Simon M. Sullivan

ISBN 978-0-521-17384-1 Paperback

To my muses, on Earth and elsewhere

CONTENTS

PREFACE TO THE PAPERBACK EDITION

SINCE *THE LIVING COSMOS* was first published in 2007, astrobiology has been in a ferment of activity. The search for life in the universe is highly interdisciplinary, and one of the best signs of the subject's rude health is the vigorous debate and questioning (with touches of incomprehension) between astronomers, geologists, chemists, and biologists at any major meeting. A young cohort of researchers who brand themselves as astrobiologists is making its way through the ranks, and is already making its mark. The research community is becoming steadily more international. Whether the topic is the exploration of Mars, extremophiles on Earth, exoplanets, or SETI, astrobiology has shown the power to capture the public's attention and fuel its imagination.

And yet, we still only know of one place in the universe with life. Does this mean astrobiology is a failure? What's the status of this young and exciting field?

The greatest progress has been made on exoplanets. The detection limit has steadily marched downwards in mass to approach the mass of the Earth. Very few of the super-Earths found so far are habitable, in a traditional sense of being able to have liquid water on their surfaces, but the Kepler satellite is poised to deliver the census of Earths. The number of exoplanets has more than doubled since this book first appeared. Exoplanet research has moved beyond counting bodies to characterizing planets and comparing observations to sophisticated models of geology and atmospheric chemistry. In little more than fifteen years, the advances have been breathtaking.

By comparison, the Search for Extraterrestrial Intelligence (SETI) has celebrated its 50th anniversary without any signal being detected. It might seem strange to compare the two fields—one announcing a new discovery almost every week, and the other suffering through a continuation of the "Great Silence." But both searches are addressing central issues of our relationship to the universe, and both are paced by advances in technology. SETI may always be vulnerable to its untested and anthropocentric assumptions, but with the increased power of radio and optical methods, the search may just be getting interesting.

I'm grateful for the tranquil but intellectually stimulating environment of the Aspen Center for Physics, where revisions for this paperback edition were completed. I'm grateful to my editor at Cambridge University Press, Vince Higgs, for his guidance on this and our other astrobiology projects. Special thanks to Dinah Jasensky for entering my life in a most delightful way.

Readers of this book may also be interested in my recently published *Talking About Life*, also published by Cambridge University Press, which contains interviews with dozens of the scientists featured in *The Living Cosmos*. One day, I expect to give this book the most dramatic update possible.

Chris Impey
Tucson, Arizona
August, 2010

PREFACE

IT'S QUITE PECULIAR to be human. Our lives are filled with event and episode, with work and recreation, with the ebb and flow of friends and family. Seen from above, our actions would seem as purposeful as the activity of bees in a hive or squirrels in a forest. Yet we each house the awareness that we're living, conscious entities. We reflect on our existence. We know that we will die. Perhaps we share self-awareness with a few other species on Earth, but no other creature has gained knowledge of its place in the largest landscapes of time and space.

The history of astronomy has been a steady march of awe and ignominy: awe at the prodigious size and age of a universe with tens of billions of galaxies, ignominy that we have no special place among those galaxies and their countless trillions of stars. Stars and nebulae and planets are the result of natural forces. Rocks and clouds weren't created for our pleasure or benefit. The last bastion of specialness is our existence. Surely life has purpose and meaning. As the poet Stephen Crane has written,

> A man said to the universe:
> "Sir I exist"
> "However," replied the universe,
> "The fact has not created in me
> A sense of obligation."

The final step in the Copernican revolution would be the revelation that we live in a biological universe. As it stands, we know of only one planet with life: Earth. But that's not a very strong statement. We've literally just scratched the surface of interesting sites for life in the Solar System like Mars and Titan. We know very little about the properties of the thousand or so planets in other solar systems or whether those solar systems also contain Earth-like planets. And our attempt to listen for signals from intelligent aliens in the vastness of space has been met with a great silence.

Astrobiology is the study of life in the universe. It's a young subject, the domain of researchers drawn from the full spectrum of biological and physical sciences. It's not immune from criticism—that it's a subject with no subject matter, that astrobiology can only subsist for so long on hopes and promises. Yet the sense of expectation is palpable. The technological revolution that powers computers and consumer appliances has also transformed our ability to gather distant light and send sophisticated probes into space. There's every reason to believe that we'll find out within a few decades whether or not our biology is unique.

This book is a survey of the state of the art in astrobiology. It begins with the history of how we've come to know our place in the universe. Then it turns to what we know about the evolution of life on Earth and what we can learn from the diversity and robustness of terrestrial fauna. Next comes discussion of the prospects for life elsewhere in the Solar System. This is followed by exciting new research on distant planets, and the book closes by considering the potential for intelligent life elsewhere in the universe. Our knowledge is very modest, so some material is speculative. The universe has surprised us before, and it will surprise us again.

The Living Cosmos is designed for a reader with no background in astronomy. Curiosity is the trait that unites us all. Perhaps you've wondered if art and music and computers and commerce are purely human constructs, or if they have emerged in a recognizable form elsewhere in the universe. Perhaps you've wondered if evolution on other planets makes creatures similar to us in function and form or organisms so wildly different that they're beyond our imagination. The language of the book is nontechnical, and details are confined to endnotes. A reading list and set of web links is provided for further exploration. Finally, a large amount of enrichment material—including podcasts, video clips, interviews, news stories, color images, and original artwork—is available on a companion web site at http://www.thelivingcosmos.com.

Working on this book has been engrossing and at times thrilling, because it has taken me far beyond my original training in physics and astronomy. I've benefited from the expertise of many professional colleagues, but all errors, omissions, and inadvertent misrepresentations are my responsibility alone.

At the University of Arizona, I'm particularly grateful to Jonathan Lunine and Nick Woolf for filling in many gaps in my knowledge. The following people carefully read sections and provided valuable advice and feedback: Mark Bailey, Steve Benner, Nick Bostrom, Roger Buick, Guy Consolmagno, Richard Gott, David Grinspoon, Roger Hanlon, Ray Kurzweil, Geoff Marcy, Chris McKay, Simon Conway Morris, Carolyn Porco, Richard Poss, Lynn Rothschild, Woody Sullivan, Jack Szostak, and Jill Tarter. Many of them also feature in the book; their ideas and enthusiasm explain better than I ever could why being a scientist is so much fun. I also warmly acknowledge George Coyne, S.J., a close friend and mentor since I was a fledgling graduate student.

To gather information for this book, I talked with and formally interviewed many scientists and deep thinkers about life in the universe, some of whom have already been mentioned above. They've each helped to shape my understanding of astrobiology, and I'm grateful for all their insights: John Baross, Ben Bova, Chris Chyba, Carol Cleland, Steven Dick, Ann Druyan, Timothy Ferris, Debra Fisher, Iris Fry, Rose Grymes, Bill Hartmann, Joe Kirschvink, Andy Knoll, Laurie Leshin, Frank Lin, Mario Livio, Renu Malhotra, Laurie Marino, Vikki Meadows, Jay Melosh, Mike Meyer, Steve Mojzsis, Hans Moravec, Pinky Nelson, Norm Pace, Richard Poss, Sara Seager, Peter Smith, Dava Sobel, Neil Tyson, Diana Wall, and Larry Yaeger.

I acknowledge the support of the Templeton Foundation for the "Astrobiology and the Sacred" project at the University of Arizona. The visitors and lectures resulting from this project broadened my scientific horizons and acquainted me with many of the people whose work is featured in this book.

Writing is a solitary activity, but no book emerges without help and support. My deepest thanks go to Katherine Larson for editing several chapters and giving excellent suggestions on the whole project and for inspiring me never to limit my imagination. I'm grateful to my agent, Anna Gosh, for her attentiveness and feedback. I thank Catherine, Ben, and Paul for their support during the initial writing of this book.

T·H·E L·I·V·I·N·G
COSMOS

1.

THE UNFINISHED REVOLUTION

There are infinite worlds both like and unlike this world of ours. . . . [W]e must believe that in all other worlds there are living creatures and plants and other things we see in this world.

—Epicurus (341–270 B.C.E.), letter to Herodotus

The young scholar clutches the book to his chest as he works his way through the crowd. Campo dei Fiori is packed; it's a jubilee year, and Rome teems with pilgrims, beggars, and pickpockets. He edges forward, brushing aside the vendors who tug at his sleeve. Days earlier, a small item in a local broadsheet caught his eye. A Dominican monk from Nola was to be put to death, having exhausted the patience and goodwill of the authorities. The scholar sighs. His heart is heavy at the prospect. It is not yet a century since the death of Leonardo, but enlightenment has dimmed so much that it seems like eons.

With difficulty, the scholar climbs scaffolding behind a merchant stall so he can see over the heads of the mob. Yelling at the far side of the square tells him that Bruno has arrived, having been paraded naked through the streets of Rome. He is bound to the stake with thick rope while a local functionary reads the charges. The scholar can only catch fragments: "impenitent heretic . . . failure to recant . . . persistent follies."

A soldier drives a nail through Bruno's tongue and into his jaw to stop him from speaking. As a token of mercy, the soldier hangs a bag of gunpowder around his neck to speed the end of his suffering. Bruno shakes his head as the crucifix is offered to him. Shouts fill the air; lit torches are raised and then lowered. The scholar cannot bear to watch; he pushes his way out of the square.

• • •

THE BOOK IN THE HAND of the young scholar was *On the Infinite Universe and Worlds*, written by Giordano Bruno in 1584. Bruno was a mystic and a philosopher. He had no formal training in science, and he never made astronomical observations. Yet his vision of the universe was strikingly modern and, for its time, dangerously bold.

Bruno was condemned for heresy—violation of the teachings of the Catholic Church. He wasn't put to death specifically for his astronomical ideas, but they were audacious. Decades before Galileo turned his simple telescope to the stars, Bruno was dreaming of other worlds. He thought it ludicrous that the Earth should be the center of the universe. The stars, he imagined, were huge balls of glowing gas just like the Sun, appearing faint only because they were so far away. He speculated that those stars would also have planets orbiting them. With a multitude of planets flung through space, surely there were some that hosted living creatures.

Bruno could only imagine, but we're on the verge of being able to know. You're about to read a survey of the frontiers of astrobiology: the study of the origin, nature, and evolution of life on Earth and beyond. In the past twenty years, we've pieced together important aspects of the origins of life on Earth and discovered a dizzying array of microbes. We've sent spacecraft to all of the major planets and moons in the Solar System. We've discovered more than a thousand planets orbiting other stars. So far, we know of life on only one planet: Earth. But we live in a time of tumultuous scientific and technological change. If we find that terrestrial biology is not unique—that this is a living cosmos—it will be a discovery as profound as any in human history.

This book is framed by three questions. Each begins by looking inward but then turns outward to ask about our place in the universe. *Is the Earth special?* Astrobiology turns this into the question, How many habitable worlds are there? *Is life special?* In astrobiology, this becomes, Is biology unique to the Earth? *Are we alone?* That last question may be the most profound, and astrobiology frames it this way: Are there any intelligent, communicable civilizations out there? In this chapter, we'll see that these questions were considered by the first scientists over two millennia ago, and we'll see how the science of astrobiology emerged.

THE AUDACITY OF THE GREEKS

THE JOURNEY THAT BRINGS US to this point began 2,500 years ago on the coast of Asia Minor. For thousands of years, large and complex civilizations had existed in Egypt and Mesopotamia without developing the means to investigate what lay beyond the edge of the sky. When later scholars decoded the artifacts from these civilizations, they found mostly long lists of land and property: the bureaucratic baggage of everyday living. They left us no theories of the universe. The Greeks were different. As members of a small maritime culture, they lived by trade and their wits. They were open to ideas and to new ways of looking at the natural world.

THINKING DEEPLY ABOUT NATURE

In the age before science, people had no mental constructs for interpreting nature, so they generally accepted the world as they found it. A rock was a rock, a flower was a flower, and a star was a star. Each had its own immutable nature. Humans were clearly special, the preeminent inhabitants of the world. The dawn of science meant that simple acceptance could give way to inquiry. Science accepts the challenge of looking below the surface for deeper meanings. Its goal is to answer the question of why things are the way they are.

Starting in the sixth century B.C.E., a series of philosophers made bold speculations about the natural world. Thales supposed that the source of the universe was water, the substance from which all materials emerged. His student Anaximander extended this idea, but in his version the primal element was an infinite substance called *apeiron*. Since everything formed from one material and would return to it, constant recycling allowed for the possibility that other worlds might have existed at other times.

Meanwhile, Pythagoras and his followers were experimenting with numbers and inventing the foundations of geometry. Pythagoras saw mathematics as a powerful tool to understand music—harmony resulted from the ratio of lengths of a plucked string or of air columns in an open flute. He extended this idea of mathematical perfection to the heavens. The Sun, Moon, planets, and stars were carried overhead on crystalline spheres, and an enlightened person might even be able to hear their "harmony." Pythagoras knew that the Moon shone by reflected light, and its phases could be explained only if it was a sphere. The arcing motions of the stars overhead, and the fact that new stars appeared as one traveled south, meant that the Earth, too, was a sphere. We can understand why Plato inscribed "Let Only Geometers Enter Here" above the entrance when he founded the world's first university in an olive grove outside Athens.[1]

FROM ATOMS TO WORLDS

Another Greek idea with profound implications was atomism. Initially proposed by Leucippus, the idea was developed more fully by his student Democritus. Suppose you cut a stone in half with a sharp knife, then in half and in half again. Eventually, it'll be reduced to a grain of sand and then become too small to see or too small to cut. Democritus found it implausible that this process could continue infinitely, so he proposed tiny, indivisible units of matter called atoms. It's a moniker that survives today: everything is made of atoms, and the atoms are in constant motion. All the familiar aspects of matter—color, smell, taste, texture—are secondary properties of collections of atoms; the atoms themselves have none of these attributes.[2]

Atomism gave new impetus to speculations about life beyond Earth. In the theory, everything on Earth and in the heavens was made of indivisible atoms, and there were an enormous number of them. The Greek idea of elements was rudimentary; there were only four: earth, air, fire, and water. Anaxagoras thought celestial bodies were made of the same elements as the Earth and suggested that the Sun was a flaming rock as large as Greece. This was brave indeed, to suggest that the world was not unique.

Democritus went even further, speculating that the Moon had mountains and valleys and that the Milky Way was an aggregation of stars. He postulated space as infinite and occupied by atoms with pure void in between. This is strikingly close to modern cosmological views. He had no trouble imagining the variety of worlds that an infinite number of atoms might provide: "On some worlds there is no Sun and Moon, others are larger than our world, in some places they are more numerous. . . . There are some worlds devoid of living creatures or plants or even moisture." Democritus was known as the laughing philosopher, content to think about puzzles of matter and space. He said, "I would rather discover a single cause than become king of the Persians."[3]

This is the birth of the "many worlds" concept, which holds that the Earth isn't special. It sits in opposition to the geocentric view. The Earth is just one world among many—perhaps an infinite number—scattered through space. And if the Earth is littered with diverse forms of life, why should other worlds be barren?

Radical ideas are risky—or, rather, the act of questioning everything is radical because it threatens the social order, as Socrates had found out. Pythagoras and his followers were hounded from the Greek mainland for operating a cult with mathematics as its secret language, and Anaxagoras was banished for impiety in daring to suggest that the Sun was as large as a country. Hypatia the geometer engaged in political intrigue and was torn apart by a mob in Alexandria. It would be recasting history to present Giordano Bruno as an archetype of science in conflict with religion; his writings had no coherent explanatory framework. But he collided with authority over ideas that are uncontroversial today and paid the ultimate price.

THE MAN WHO DISPLACED THE EARTH

Viewed through the mists of time, the Greek philosophers are enigmatic. We know very little about the man who anticipated Copernicus by nearly two millennia. Aristarchus lived on the rugged island of Samos, a wealthy city-state that had been run by the tyrant Polykrates during the time of Pythagoras. Aristarchus wrote many commentaries on mathematics and natural philosophy, but only one survives. He was one of the strange breed of men who thought deeply about the heavens—like the earlier philosopher who fell into a ditch and was mocked by a servant girl because he cared more about the things above his head than he did about the things under his feet.

Aristarchus was the first philosopher we know of to make actual observations. Presuming only that the Moon shone with reflected light from the Sun, he used the curved shadow of the Earth during a lunar eclipse to measure the relative sizes of the Moon and Earth. He then used timing of lunar phases to argue that the Sun was much farther away from us than the Moon. Combining the observations, he showed that the Sun was much larger than Earth (Fig. 1).

To Aristarchus, the idea that the larger Sun could orbit the smaller Earth was as nonsensical as a hammer thrower spinning a hammer that exceeded his weight. He proposed a Sun-centered, or heliocentric, cosmology, which was a radical idea at a time when to most people the Earth *was* the universe. But Aristarchus still had to explain why the stars did not change their relative positions or apparent brightness as the Earth moved in its orbit. He guessed correctly that the stars were so far away that these effects were too small to detect. His universe was one billion miles across, a phenomenal number in an age when most people never ventured far from where they were born.

This glimpse into the true nature of the Solar System was a cul-de-sac; the tradition established by Plato and Aristotle was to define astronomical thought for two more millennia. Aristotle dismissed the notion that the stars moved

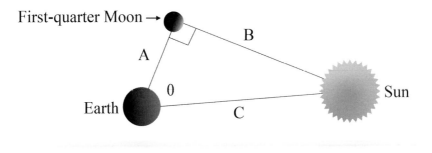

Figure 1. Aristarchus knew that the Moon was illuminated by the Sun's reflected light. By imagining the geometry of a quarter Moon, he realized he could use the triangle to calculate the relative size of the Sun and the Moon. He also used the fact that the Sun and Moon appear to be the same size during an eclipse. Note: object sizes and distances are not to scale.

overhead because the Earth was spinning. If that was true, he argued, we should be moving at nearly one thousand miles per hour—a speed we would certainly feel. He dismissed the heliocentric model. If the Earth was moving around the Sun, Aristotle reasoned, the stars should alter their alignment and apparent brightness over the course of a year, just as we know that nearby objects appear to move against a distant backdrop when we look out of a car window. This phenomenon is called parallax. Aristotle's universe was a cozy million miles across, and its outermost crystalline sphere shut out any thought of distant worlds.

Aristotle also argued against atomism because he believed that each element had its own natural tendencies of motion. Earth and water moved naturally to the center of the universe—the center of the Earth. In our world (in Aristotle's mind it was The World), everything was composed of earth, air, fire, and water. The celestial realm was made of utterly different material, an ethereal substance called quintessence.

Greek thinking ran far ahead of Greek technology. They simply didn't have the tools to test their hypotheses. However, their early instinct that the universe had an underlying unity described by mathematics has proved to be uncannily accurate.[4]

The brilliant mathematician Archimedes even used the Aristarchan Sun-centered model to estimate the amount of matter in the universe. His work *The Sand Reckoner* is a remarkable work designed to impress his sponsor, King Gelon II, with his mathematical prowess. In it, he estimates that the universe is several trillion miles across and calculates that it would take a staggering 10^{64} grains of sand to fill it. If we imagine that these grains are clumped into planets and spread over the much larger volume of the modern universe, we can calculate the number of Earths it would contain: 10^{33}, a billion trillion trillion.

WITNESSING THE BIRTH OF SCIENCE

The scene is Asia Minor. The year is 584 B.C.E. We can imagine the scene as two Greek tribes are hacking away at each other with clubs and swords on a rocky plain near the shore. It's near noon, but the air chills, and the sky darkens. Dazed and confused, the warriors drop their weapons and wander from the battlefield. History veers slightly in its course. A hundred miles away, according to Herodotus, Thales has used knowledge of astronomy to predict this event.[5] He knows that solar eclipses are part of the rhythm of the heavens and not omens from vengeful gods. It's a pivotal moment in history—the first recorded time that humans use sheer intellect to make sense of the cosmos.

Consider this for a moment. At the end of the classical Greek era, most people traveled no more than a total of fifty miles in the course of their lifetimes, yet the average educated person knew that the Earth was round and twenty-five thousand miles in circumference. They had no atom smashers or telescopes, yet they suspected that matter was made of invisibly small atoms and that the universe was millions of times larger than the Earth. While most people saw the objects of the night sky as mysterious and supernatural, the Greeks knew they were subject to rational inquiry. Armed only with logic and rudimentary math, they gave birth to science.

A brave few even ventured the questions that form the heart of astrobiology. They imagined there were many other worlds in space and that life wouldn't be confined to our realm. At the end of the Greek era, Lucian of Samosta even asked the third question: are we alone? His work *True Histories* is a precursor of modern science fiction, extremely speculative but written in the style of a travelogue or a historical narrative. He wrote of trips to the Moon and interstellar warfare. Everything about his work was designed to make the reader think "what if . . . ?"

HOW WE KNOW WHAT WE KNOW

THE STUDY OF ASTROBIOLOGY takes us to the edge of knowledge. Understanding the range of diverse conditions under which life can exist on Earth takes us to the limit of exploration of our own planet. Exploring the Solar System for life takes us to the limit of space technology. Looking for life on planets around other stars takes us to the limit of the telescope. Conjecture can fill the sails, but observations are the ballast that keeps the ship of science on course. To critically examine astrobiology, we first must answer the question, How do we know what we know?

Scientists aren't prone to introspection about what they do; they just get on with it. But scientists in all fields use the same method to create knowledge. This method is the source of all technological innovation—just try to imagine the world without air travel or medicine or electronics.

The scientific method centers on evidence. Evidence separates the factual from the fanciful. It's the reason scientists think there may have been life on Mars but don't think UFOs are alien visitors. It's the reason they think the dinosaurs died sixty-five million years ago even though nobody was there to witness the event. It's the reason they believe that stars in distant galaxies are made of the same stuff as the Sun. Painstakingly gathered evidence is fashioned into nuggets of knowledge, which form the bedrock on which scientists build castles of theory and speculation. Science is exciting because we don't know where it will end or how far it can take us.

WE ARE ALL SCIENTISTS

Everyone is naturally born a scientist. Babies are endlessly curious; their freshly minted senses eagerly absorb every aspect of the surrounding world. The plasticity of their brains enables them to forge new connections every day. At some point in the first six months, a baby learns the power of abstraction. Before that, when an object is held in front of its gaze and then removed, it's lost from the world: out of sight, out of mind. But at some stage, the baby can hold the idea of a toy or a doll even when it's removed from plain view. The ability to use an idea as a placeholder for something concrete is the basis of mathematics.

Science begins with the recognition of patterns in nature. We can use playing cards as an analogy of this process. Looking at Figure 2, you'll see four sequences of numbered cards. The first pattern is trivial: a simple pattern of increasing numbers. The second shows no obvious numerical pattern, but

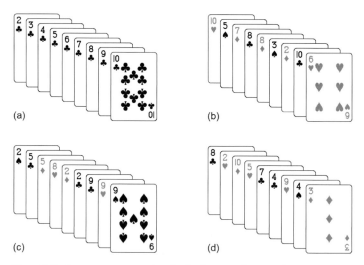

Figure 2. Card sequences as an analogy for the discovery of patterns in nature. Sequence (a) is trivial, while sequence (b) is simple only when you realize what color is associated with each suit and that card value is irrelevant. In (c), each successive card matches either the color or the value of the card before it. In (d), even-numbered cards are followed by any red card, while odd-numbered cards are followed by any black card. This black-and-white image removes an important visual cue to the suit and shows how important the hidden or "coded" information of color is in the analysis.

once you recall that hearts and diamonds are red while spades and clubs are black, it can easily be identified as alternating red and black cards. What about the third and fourth examples? Without reading the caption, can you think of a simple rule that describes how the cards have been laid down?

In nature, a pattern is rarely as simple as a numbered sequence. It can be very complex, like the three billion base pairs in the human genome, or imperfect, like layers of rock that have been jumbled by geological activity. Our innate drive to recognize patterns is so strong that we sometimes see patterns that aren't there. If you shuffle a deck of cards and lay out sequences as shown in the image, you might find a few where you could come up with a rule that explains them. Is this evidence of a deeper meaning or pure delusion? On the plains of Africa hundreds of thousands of years ago, there was adaptive advantage to our skill at recognizing patterns. If you saw a leopard hiding in the dappled grass of the savannah when none was really there, you would be spooked; if you missed the leopard, you'd be lunch!

FROM PATTERNS TO UNDERSTANDING

The card analogy demonstrates other aspects of the scientific method. In the top right panel of Figure 2, the pattern is determined by color. If we had no sense of color or inferred by number instead, we'd see no pattern. The senses through which we absorb information are important, because they're intimately tied to cognition. And not all the information is equally relevant; in this case, color is more important than number. At a first glance, the two lower panels appear inscrutable. Yet the rule that describes each sequence can be simply stated. Are these the only rules? There's often more than one hypothesis that describes the data, which is one reason scientists can disagree.

We also see why science is such a data-hungry enterprise. The sequence on the top left can be recognized after only three or four cards. It takes more cards to identify and confirm the alternating color pattern on the top right, mostly to be sure card number is irrelevant. But decoding the lower two sequences requires even more data, because the patterns aren't as obvious. Scientists are always pressing for more experiments and better observations because the patterns in nature are so subtle and profound.

However, pattern recognition does not imply understanding. It's merely a first step. Our ancestors observed seasons and eclipses and planet motions for thousands of years and thought it was a complex shadow play orchestrated around us. They had no way to know that these phenomena acted in a space that dwarfed the Earth.

Imagine a deck of cards with one-third of the cards randomly removed. If you tried to lay the cards out in rows, from ace to king, one row per suit, the gaps would be scattered across the sequence. In 1869, Dmitri Mendeleev used this kind of arrangement of elements to discover the periodic table. He didn't know about the role of electrons in chemistry, but he could see patterns in chemical behavior, and he used the placement of the gaps to predict properties of elements that hadn't yet been discovered.[6]

In another example, the ancient Egyptians resurveyed their rich alluvial delta every year after the Nile flooded, using huge loops of rope knotted at intervals. This let them lay out the land with right-angled triangles. Although they knew sets of numbers that had this useful property, like 3, 4, and 5 ($3^2 + 4^2 = 5^2$) and 5, 12, and 13 ($5^2 + 12^2 = 13^2$), it took Pythagoras to figure out the general case that applies for any right-angled triangle. His equation gave him an algebraic "net" where the Egyptians had just caught a few fish. He was sufficiently impressed by his aha moment that he sacrificed one hundred oxen to the gods, and as we all know from the Scarecrow's rapturous recapitulation in *The Wizard of Oz*, the Pythagorean theorem is the definition of braininess.

THE TOOLKIT OF SCIENCE

The foundation of the scientific method was invented by Greek philosophers. Modern scientists inherited two ways of looking at the world. From Plato, we acquired rationalism: the idea that nature can be understood by the power of thought alone. Plato disdained observation, as he considered senses to be flawed. This thread continues today in the almost mystical power of mathematics to describe the natural world. Aristotle, Plato's student, was by contrast an empiricist who thought there could be no real understanding without observation. Science today is driven by observations. Assertions must be backed up by evidence that's shared and verified by other scientists. That's why scientists don't believe in ghosts and psychic powers and other ideas that have continuing traction in the popular culture.

On the other hand, data alone are mute to meaning. Scientists are known for being fanatical counters and classifiers—obsessive to the point of being slightly scary. Methods like these are essential to progress. But without theories, a rock is just a rock, a flower just a flower, and a star just a star.[7]

When a scientific field is healthy, there's close interplay and sometimes tension between theory and observation. Speculation unconstrained by evidence descends into intellectual exhibitionism, and a pile of data without a conceptual framework in which to interpret it doesn't advance knowledge. Astrobiologists must lean on speculation to an uncomfortable degree. Since we know of only one planet with life, our sample size is small—all biology is based on

Earth's example. Evidence for giant planets is overwhelming but we've not yet found Earth "clones" around other stars. We've no idea how often evolution leads to intelligence and technology. Part of the excitement of the field comes from the dazzling array of possibilities and the sense that we're close to erasing our state of ignorance about them.

BETTER LIVING THROUGH LOGIC

Logic is another fundamental tool of science. Aristotle invented the framework for deduction, which today forms the nuts and bolts of scientific progress. Deduction gives scientists the way to combine arguments and draw conclusions that can be tested. For example, we know that some forms of life today can withstand extreme heat and chemical environments. We also know that our planet was hot and toxic just after it formed. Together, these pieces of evidence suggest that life could have survived in the rugged environment of the early Earth. We might conclude by deduction that life began early in the Earth's history. As we'll see, this supposition is very difficult to test.

The modern heir to Plato and Aristotle was Bertrand Russell, whose master-work was the *Principia Mathematica,* three brick-sized volumes filled with dense, abstract reasoning that beautifully connected logic in the form of words and logic in the form of symbols. Russell was a heroic figure in the history of thought: so eloquent that he won a Nobel Prize for literature, so dedicated that he did original work into his eighties, and so principled that he wrote his best work while in jail for antiwar activities during World War I. Extreme powers of perception can be uncomfortable; Russell wrote that it was only his desire to learn more mathematics that kept him from suicide.

The complement to deduction is induction. Scientists always have to work with limited information, but when they boldly generalize from a specific observation to the general case, that's induction. Isaac Newton provides the best example of its power. His law of gravity was formulated to explain the motions of the Moon and the planets. Yet he called it a universal law of gravity, because he expected it would also apply to stars and planets beyond the Solar System. Newton's friend Edmund Halley used it to confidently predict the return of the comet that bears his name.

The view that chemistry is universal is another form of induction. Astronomers haven't verified the composition of all regions of space, but everywhere they've looked so far has the same periodic table and the same chemical reactions as are found on Earth. We'll see that astrobiology uses induction to form its core premise: if life exists on a planet around one unremarkable star, then it should exist on similar planets around other stars.

Induction is powerful, but it must be used with care. A classic trap is to generalize from insufficient data.[8] You might be tempted to conclude that all people you meet who have tattoos are jerks, until you meet the tattooed person who becomes your spouse. Science doesn't have the comfort of certainty; at its best, induction conveys a high probability of correctness, no more. A theory is always on its mettle, potentially rejected by discordant data. But interpreting observations that don't fit is rarely a straightforward process, and that's part of the fun. When unexplained wiggles were found in the orbit of Uranus in the mid-nineteenth century, they led to the prediction of a new planet, Neptune, and confirmation of Newton's law of gravity. But when analogous deviations were found in the orbit of Mercury, it was the first step on a path that eventually led to a new concept of gravity: Einstein's general theory of relativity.[9]

Logic is intoxicating once you get the hang of it. You can use it in your everyday life to deconstruct the arguments of charlatans and hustlers and pyramid sellers. You can use it professionally to make the workplace run more smoothly. Be wary of applying it in your personal life, however, or you may find your friends slipping away one by one.

DOES SCIENCE HAVE LIMITS?

It sometimes seems that science has no bounds. We've learned to split the atom and manipulate genetic material. We can describe places and times as far from human experience as the crust of a neutron star and a microsecond after the big bang. Einstein once said that the amazing thing about the universe is that it's understandable at all. Yet there are limits. Parts of the universe are receding from us so quickly we'll never see them, and some places will always be enigmatic, like the interior of a black hole. Science may never explain why the universe exists, how stock markets work, or why people fall in love.

Much of astrobiology isn't done in the lab, so scientists don't have a controlled environment in which to test the range of conditions under which life begins or explore the possible consequences of evolution. We might think biology elsewhere is inevitable because ingredients like carbon and planets and stars exist throughout the universe, but this inductive reasoning is not secure. If life on Earth is a fluke, all bets are off.

Science begins and ends with curiosity. A humble or introspective species might not care whether or not it was alone in the universe. But we're apes—full of piss and vinegar. We've always wanted to poke a stick in the beehive, venture into the dark cave, ride the wild beast. After four billion years of life on Earth, one type of primate learned how to explore the heart of the atom and galaxies at the edge of time. Science is the itch that demands to be scratched.

THE COPERNICAN REVOLUTION

A EUROPEAN OF THE fifteenth century would have been unaware of the achievements of the Greek philosophers. Science and technology made halting progress in the intervening two millennia. The majority of the population was illiterate and innumerate and believed the Earth was flat. Anyone who ventured too far in a ship would encounter sea monsters and demons. Outer space was beyond comprehension (Fig. 3).

Someone with education would have known the Earth was round but clearly immobile and planted firmly at the center of the universe. The astronomical knowledge of antiquity was summarized in a thirteen-volume epic work from the second century written by Ptolemy and modestly titled *The Greatest.* In it, the Sun, the Moon, and five naked-eye planets whirled overhead on translucent spheres, nested like Russian dolls. However, Ptolemy had been forced to add off-centered spheres to the model to explain the nonuniform and occasionally reversing motion of the planets. In the thirteenth century, as Spain's King Alfonso X watched his court astronomers laboriously calculate planetary positions, he dryly remarked that if he'd been present at the Cre-

Figure 3. A woodcut that appeared in a work by the French astronomer Camille Flammarion in 1888. The image is a metaphor for thinking about the universe beyond the celestial spheres of the Greek philosophers and beyond flat-Earth ideas that dominated in medieval times.

ation, he would have suggested a simpler arrangement. Ptolemy's contraption was capped with a last sphere carrying the fixed stars, which spun overhead at a blistering million miles per hour.

In accord with the Christian theology of Thomas Aquinas, the static hierarchy of the heavens mirrored a static hierarchy on the Earth. Medieval elaboration of Ptolemy's model included angels and archangels, with God as the animator beyond the last sphere. The poet Dante Alighieri drew spheres carrying celestial objects as counterparts to the layers of the Earth encountered in the descent to hell (Fig. 4). Life's hierarchy was defined by "The Chain of Being" (Fig. 5). The oak was nobler than the bramble. The dog was nobler than the

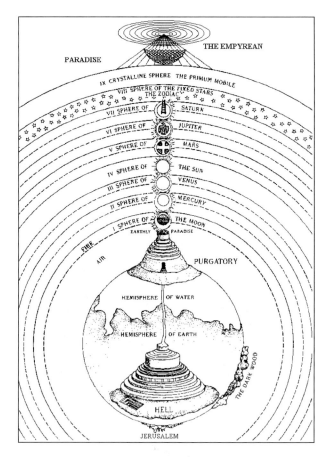

Figure 4. From an early edition of Dante Alighieri's *Divine Comedy*, written in the fourteenth century. This medieval cosmology has planet spheres ordered according to astrological principles, with heaven above and hell below to reflect the Christian theology of the times.

shrew. Man was nobler than all the other creatures because he'd been made in God's image.

This arrangement cemented a worldview in which the science of astrobiology couldn't exist. It took later innovations of biology to show the molecular commonality among all creatures; humans aren't special. And it took the advance of astronomy to show that we have no privileged position in the universe.

Figure 5. From Robert Fludd's *Utriusque Cosmi Maioris* (1617), still showing a traditional medieval view even at the end of the Renaissance. Humans are at the pinnacle of a static ordering of creatures on Earth that mirrors the static ordering of the celestial bodies. The Copernican Revolution took a long time to penetrate the popular culture.

THE RELUCTANT REVOLUTIONARY

Copernicus was an unlikely revolutionary. An introverted mathematician who also worked as a minor cleric in the Catholic Church, he spent years studying different models for the layout of the Solar System. He was motivated by a complex set of aesthetic and theological criteria. He decided that it would be simpler to predict planetary positions if the Sun was at the center and the Earth was just one of the orbiting planets—simpler, but not more accurate. Copernicus was unwilling to abandon the Greek preference for spheres and circles, so his model didn't fit the elliptical motions of the planets any better than the geocentric model it replaced.[10]

The conceptual leap from a geocentric to a heliocentric model is huge. Earth plummets from its position as the pinnacle of Creation to become just one of the planets orbiting the Sun. Copernicus dragged his feet for years before preparing his ideas for publication. He was on his deathbed when the book finally appeared. The publisher inserted an apologetic preface in an attempt to head off controversy. Fewer than four hundred copies were printed, and only a small fraction of the book touches on the heliocentric hypothesis. Yet *On the Revolutions of the Celestial Spheres* changed the world. Our modern word "revolution"—meaning a time of dramatic social and political upheaval—dates from Copernicus.

THE MYSTIC OF MANY WORLDS

In one sense, the Copernican Revolution was mild. Copernicus retained the outer sphere of fixed stars and simply rearranged the order of the Sun and the planets within it. His universe still had a cloistered hierarchy with the Earth fairly close to the center.

Giordano Bruno was more audacious. He proposed that the stars were enormous suns like our own. He postulated that the universe was infinite, that blazing stars with orbiting planets were scattered through the void without end, and that these planets hosted all sorts of creatures, perhaps even creatures like us (Fig. 6).

Bruno didn't improve the heliocentric model, being foggy on its details. He didn't even invent the idea of "many worlds"—philosophers had speculated about life beyond Earth since the third century B.C.E. But he was an influential visionary, daring people to stretch their imaginations beyond the scope of our Solar System and into the vastness of the universe. Luckily, we live in more accepting times. Bruno foreshadowed another Italian scientist, working in America in the last century, who used similar logic. Enrico Fermi asked, "Where are they?" and his provocative question will be addressed later in this book.

CEMENTING THE HELIOCENTRIC MODEL

In 1600, the same year that Bruno was put to death, Danish astronomer Tycho Brahe hired a young mathematician named Johannes Kepler. Brahe was a colorful, turbulent character who lived a charmed life. He was born into poverty, but his life had improved greatly when his uncle, who had no male children, "bought" him from his father. It took another turn for the better when his uncle happened on an accident involving the king of Denmark, rescued him, and was rewarded with a large estate. Brahe was vaulted into the aris-

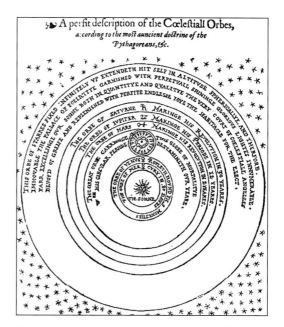

Figure 6. In this picture from *A Perfect Description of the Celestial Orbs*, Thomas Digges (1576) makes a small but profound extension to the Copernican idea by speculating that the stars are distributed without limit beyond the outermost sphere of the planets. The same idea was later propagated by Giordano Bruno.

tocracy. Bankrolled by the king, he set up an observatory on an island off the coast of Sweden, where, for twenty years, he made painstaking naked-eye observations of the motions of the planets.

Brahe's island observatory conjures visions out of a gothic novel. Imagine Uraniborg at night, its coastline shrouded in mist, the observatory towering from the highest point. Within the ramparts, a cluster of metal shapes reach into the darkness—circles and armillary spheres for tracking objects' positions in the sky. Dwarves scurry around making adjustments; they've been hired specifically for this purpose since they can fit under gaps in the machinery. Presiding over it all like a monarch, Brahe sits in a chair that barely contains his girth and barks orders at the dwarves. His nose glints; its metal tip is the relic of a dueling injury sustained as a student, in an argument over who was a better mathematician.

Although he didn't know what it was at the time, as a young man Brahe had witnessed a dying star, so he knew that the Greek idea of an unchanging cosmos was wrong. He had also tracked the path of a comet across the orbits of two planets; apparently there were no crystalline spheres. His data showed

clearly that the planets moved in ways that were inconsistent with circular orbits. Another cherished Greek idea was in jeopardy.

But Brahe was a hands-on guy, an observer, not a theorist. So he grudgingly
let Kepler work with him. There was one final twist in his colorful life. Just a few
months after changing his will to let Kepler inherit his data, Brahe was dining
with a nobleman and taking his usual copious fill of food and beer. Following
custom, he didn't leave the table until his host was finished. Uncomfortably
bloated, he staggered to his carriage. On the way home, the jolting ride rup-
tured his bladder, and he died a few days later.

Kepler used Brahe's data to derive three laws of motion that explain the motions of all the planets. Their elegance and simplicity left him in no doubt that
Copernicus was correct: the Sun was indeed at the center of the Solar System.
Kepler also hinted at some remarkably modern ideas in *Somnium*, a work of science fiction that inspired Jules Verne and H. G. Wells centuries later. The book
describes the voyage of an Icelandic fisherman, Duracotus, to the Moon. There
he finds giant serpentlike creatures with porous skin who grow to a huge size
but live short lives in the harsh environment. Kepler anticipated Charles Lyell
and Charles Darwin in his idea that life-forms are shaped by their environment.

Kepler used his mind to escape the confines of what was often a difficult life.
All his brothers and sisters suffered from physical or mental handicaps, and Kepler himself was bowlegged, severely myopic, and often covered with large
boils. He spent five arduous years defending his mother from the charge of
witchcraft. He bolstered his belief in the Copernican system by correspondence
with Galileo Galilei, to whom the revolutionary torch now passed.

WORLDS IN COLLISION

Galileo was working at the University of Padua when he heard that opticians in
Holland had discovered that they could combine the lenses of eyeglasses to obtain greater powers of magnification. He quickly improved on this idea and assembled a primitive telescope.[11] His observations not only provided strong
support for the heliocentric model but also cemented the idea of "many
worlds." When Galileo saw mountains and valleys on the Moon, he realized
that it wasn't a perfect sphere but a complex geological world like the Earth
(Fig. 7). He discovered four moons orbiting Jupiter—further proof that the
Earth was not the center of all motion. He resolved the blur of the Milky Way
into the pinpoint lights of individual stars and recognized that the stars visible
to the naked eye were just the brightest examples of legions that extended into
the depths of space.

The insights afforded by the telescope were unwelcome to some. When

Galileo took the Venetian city fathers to the top of the campanile in St. Mark's Square to point out the mountains on the Moon, some of them denied the evidence of their own eyes because the information did not fit their worldview. Galileo was savvy enough to sell the idea of the telescope based on commerce rather than science. Someone with a telescope could see which ships in a merchant fleet had been lost at sea hours before the fleet landed, then go around the city buying up goods that would soon be in short supply. A more legitimate complaint was that early telescopes produced flawed images that were hard to interpret.

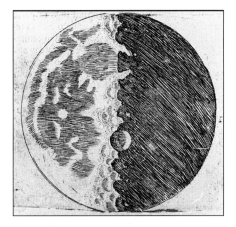

Figure 7. Galileo's rendering of a quarter-moon phase from 1610 is the first published drawing of the Moon as seen through a telescope. The drawing makes it clear that the Moon is a geological world with mountains and valleys, like the Earth.

Galileo was brilliant, but he was also willful. After the death of Bruno, he could have been in little doubt of the risks he was taking by pursuing his radical ideas. As early as 1597, he had written to Kepler, "Like you, I accepted the Copernican position several years ago. . . . I have not dared until now to bring my ideas into the open." Yet he presented the heliocentric model as a fact rather than a hypothesis, lecturing in Italian rather than the more scholarly Latin to reach a wide audience. He delighted in baiting his adversaries. A scene in one of his books had a debate in which the geocentric view held by the Catholic Church was espoused by a simpleton. He resisted the entreaties of colleagues who begged him to work in the more tolerant climate of Holland or England. A collision course was set.

Galileo was an old man by the time he was ordered to face the Inquisition in 1632. He was shown instruments of torture. The message was unmistakable; they were oiled and ready for use. Faced with the certainty of imminent physical harm, he recanted and spent the last decade of his life under house arrest, blind from careless observations of the Sun.

The trial of Galileo was a desperate attempt to defend a dying idea. Science and religion were to continue to tussle over man's place in the universe—skirmishes continue to the present day—but Galileo let the genie out of the bottle. By the mid-seventeenth century, people had to face the possibility that God had created a universe where man and Earth were not preeminent. The universe was much larger and more complex than anyone had guessed. It had

Figure 8. Isaac Newton (1643–1727). Perhaps the greatest scientist who ever lived, Newton made essential contributions to the theories of mechanics, light, optics, and gravity. His laws of motion and understanding of the application of force set the stage for the Industrial Revolution, and his theory of gravity applies to parts of the universe that were undreamed of when he was alive.

been comforting to think that humans were the main actors in this cosmic drama, but people began to suspect that things might be very different. The new paradigm was intimidating: a "clockwork universe" running according to physical law. For this we thank Isaac Newton.

MASTER OF THE UNIVERSE

Born in the mid-seventeenth century, Newton is a titan in the history of science (Fig. 8). Against a backdrop of turmoil during the English Civil War, his early years weren't promising: frail at birth, he came from a rural background and had to suffer village schooling and an abusive stepparent. Although he was surly and inattentive as a child, his intellect blossomed as a teenager, and when he went to study at Cambridge he eclipsed his tutors and professors. In one extraordinary year, while southern England was locked down with the plague, he invented calculus, discovered the law of gravity, defined the properties of light, and made improvements to the design of the reflecting telescope.

Although we think of Newton as a pure scientist, his ideas unlocked technology. His systematic investigation of motions and forces led to the invention of new machines. The spinning jenny and the steam engine were just the first in a series of innovations that transformed the economic landscape of England and then the world.

Newton's crowning achievement was his theory of gravity. He removed the understandable but arbitrary distinction the Greeks had made between the terrestrial and the celestial. The story of the apple falling on his head is apocryphal, but there is at least an apple orchard behind his childhood home at Woolsthorpe in Lincolnshire. Let's see how thinking about an apple and the Moon unlocks the secret of gravity.

A falling apple is pulled toward the center of the Earth by gravity. Dropped from the hand, it falls faster and faster. The orbiting Moon also undergoes acceleration as it is kept in orbit by the Earth's gravity. Newton calculated the distance an apple or any object falls in one second under the action of gravity. He was able to relate it to the distance the Moon is deflected toward the Earth in a second using the inverse-square law of gravity. Motion isn't particular to each object, as the Greeks believed. Rather, everything moves with the same acceleration due to gravity, whether the situation is terrestrial or celestial. It's a universal force.

Newton used this idea to anticipate space travel by three hundred years (Fig. 9). Fire a cannon horizontally from the top of a tall mountain, he hypothesized, and the cannonball will fall to the Earth in an arcing path. If you increase the speed of the cannonball, it will travel farther from the base of the mountain before it hits the ground. Newton speculated that if you were on a mountaintop high enough, you could escape the air that resisted the motion of objects in flight. There, if you fired the cannonball fast enough, it would curve in its flight at the same rate as the Earth curved underneath it. An orbit!

Not only are the Earth and Moon and apple united—everything is subject to this natural law, one of only four forces in the universe. Gravity applies to a tiny electron and the largest galaxy. Its reach is infinite, like Newton's austere and unflinching gaze.

The monumental figure that finalized this first phase of the Copernican Revolution was humble before nature but cantankerous with his colleagues. Newton argued bitterly with Gottfried Leibniz over who had invented calculus. He developed a rivalry with Robert Hooke that was so intense he tried to prevent Hooke's papers from being published after his death. In his honorary role as master of the Mint, he zealously prosecuted forgers and led the cheering section at their public hangings. He was inattentive to his personal hygiene, abrupt with friends, and a notoriously bad lecturer. It took years of cajoling by friends—especially astronomer Edmund Halley and architect Christopher Wren—before he published his masterwork, *Principia*.[12] For Newton it was enough that he alone knew the answer.

Newton's contribution to the awareness of our place in the universe was immense. He's

Figure 9. This theoretical experiment, which Newton could not carry out, shows how an orbit can be created. If a cannon is fired horizontally from the top of a mountain high enough to be above the Earth's atmosphere, there is a muzzle speed at which the surface of the Earth falls away at the same rate that the cannonball is pulled toward its center by gravity. The speed required is seventeen thousand miles per hour. With this concept, Newton anticipated space travel by 300 years.

buried in Westminster Abbey, with no epitaph. It isn't needed because his legacy is everywhere: our concept of infinite yet invisible time and space.

MODERN COSMOLOGY

THE CONTINUING COPERNICAN REVOLUTION is a story of ignominy. We've been steadily displaced in importance as we learn more about the universe. Apart from the blows to our self-esteem, this progression forms a backdrop for today's arguments over whether or not life in the universe is common.

Like Bruno, Newton thought the universe was an infinite sea of stars. He could think of no logical reason to give it a limit, and he calculated that if it did have a limit it would have to collapse. Unfortunately, this cosmology is fatally flawed. Gravity has an infinite reach. The force of gravity does decline as the square of the distance, and so as the distance gets very large the force gets very small, but it never goes to zero. So an endless universe filled with objects whose gravity stretches forever will have an infinite amount of gravity. Newton never addressed this conundrum.

Cosmology describes the extent of the canvas on which life in the universe might be painted. In 1750, Thomas Wright used a curious mixture of logic and theology to estimate the number of habitable worlds in the universe, which was just the Milky Way before the other galaxies were discovered (Fig. 10). His conclusion: "There cannot possibly be less than 10,000,000 Suns, or Stars . . . admitting them all to have an equal number of primary Planets around them, . . . and if to these we add those of a Secondary class, such as the Moon . . . in all together then we may safely reckon 170,000,000."[13] Such a huge number spurs a question based on probability: how unlikely would life have to be for there to be no companionship on any of those habitable worlds?

HERSCHEL SCANS THE SKIES

Does the universe have a limit? This empirical question was attacked by a series of telescope pioneers, the counterparts to the navigators and explorers who mapped the surface of the Earth in the eighteenth and nineteenth centuries. The best of the celestial cartographers was William Herschel (Fig. 11). A German deserter from the Seven Years' War, Herschel came to England, where his skill as a musician found him easy employment. He also built furniture and crafted his own cellos and oboes with the same care and attention to detail that he brought to his maps of the night sky. But Herschel's true passion was as-

Figure 10. This print from Thomas Wright's 1750 book *An Original Theory or New Hypothesis of the Universe* shows a universe where the stars are clustered into many different structures, each centered on an "eye of Providence." Wright suffused his writing with theological speculation, but his estimate of the number of habitable worlds was close to the mark.

tronomy. With his sister, Caroline, as an able partner, he conducted "sweeps" of the sky every clear night, often rushing home during the intermission of a concert he was playing in to squeeze in a few extra observations.

Herschel achieved notoriety after spotting Uranus in 1781, the first new planet to be discovered since the earliest humans looked at the night sky. He used his fame and a salary from King George III to build a series of larger and larger telescopes. Each was a work of art, with exquisite inlays and brass fittings; owning a Herschel is like owning a Stradivarius, but there are even fewer of them to go around. Herschel figured out a way to compare the brightness of

Figure 11. William Herschel (1738–1822). Knighted by the eccentric King George III, Herschel was the greatest observer of the eighteenth century, discovering Uranus and many new nebulae and mapping out the structure of the Milky Way. With exquisite craftsmanship, he fashioned the best and largest telescopes of his time.

stars and deduce their relative distances.[14] He scanned the sky in swaths and mapped out the distribution of stars in the Milky Way. He also cataloged hundreds of fuzzy objects or "nebulae," the nature of which was unclear. Herschel's "universe" was a slablike distribution of stars, with us embedded in it, but he had no idea how far it extended.

Herschel's biggest telescope was forty times larger than Newton's reflector and was not surpassed in size for sixty years. Then William Parsons, the third earl of Rosse, began construction in Ireland of a leviathan with a mirror six feet across and a tube sixty feet long (Fig. 12). But Parsons didn't have an ideal setup for clear viewing. His beast needed ten assistants to control its cables, pulleys, and cranes. It was so sensitive that the image shook every time a horse and rider passed by. And Ireland is famously green for a reason. Parsons used to joke with friends

Figure 12. The "leviathan" was built by Irish aristocrat and amateur astronomer William Parsons, the third earl of Rosse. Parsons used it to discover several hundred star clusters and nebulae. It was not surpassed in size for seventy-five years, until the Hooker hundred-inch reflector was built on Mount Wilson in southern California.

that if you could see his huge telescope, it was going to rain, and if you couldn't, it was raining already.

BETTER TELESCOPES, BETTER TECHNOLOGY

In astronomy, size is not everything, but it's very important. Lord Rosse's telescope gathered enough light to see details in Herschel's nebulae. Some had spiral structures, and Rosse revived a speculation from the German philosopher Immanuel Kant that the nebulae were "island universes" or vast collections of stars beyond the Milky Way.

The issue wouldn't be decided for another sixty years. Meanwhile, the mid-nineteenth century saw two critical innovations. Improved telescope optics sharpened the sizes of images to the limit imposed by the blurring of the atmosphere.[15] This in turn allowed stellar parallax to be measured for the first time. Parallax is the tiny angular shift of nearby stars against a backdrop of more distant stars caused by the Earth's orbit of the Sun (Fig. 13). Hold a finger at arm's length and stare at it with one eye, then the other; the displacement of your finger against a distant backdrop is a parallax shift.

It is important because until then astronomers had only crude estimates of the distance to any star. How bright a star appears gave them little idea because there are stars that burn fiercely and stars that burn feebly—a one-watt flashlight bulb seen up close can appear as bright as a hundred-watt lamp far away. The apparent size of a star gave them no idea because stars are all blurred the

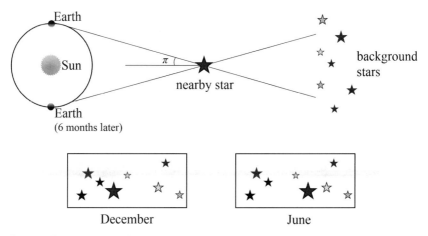

Figure 13. Parallax is the angular shift of a nearby star as seen against a backdrop of more distant stars, caused by the change in perspective as the Earth moves in its orbit of the Sun. By direct application of geometry, the distance to the nearby star is measured. The effect is subtle and wasn't detected until 1838.

same amount by the Earth's atmosphere, whatever the distance. Parallax allows distance to be measured by triangulation, just as a surveyor uses geometry to calculate the distance to a remote landmark.

It took a long time for astronomers to measure parallax because the effect is tiny; the star shifts only a hair's breadth. But by measuring this shift, simple geometry showed the nearest stars to be several trillion miles away. Herschel had already showed that stars extended thousands of times farther than the nearest and brightest ones. The Milky Way was a million billion miles across—staggeringly huge.

The second innovation was photography. When Galileo recorded what he saw through his simple spyglass, he made delicate watercolor drawings (Fig. 7). When photography was perfected in the 1840s, it was immediately adopted by astronomers. Now there was a way to make a permanent record of the night sky. Photography also enabled much deeper images by allowing light from faint objects to be gathered and built up for minutes or hours in a single exposure.

HUBBLE AND THE NATURE OF THE NEBULAE

In the early twentieth century, a new telescope took shape on a mountain north of the sleepy town of Los Angeles. A young man waited impatiently for its completion. He'd been recently hired at the Mount Wilson Observatory and quickly decided to tackle the enigma of the nebulae. Were they swirling regions of star formation nearby in the Milky Way, as most astronomers guessed, or were they remote systems of stars, as a few had speculated? Edwin Hubble had the perfect tool for the job: the world's largest telescope.

Hubble was endlessly talented. Before moving to Mount Wilson, he'd been an athlete, a boxer, an attorney, and a Rhodes Scholar. He possessed an imperious air and a self-confidence that bordered on arrogance (Fig. 14). By the early 1920s, Hubble was taking nightly photos of nebulae with the new hundred-inch telescope. Hubble's assistant, Milton Humason, had his start as a mule driver on the road up Mount Wilson, moved on to construction work on the telescope, and eventually became a night assistant as the telescope began operation. Humason never graduated from high school but later achieved recognition as a scientific staff member of the observatory. In their partnership, Hubble was the ideas man. Photographic plates taken by Humason have images that are crisp and round, while Hubble's plates are usually trailed or fogged or otherwise marred.

As Hubble stared at a sequence of images of the Andromeda nebula, he noticed stars within the nebula that varied in brightness. The stars seemed similar to Cepheid variables, a particular kind of star in the Milky Way with a well-regulated brightness.[16] By comparing the brightness of the variable stars in Andromeda to variables at known distances in the Milky Way, he deduced that Andromeda was about a million light-years away.[17] The universe was hundreds of times larger than Herschel had imagined. Hubble wasn't done yet.

Figure 14. Edwin Hubble (1889–1953) made two momentous contributions to our understanding of the universe. The first was the demonstration that many of the spiral nebulae are "island universes" or systems of stars millions of light-years from the Milky Way. The second was the discovery that the light from most galaxies is redshifted by an amount that increases with increasing distance. This indicates cosmic expansion.

WELCOME TO THE EXPANDING UNIVERSE

Early in the twentieth century, Vesto Slipher measured spectra of a few dozen spiral nebulae—thinking he was studying swirls of planet formation nearby in the galaxy—and noticed that their spectra were mostly shifted to red wavelengths compared to spectra of stars near the Sun. In the nineteenth century, the Austrian mathematician Christian Doppler had explained a familiar wave phenomenon: the pitch of a siren rises as it approaches, then falls as it moves away. As an ambulance rushes up to us, it catches up with its own sound waves, compressing or shortening the wavelength. As it rushes away, the opposite effect happens and the wavelength is stretched or lengthened.

The Doppler effect applies to any waves, not just sound waves. So when a light source moves toward us, its waves will be compressed in a blueshift, and when it moves away from us its waves will be stretched in a redshift. (Blue light has a shorter wavelength than red light.) The fractional wavelength shift is the speed of motion as a fraction of the speed of light.

Let's get back to Hubble. He took Slipher's velocities, added some of his own, and then did the painstaking work of measuring distances to the nebulae by monitoring Cepheid stars in each one. He reasoned that if the light waves from galaxies are stretched by the Doppler effect, those galaxies must be moving away from us, too, at speeds of tens of millions of miles per hour.[18] That wasn't the only surprise. As he accumulated spectra, Hubble saw almost all redshifts

Systematic Expansion

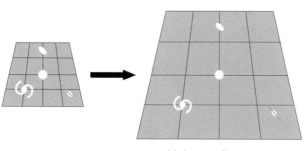

Universe 2x larger
Galaxies are 2x further apart

Figure 15. Universal expansion means that we are not the center of the universe, even though all galaxies are moving away from us. An observer in another galaxy would see the same thing. There's no preferred position; the expansion has an origin in time 13.7 billion years ago but no center in space.

and very few blueshifts. If galaxies had just been milling around aimlessly in the vastness of space, there would have been roughly equal numbers moving toward us and away from us and similar numbers of blueshifts and redshifts. But the galaxies were almost all receding.

At first glance, it seems like a reversal of the Copernican principle. If all galaxies are moving away from us, surely we're the center of the universe. In fact, any intelligent creatures with telescopes in other galaxies would measure exactly the same thing (Fig. 15). Every galaxy moves away from every other galaxy. No galaxy, including our own, is special. Hubble was cautious about interpreting what he observed and knew little of Einstein's work on general relativity, but he had detected the expansion of the universe.

The expansion that Hubble observed has a particular property: the larger the separation of any two galaxies, the faster they're moving apart. Astronomers have used modern telescopes to detect galaxies that are billions of light-years away and moving away from the Milky Way at more than 90 percent of the speed of light. So our view of the universe has a limit. Returning to sound waves, imagine a lot of people rushing away from you but chattering back over their shoulders. Nearby, they're moving away slowly so you can hear them, their voices Doppler-shifted to a lower pitch. Farther away is a distance at which they flee at the speed of sound or faster, so their sound waves never reach you. If you had only sound to go on, you'd never know they exist. Similarly, the universe has regions that are receding from us faster than the speed of light, as expanding space has no speed limit in general relativity defining the limit of our vision.

We can now answer the question posed by Archytas, the Greek general and leader of Tarentum in the fourth century B.C.E. Archytas rescued his mentor Plato from Dionysius, the tyrant of Syracuse. He was an accomplished mathematician, but he is best known for this cosmological speculation: "If I arrived at the outermost edge of the heaven, could I extend my staff or hand into what is outside or not?" Does the universe have a limit? In modern cosmology, the universe has no edge or boundary to bump into, and it doesn't just run out (for how can you run out of empty space?). Rather, there is an information limit set by the expansion and the speed of light.

No center, no edge; it sounds very disorienting. There is, however, an origin in time. The diffuse current state—galaxies separated by voids dozens of times larger than their sizes—has not always been the case. If we ran a movie of the universe backward, effectively reversing the expansion, the galaxies would get closer together. Like any gas undergoing compression, the infant universe was hotter and denser than it is now. Some time in the distant past, everything was on top of everything else.

We're bit players in a cosmic pageant. The lives of stars and galaxies so dwarf our own brief life spans that it's easy to believe the universe has lasted forever. This much time and space offers abundant opportunity for life beyond Earth. But the pageant isn't eternal. It began nearly fourteen billion years ago when the universe was very much smaller, in conditions of unimaginable heat and pressure.

EINSTEIN GIVES SHAPE TO GRAVITY

The birth and evolution of the universe can't be understood without a theory of gravity. Newtonian gravity is fairly easy to visualize. Objects have their places in smooth, infinite, three-dimensional space. Time flows linearly and imperturbably forward. Now try and imagine a universe where objects dictate the shape of the space that contains them, a universe where time and space are malleable. Imagine your living room if the walls and floor bulged and warped depending on where you put the furniture or where you were standing. Now you're getting the flavor of general relativity.

The young man who revolutionized our concept of gravity had been told by his classics teacher that he'd never amount to much. He had been a mediocre student, had failed his college entrance exams, and hadn't even been accepted into a teacher-training school. He passed several years reviewing patent applications and dreaming about space and time. Albert Einstein's theory of general relativity rocked the science world when it was published in 1916, the same year the arts world was staggered by the genius of Igor Stravinsky, James Joyce, and the Postimpressionists.

General relativity banishes the linear space and time of Newton. In its place, the universe becomes an undulating fair ride. Any object like a star or a galaxy gathers space in around itself slightly. The denser and more massive an object is, the tighter it grabs and pinches space. Light traveling in the universe follows the undulations of space and is deflected just like a marble rolling over an uneven floor. Usually gravity is weak, so the deflections are slight, as in a funhouse mirror. But where gravity is intense, the curvature can be substantial, and in the extreme environment of a black hole space is pinched off entirely and light is trapped.[19]

The mathematics of general relativity are wickedly hard—there are probably only one hundred people in the world who understand it at a gut level—but we can get the gist of how it works with two basic ideas. They both came to a man who had little formal training in physics and who was working outside the halls of academia.

Einstein was struck by a strange coincidence. The resistance of an object to a change in its motion, such as when you push it, is called its inertial mass. The mass that dictates how an object moves under the action of gravity is its gravitational mass. These two masses are identical, but why? Einstein's answer is the essence of "relativity," the idea that acceleration due to gravity is no different from acceleration due to any other force. Imagine you are stuck in a stationary elevator. There's no way you could tell the difference between that and the more ominous situation of being trapped in an elevator being accelerated "upward" through space at 9.8 meters per second squared. Now imagine you're floating weightless in an elevator in deep space. Einstein realized there was no way you could tell the difference between that and the perilous situation of being in an elevator that was plunging to the ground after the cable snapped.

The second idea uses Einstein's earlier insight that mass is interchangeable with energy, according to the iconic equation $E = mc^2$. So if anything with mass is subject to gravity, and energy and mass are equivalent, then light has mass, too, which makes it subject to gravity! Let's return to the elevators. Suppose you are in the elevator being accelerated through space at 9.8 meters per second squared. Your weight is normal, and you have no inkling of your true situation. You shine a flashlight across the elevator. The elevator is accelerating while the light travels, so the spot of light is deflected slightly in its path across the elevator. But relativity says that this situation is indistinguishable from the elevator sitting stationary on the Earth's surface. So the light beam is deflected in that situation, too. Einstein's theory makes a surprising and profound connection between matter and the shape of space.

General relativity describes the history of the universe's expansion. Mass curves space on a small scale—leading to the extreme curvature of a black hole—and on a large scale, leading to a gentler curvature of the entire universe.[20] Galaxies are all pulled apart by the expansion, like dots drawn on an inflating balloon. After billions of years of steady expansion, the universe is huge, dark, nearly empty, and essentially flat.

BIG BANG

The Belgian priest Georges Lemaître, a contemporary of Einstein, first came up with the idea of the big bang: a "day without a yesterday." By the 1940s, physicists had calculated that the universe should be filled with a sea of photons left over from the early hot phase of the universe. These waves should have been stretched by cosmic expansion into low-energy microwaves. Unfortunately, technology to detect microwaves didn't yet exist: it was twenty years before Bell Labs engineers blundered onto this signature of the big bang. Images of the microwave sky give us baby pictures of the universe, when it was a thousand times smaller than it is now (Fig. 16).[21]

The idea of a hot origin solved another puzzle. The universe is one quarter helium, which is too much helium to be manufactured by all the stars in all the galaxies, even if they churned it out for fourteen billion years. But a few minutes after the big bang, the entire universe was as hot as the core of a star, and, just as in the Sun, that kind of temperature is hot enough to fuse hydrogen into helium. The big bang created a lot of helium and tiny amounts of deuterium and lithium, in amounts that agree beautifully with calculations of fusion a minute or two after the big bang.

Despite its audacity, the big bang is actually one of the most robust ideas in all science, with experimental verification going back to a microsecond after the creation event. It's very difficult to explain the cosmic expansion and the microwaves that bathe us in any other way.

ENIGMAS OF THE UNIVERSE

Despite the successes of cosmology, there are still some mysteries to explain. In the 1970s, it became clear that galaxies would fly apart unless they were held together by some kind of invisible material. Think of it as an accounting problem: on one side of the ledger is all the visible mass in the form of stars and gas and dust, and on the other side is the mass required to explain the motions of stars within galaxies and galaxies within huge galaxy clusters. On every scale, and by every measure, only 15 percent of the mass is visible. The rest is called dark matter; it has gravity but emits no light. Very strange.

Dark energy is even stranger. In the mid-1990s astronomers found that dis-

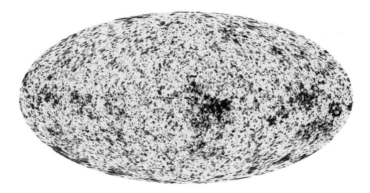

Figure 16. A baby picture of the universe, taken when it was less than 0.003 percent of its current age. This sky map was made with the Wilkinson Microwave Anisotropy Probe (WMAP), a satellite sensitive to millimeter wavelengths of radiation. The shading from gray to black conveys a very small interval around the average temperature of 2.7 degrees above absolute zero. Overall, the temperature of this radiation from the big bang is uniform to a tiny fraction of a percent. The speckles are quantum fluctuations greatly expanded in the epoch of inflation; they then become seeds for galaxy formation.

tant galaxies were fainter than anticipated, meaning that they were farther away than the big bang model would have predicted. Several billion years ago, the decelerating expansion that occurs as galaxies tug on one another by gravity was overcome by some new kind of repulsive force. As a result, the expansion rate is accelerating, and galaxies are fleeing one another ever and ever faster. The best guess is that the shift from cosmic brake to cosmic accelerator was caused by energy from the vacuum of space—a form of dark energy anticipated by Einstein, though he rejected it as an outrageous concept.

So modern cosmology adds several bizarre twists to the Copernican tale.[22] All of the atoms in our bodies and in all the planets, stars, and galaxies make up only 4 percent of the universe. Dark matter is 23 percent, and dark energy is the remaining 73 percent. Galaxies are bright jewels responding to an undulating sea of dark matter as they are thrust apart by dark energy. As we watch this amazing pageant, we're not even made of the stuff that most of the universe is made of (Fig. 17).

The standard big bang model doesn't explain the smoothness and flatness of the universe, so it's been embellished by an additional component: inflation. A minuscule fraction of a second after the big bang, the universe was propelled into an exponential expansion that increased its size from a proton to a grapefruit. Evidence for inflation includes the almost perfect flatness of space and the special properties of the ripples in the microwaves from creation. Since the early universe expanded much faster than the speed of light, there are regions of space that have been removed far from view. As a result, the physical universe—all there is—is very much larger than the observable universe, all we can see.

Figure 17. A pie chart of what the universe is made of shows that all the 10^{80} particles in normal atoms compose only 4 percent of the universe. (There are also 10^{88} photons in the microwave background radiation, but their equivalent mass, calculated by $E = mc^2$, is far less.) Normal particles are exceeded by a factor of six by dark-matter particles, the nature of which is unknown. The contribution from dark energy is three times larger still, and its basis is still a deep physical mystery.

Moreover, if we accept the theory that the universe emerged from a quantum seed and exponentially expanded in the big bang, there is the possibility that other regions of space-time exist, remote in time or space from our universe. Due to the random nature of quantum processes, these parallel universes could have wildly different properties. This extravagant concept is called the multiverse.

Meanwhile, the work of Herschel and Hubble continues. In the regions of the observable universe we *can* measure, powerful telescopes have been used to project a census of all galaxies (Fig. 18). To the limits of our vision, there are roughly sixty billion galaxies. The number of stars contained in those galaxies is 10^{22}, or ten thousand billion billion. In a nutshell, that's the basis for the awe-inspiring potential of astrobiology. Imagine how unlikely it would be, on all the planets around those ten thousand billion billion stars, for life on Earth to be unique.

ENHANCING OUR SENSES

SCIENCE RELIES ON TECHNOLOGY to extend the capability of our senses. The universe is a near-perfect vacuum in which no sound can travel. It's made of 99.9 percent hydrogen and helium, which are odorless and tasteless. Even vision—our most developed sense—has limits. No physicist has even seen an electron or a magnetic field, but they have tools that leave them in no doubt of their existence. No astronomer has ever seen the fusion core of a star or the heart of a distant quasar, but they've developed well-tested theories of these and other exotic environments.

Vision is so central to our existence that we rarely stop to think about it. Almost all animals, from spiders to blue whales, use some form of eye to gather light and learn about their environments. But evolution does provide other means to adapt and survive. Bats use sonar, and deep-sea fish sense infrared radiation. Some microbes can sense heat, and others sense magnetic fields. On

Figure 18. The Ultra Deep Field is the deepest image of the sky ever made. The Hubble Space Telescope stared at a small patch of sky for several weeks, reaching ten billion times fainter than the unaided eye can see. In this region, which is representative of any direction in the sky, several thousand galaxies are present, each a fuzzy white dot, most of which are five to eight billion light-years away. From this sample, the stellar content of the universe is estimated.

the surface of any planet with a thin atmosphere, visible light is abundant, so biology will likely make use of its energy and information. Perhaps eyes are even universal. But in the diverse astronomical habitats where life might exist beyond the Earth, the senses might be wildly different. When we come to ask the question "Are we alone?" the problem may be that alien life experiences its world so differently that there's no way to communicate with it.

EXTENDING THE REACH OF THE EYE

The telescope and the microscope were invented within a single decade early in the seventeenth century. After millennia of naked-eye observations, scientists could study the unimaginably large and the invisibly small. By today's standards, this was a modest revolution; a cheap pair of binoculars can see better than Galileo's best telescope. The first telescopes were refractors that used lenses to collect light and bring it to a focus. However, as lenses are made larger, the weight of the glass causes them to sag and distort. Worse, light of different colors comes into focus at different positions, so the red light in an image is in focus while the blue light is slightly blurred and vice versa. The largest refractor ever built, at Yerkes Observatory near Chicago, is just one meter across.

All modern research telescopes are reflectors. They follow the design established by Newton, with minor variations. Basically, a large, curved primary mirror reflects the incoming light upward, then a secondary mirror bounces the light back down through a hole in the primary to a focus. Unlike mirrors in your house, the aluminum coating of a telescope mirror is exposed to the elements every night, so it must be resurfaced every year or so. This is a complex and dangerous process, since the fragile mirror must be removed from its cell.

The largest single mirrors in the world are eight meters across, larger than most people's living rooms. The twin Keck telescopes in Hawaii reach ten meters aperture by combining and aligning individual hexagonal mirror segments. Astronomers worldwide are planning for a new generation of monster scopes twenty to forty meters across.[23]

But size doesn't come cheap. A decent starter telescope for an amateur might cost one thousand dollars, and the fanciest, complete with dome, around fifty thousand dollars. Meanwhile, a state-of-the-art research facility with a telescope six to eight meters costs more than one hundred million dollars. Taxpayers get to pick up most of the tab, but philanthropists get the bug, too, so telescopes have been funded by industrialist Andrew Carnegie, oil man Bill Keck, and software mogul Paul Allen.

The Hubble Space Telescope is the most powerful telescope ever built (Fig. 19). After a rocky start (when it was perfectly crafted to the wrong specification), it has taken exquisite images of planets, nebulae, and galaxies. As an analogy of its power, Hubble can read a book three miles away, or detect a hundred-watt lightbulb twenty-five times farther away than the Moon. Its deepest image reaches ten billion times fainter than the eye.

Where does this enormous factor of extra light grasp come from? Two orders of magnitude are due to improvements in detectors. The eye's retina converts photons into electrical signals with 1 percent efficiency. However, modern digital detectors or CCDs—high-octane versions of the CCDs in camcorders and cellphones—detect almost every photon. With these nearly perfect detectors, astronomers can collect more light. The next factor is the gain in collecting area going from a fraction of an inch for the eye to ten meters for the biggest telescope, another four orders of magnitude. The last factor of ten thousand comes from the fact that the eye collects information for a fraction of a second before delivering it to the brain to give vision with continuous motion. As-

Figure 19. The Hubble Space Telescope during a routine servicing mission with the Space Shuttle. Astronauts work on the telescope while tethered to a robotic arm. Even though the 2.2-meter aperture of the HST is far exceeded by a dozen telescopes on the ground, the sharp images and low background levels of sky emission of the space environment ensure it remains the most powerful tool in astronomy.

tronomers, on the other hand, can track a target across the sky and gather light for hours, giving a much deeper image.

DETECTING INVISIBLE WAVES

Imagine there were millions of books in a vast library, but you were allowed to read only one. Or that you were blessed with eyes that could discern incredible subtleties of color and hue but were forced to wear glasses that turned the world monochrome. Such was the situation of astronomers until fifty years ago. The universe is filled with invisible radiation, but we were unaware of its existence.

Hints of this wealth of information came in 1800, when Herschel dispersed the Sun's rays with a prism into the familiar rainbow colors from long-wavelength red to short-wavelength blue light, a sequence tethered by the mnemonic "roygbiv" for red-orange-yellow-green-blue-indigo-violet; the acronym dates back to Newton, who picked seven colors to match the octave of a musical scale. Herschel wondered if the reddest color meant the end of the spectrum or just the end of our ability to see light, so he placed a thermometer past the red edge, where it was dark. The temperature rose, showing there was invisible energy beyond the red end of the rainbow: infrared!

The next year, prompted by Herschel's discovery, German chemist Johann Ritter put paper soaked in silver chloride (a precursor of the method that would later turn into photography) beyond the bluest color in the spectrum from a prism. The paper darkened, revealing invisible energy too short for our eyes to see. This is the invisible UV part of the spectrum. The energetic ultraviolet (UV) rays from the Sun are mostly absorbed by the Earth's atmosphere. Contrary to the tradition in music and art, where blue is "cool" and red is "hot," spectral color goes the opposite way because longer red waves carry less energy than shorter blue waves.[24]

Discoveries a century later pried the spectrum open. In the late 1880s, Heinrich Hertz made sparks fly between metal rods in a scene worthy of a mad scientist's lab. In addition to making his students' hair stand up, he'd created radio waves. "They're of no use whatsoever," he commented, but an Italian teenager vacationing in the Alps read about his discovery and thought that there might be practical applications. Guglielmo Marconi was soon sending radio waves through people and buildings and launching the age of global communications. Radio astronomy has used invisible waves the longest; ham-radio operator Grote Reber built a radio dish in his backyard in 1937. For ten years, he was virtually the only person studying enigmatic radio waves from the sky.

Not long after Hertz's experiments, the German physicist Wilhelm Roentgen was startled when an electrical discharge tube in his lab created radiation that made an image of the bone in his hand, as if the flesh had been stripped away. In the case of X rays, the practical applications were immediately clear, and they quickly became an essential tool of medicine. X rays from the cosmos can't penetrate the Earth's atmosphere; astronomers had to wait until satellites in the 1960s to begin looking at the high-energy universe.

All these are manifestations of the same fundamental phenomenon: oscillating electric and magnetic fields that carry energy through space at a speed of 186,000 miles per second. The visible spectrum from red to violet spans a factor of two in wavelength, while the rich phenomena of the universe span a factor of several trillion, from radio waves the size of a car down to gamma rays smaller than an atomic nucleus (Fig. 20). Using a sound analogy, the frequency range from the bluest blue to the reddest red is just one octave, while the electromagnetic spectrum is thirty-five octaves. One octave is enough to make a basic tune, but think of the music you could make with thirty-five octaves!

Astronomers have detectors that can "see" all these varieties of radiation. The Earth's atmosphere is transparent to visible light, radio waves, and the shorter waves of infrared radiation; specially designed telescopes in space are needed to observe the rest. Busting open the electromagnetic spectrum has had

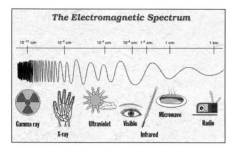

Figure 20. Electromagnetic waves range in size from meters to smaller than the nucleus of an atom, yet they are all aspects of the same phenomenon. The visible spectrum of light forms a tiny sliver at the center of this vast range. All these types of radiation travel at the same speed, 186,000 miles per second.

practical benefits. Modern medicine wouldn't be possible without X rays, and infrared sensing allows us to monitor the energy usage of our industry and our planet.

EXOTIC MESSENGERS

Humans are frail. Several hundred of us have hugged the Earth in tight orbit, but only a dozen have walked on the nearest rock in space: the Moon. However, our robotic emissaries have taken close-up looks at all of the planets, most major moons of the giant planets, and several asteroids and comets. These robots are our sense extenders, exploring places too hot or too cold or too toxic for us. (Hypothetical aliens will be adapted to their planets, too, so we imagine some will also advance sufficiently to invent machines to do their bidding.) Planetary exploration appeals to a new generation in part because it's like a video game—joysticks control machines as they roam over strange worlds. Space exploration is the projection of life beyond the Earth, but without new technologies we'll be limited to our backyard, the Solar System.

In recent years, astronomers have learned how to detect messages from space that aren't based on electromagnetic radiation. It's like being used to your normal five senses and finding out you have several more, never yet used.

Nature contains places of unspeakable violence: the blast wave of a dying star, the perimeter of a black hole, the place where beams of radiation from active galaxies hit the diffuse gas between galaxies. In these situations, subatomic particles or "cosmic rays" are accelerated to amazing energies—think of a proton careening into space with the power of a well-hit tennis ball. When cosmic rays reach the top of the Earth's atmosphere, they produce a cascade of lower-energy particles that are detected with arrays on the ground. Astronomers also study neutrinos, wraithlike subatomic particles created by solar fusion. They interact so sparingly that trillions from the Sun pass through our bodies each second without us feeling a thing. To snare them, scientists lower strands of detectors into holes melted into the Antarctic ice, using the ice pack as a physics experiment. Another new and ambitious project will detect gravity waves, which are ripples in space-time created anywhere that matter undergoes a dramatic change or cataclysm.[25]

The outcome of this technological innovation is a truly incredible expansion of our senses. Astronomers can use microwaves to see the universe when it was less than 0.003 percent of its current age. They can use radio waves to see the center of the Milky Way, where light cannot penetrate. They can use neutrinos to see the core of the Sun. They'll soon even be equipped with gravity eyes. If Galileo could see us now. . . .

Even our familiar friend light has gained a new capability thanks to the largest telescopes: time travel. Time and space are inextricably linked. Light moves so fast that we never notice it on Earth. But the waves that fly from New York to Los Angeles in one sixtieth of a second take millions of years to cross the gulfs of intergalactic space. The Hubble Space Telescope sees objects so far away that light left them when the universe was 5 percent of its current age. Distant light is old light. Telescopes are time machines.

OUR PLACE IN THE UNIVERSE

WE SPEND OUR LIVES embedded in a comfortable illusion. No wonder ancient cultures, with the exception of wild-eyed dreamers, accepted the Earth as an entirely self-contained universe. Rooted by solid ground underfoot, we find security in the sheltering sky, the predictable outcomes of gravity, and the reliable unwinding of time. The planet seems to have been put here for our use and pleasure. It is our dominion.

The truth is more unsettling. The notions of up and down are pure conventions of local gravity. Our slender sheath of atmosphere, thinner relative to the Earth than the skin of an apple is to the apple, shades quickly into the black of space that is a lifeless vacuum only three degrees from absolute zero. The Earth spins and orbits the Sun—when we bury the dead, they aren't laid to rest but instead whirl in a corkscrew motion at more than seventy thousand miles per hour. Even the Sun is not fixed; it travels around the center of the Milky Way, completing a circuit every 225 million years. The Milky Way is tied by gravity to Andromeda, and both great galaxies along with a few dozen of their dwarf companions are falling into the Virgo cluster at a million miles per hour. Even this huge agglomeration of matter is moving. Ten thousand galaxies and their thousand trillion stars are collectively in motion at a million miles an hour in yet another direction.[26]

"Unsettling" doesn't begin to cover it. To consider your true position in space is a trouser-staining, night-sweating, existential experience. Maybe it's just as well astronauts are disciplined and technical minded. If NASA sent artists or poets into orbit, they might be overwhelmed by the experience.

SPACE: A TOY MODEL

Intuition is useless when it comes to scales of time and space that are far beyond human experience. Take a scale model: the state map you might have in your car has a scale of one inch to ten miles or so, shrinking space by a factor of more than 600,000. The globe in a school or office shrinks space even further. We can shrink space by any factor; choosing 1:300,000,000 brings the Earth down to four centimeters—the size of a golf ball. At this scale, if the Earth were in front of you, the Moon would be a little further than arm's length and not much bigger than a pea. Mars would be the size of a gumball, eight hundred feet away at its closest approach. Jupiter would be a beach ball a mile and a half away. And the Sun would be an incandescent globe fourteen feet across at a distance of 1,600 feet.

So far, it's fairly comforting. Golf balls, peas, and beach balls are familiar objects. The entire Solar System is nineteen miles across—the size of a small town. Notice how empty space is; this small town contains a star the size of a minivan at the center, eight planets (none larger than a beach ball), asteroids, and meteors that would be no bigger than specks of dust, and nothing else. Wait, we're just getting started.

At a 1:300,000,000 scale, the nearest stars are eighty thousand miles or more distant. In other words, if each solar system is a "town," space is so empty that the Earth's surface would contain only one town.

The scale we've chosen is not arbitrary. In this model, speed is also reduced by the same factor. Light, which moves so fast in everyday life that its departure and arrival cannot be distinguished, is slowed to a manageable one meter per second, a slow walking speed. It takes eight minutes for light to stroll from the Earth to the Sun and five hours to walk across the Solar System. But if you set out for the nearest star at light speed, it's like walking around the Earth three times. With no rest breaks and no oceans to stop you, it would take several years. We see the Sun as it was eight minutes ago, Pluto as it was five hours ago, and the nearest stars as they were several years ago. It makes no sense to ask what the universe is like "now" because light is the fastest messenger we know of. Also, we're limited in our vision by light's speed, able to see only regions from which light has been able to reach us in the age of the universe.

One huge factor of reduction doesn't bring most of the universe into view, so we apply a second factor of 30 billion. Our scale is now $1:10^{19}$. The Sun reduces to the size of an atom, and other stars shrink to microscopic scales. The typical distance between them is three millimeters. The Milky Way is a twisting spiral of stars about three hundred feet across—the size of a large supermarket. Within this "store" are four hundred billion stars. The nearest other building is a mile away. Now we can visualize the extragalactic universe—where building-like galaxies, each containing billions of stars, are sprinkled across a vast landscape—like farmland grain silos on the Great Plains of the United States.

In this analogy, the observable universe is about the size of the Earth. But here the analogy breaks down. In our universe, we don't see the contents of the universe laid out before us as a snapshot in time. We see the outer edge of the Milky Way as it was twenty thousand years ago. For example, if we ever received a signal from a species on the far side of the galaxy, it may already have gone extinct by the time we can reply. We see the nearest galaxies as they were three or four million years ago and the most distant galaxies as they were eleven or twelve billion years ago. There are regions of the universe we've not yet seen and some we'll never see. We are bounded by time—not space.

TIME: A TOY MODEL

Another scale model reveals our place in cosmic time. Let's compress time by a factor of fifty thousand trillion (5×10^{16}). This scaling is similar but not arbitrary; it reduces the 13.7 billion years since the big bang to a calendar year.

On New Year's Day, all time and space, matter and energy, are created. By late February, the Milky Way is forming. Generations of stars are born and die through the spring and summer. The Milky Way devours a series of its smaller companions and rotates once each week. In early September, near the Orion spiral arm, a midsized star forms in a busy stellar nursery, its gas cloud nudged into collapse by the violent death of a nearby supergiant. Two days later, eight large rocky bodies have formed from dust swirling around the infant Sun. One is our Earth.

Within one or two weeks, the freshly minted planet is alive. Microbes spread across the face of the planet and deep into its oceans. The Sun converts hydrogen into helium and sends a steady stream of warming photons into space. Several times a week, at random intervals, huge rocks slam into the planet and instigate chaos. Many organisms are extinguished, but the rest adapt and diversify. In October, life invents a new way to harness solar energy, and the atmosphere begins to fill with oxygen.

In mid-December, the pace of life begins to pick up. After several months with no organism larger than a fist, new creatures proliferate in the oceans. Some move on to the land, and others learn to fly. By Christmas, dinosaurs rule the forests and swamps of the lush planet. Within three hours of the stroke of midnight on New Year's Eve, hominids appear for the first time. They are descendants of mammals, who were the successful survivors of a huge asteroid impact several days earlier. By twenty seconds to midnight, the hominids have evolved to be just like us; they invent tools and agriculture and build the first cities. The Copernican Revolution occurs at one second to midnight.

The scale model of time has described our late arrival on the scene as intelligent life-forms, able to explore space and understand the cosmos, but consider this: we may not be the first. An Earth-like planet could have formed somewhere else in the universe much earlier. Let's say it's early June. If evolution followed the same pace as on Earth, there would be an alien species attaining our level of technology in late September, just as life was first stirring on the Earth. What would that alien species be capable of now, with a four-billion-year head start on us?

We live on a cusp of exponential change. All the marvels of the modern world—computers, TV, space travel, genetics, the Internet—are crammed into the last tenth of a second of the cosmic year. The surging rate of technological change makes it impossible to confidently predict the future. Knowing our insignificance in time and space, we are both awed and unnerved by our vast potential as a self-aware species. If we aren't alone in having these capabilities, the universe must be a very, very interesting place.

THE EMERGENCE OF ASTROBIOLOGY

SPECULATION ABOUT LIFE in the universe dates back more than two thousand years. Yet a theorist like Bruno couldn't improve on the work of the ancient Greeks because he had no evidence. Astrobiology draws on the development and the expertise of many different disciplines, and it began only about fifty years ago.

CHEMISTRY IS COSMIC

Several fields of science had to mature before astrobiology became a legitimate endeavor. One was chemistry. Modern chemistry emerged in the late eighteenth century when Antoine Lavoisier showed that there were fixed components in chemical reactions that we now call elements. As a young and brilliant man, he won a prize for lighting the streets of Paris. He also married a thirteen-year-old girl who acted as his translator and illustrator. He debunked the Greek idea that earth, air, fire, and water were fundamental, and he showed that mass is conserved in chemical reactions. The chemical notation that we use today was invented by Lavoisier. He said as a student, "I am young and avid for glory," and he had a controversial career, taking credit for Joseph Priestley's discovery of the composition of air. The father of modern chemistry was an aristocrat and reviled tax collector who was beheaded at age fifty in the French Revolution.

John Dalton derived the rules by which elements combined into compounds, based on fundamental microscopic units. Dalton was from a Quaker family, so less rambunctious than Lavoisier but just as precocious; by age twelve, he ran a school in his village, and he published original theories on auroras and color blindness while in his early twenties. Dalton breathed life into the old Greek idea of atoms, discovering rules for chemical reactions and showing that elements were immutable. Within fifty years, the periodic table had been proposed as a framework for understanding chemistry.

Astronomy uses spectroscopy for sensing the chemical composition of remote objects. This innovation stems from the work of a young German scientist, someone who had a change in fortune reminiscent of Tycho Brahe's. Joseph von Fraunhofer was the eleventh son of a struggling glazier. Apprenticed to a harsh man who forbade him from going to school or even reading, he seemed destined for a life of physical labor and penury. When Fraunhofer was fourteen, his master's house collapsed, killing several people but leaving him unscathed in the rubble. The future Bavarian king happened to be one of the first on the scene and became Fraunhofer's mentor and sponsor. Within a few years, Fraunhofer was a master optician and entrepreneur. In 1814, he discovered the narrow spectral lines of hydrogen and helium that bear his name.

Astronomers soon applied spectroscopy to the Sun and other stars in order to determine their chemical constitution. They dispersed starlight into a spectrum with a grating or prism and matched the pattern of sharp lines to lab spectra of known elements. In this way, they showed that the Sun and other stars were made primarily of hydrogen and helium, with trace amounts of other elements. Helium was discovered in the Sun long before it was isolated on Earth. The development of remote sensing by spectroscopy was profound— astronomers could figure out what a star trillions of miles away was made of, even if they didn't know its mass, size, or age.

Cosmic chemistry came of age in the twentieth century. Radioactivity was understood as a natural phenomenon in which the atomic nucleus decays and elements can transmute. The predictable timescale for the decay of heavy elements gives us a way to measure the age of the Earth and the entire universe.

Even though the universe is made mostly of hydrogen and helium, the third, fourth, and fifth most abundant elements were discovered to be carbon, nitrogen, and oxygen—the life or "biogenic" elements. These elements are created in the cores of stars, ejected into space, and then incorporated into a new generation of solar systems in a pageant of birth and death that continues all around us in space. Stars in other galaxies make the same elements in similar proportions. The first stars in the universe formed a few hundred million years after the big bang, so the ingredients for chemistry existed for eight billion years before life began on Earth.

The fact that chemistry is universal is crucial for astrobiology because it acts as a guide and a limitation on unbridled speculation. The many millions of habitable worlds across the Milky Way will be broadly familiar, made of minerals and metals that might be found on the Earth. And the main reason scientists think that biology is not unique to the Earth is the fact that carbon and water exist everywhere in the universe. Raw material for life is abundant. Life doesn't *have* to be based on carbon chemistry, but the motivation to look for biology elsewhere is very strong.

THE NATURE AND EVOLUTION OF LIFE

In the second half of the seventeenth century, the invention of the microscope unlocked the hidden mechanisms of living organisms, just as the invention of the telescope and its use by Galileo had unlocked the heavens. Robert Hooke, a contemporary of Newton, wrote the first book of observations made with a microscope, called *Micrographia.* He was the first to use the word "cells" to describe the tiny structures within living things. Less than ten years later, the self-taught Dutch tradesman Anton van Leeuwenhoek single-handedly discovered bacteria, protists, blood cells, and sperm cells, launching the new subject of experimental biology for microscopic organisms.

During the Middle Ages and throughout the Renaissance, people imagined that life could arise spontaneously from nonliving matter. One seventeenth-century recipe for the spontaneous generation of mice involved putting sweaty underwear and wheat husks in an openmouthed jar and waiting twenty-one days. It was believed that tiny, fully formed humans, or homunculi, lived in every sperm, ready to unfold and grow into humans. The poet and physician Francesco Redi attacked spontaneous generation with his careful experiment showing that maggots didn't arise spontaneously in rotting meat. But it was two hundred years before the idea was laid to rest.

Meanwhile, most religions didn't accept the spontaneous generation of life, so that idea existed uneasily alongside the belief that the species were divinely created and unchanged since their simultaneous creation. Aristotle's view of the Earth and humans as centerpieces of the universe had been cemented into thirteenth-century theology by Thomas Aquinas.

Progress in the life sciences accelerated through three separate developments in the mid-nineteenth century. The real spur for biology was public health. For centuries, Europe had been ravaged by waves of infectious disease. Louis Pasteur did elegant experiments to show that microorganisms did not emerge spontaneously from simple chemicals and that disease-causing microorganisms could be killed by heat or chemical treatment. Then the Augustinian monk Gregor Mendel revealed the rules of heredity, codifying what animal breeders had understood instinctively for hundreds of years. Finally, Charles Darwin published a book that had been thirty years in the making.

The first edition of *On the Origin of Species* sold out in a day. Darwin used a pile of data to present a compelling mechanism for the diversity of species. (Strangely, the word "evolution" does not appear in the book.) In it, he proposed that any species can produce more offspring than the environment can support, and the individuals within a species vary in their degree of adaptation to the environment. The result is unequal reproductive success; the mechanism necessary for continued propagation of the species is natural selection. Darwin realized that all species must have evolved from a common ancestor, but he delicately sidestepped the question of the origin of life, noting that the evidence was probably lost in the mists of geological time.

Is natural selection universal? It's a question that we can't answer without finding life somewhere else. If evolution turns out to be unique to the conditions of the Earth, then the universe may be filled with failed life experiments or places where organisms stagnate and never develop advanced capabilities. Most scientists are quietly confident that life will always be shaped by its physical environment and that the dictates of survival will lead to new features. The restless experimentation of life elsewhere may not always lead to intelligence, but it might. What if evolution goes faster in other environments? What if it generates capabilities beyond our imaginations? What if organisms evolve to become independent of sculpting by the environment? Astrobiologists are hungry for the answers to these questions.

ASTRONOMY AND THE VASTNESS OF SPACE

Progress in astronomy was hindered by the vastness of space. While astronomers were still struggling to measure the distance to the nearest stars, philosophers were taking leaps of imagination and logic. Thomas Digges speculated about stars arrayed through infinite space decades before the invention of the telescope. In the mid-eighteenth century, Thomas Wright hypothesized that nebulae were distant stellar systems and guessed that creatures lived among the stars. By the mid-nineteenth century, the idea of other worlds took such a strong hold that crusty Scottish essayist Thomas Carlyle grumbled in *Signs of the Times*, "If they be inhabited, what a scope for folly; if they not be inhabited, what a waste of space."

The notion of cosmic evolution—life on Earth as one aspect of the evolution of planets and stars throughout the universe—got a strong push from popular books by Richard Proctor and Camille Flammarion in the late nineteenth century. These writers created public expectation of worlds with life scattered through space.

Detecting these worlds was another matter. The nearest stars are thousands of times farther from us than the most distant planets in the Solar System. The dimming of light as it travels through space makes even the largest remote planets many millions of times fainter than our planets. As for terrestrial planets, if it orbited a nearby star Earth would be like a golf ball seen at a distance of twenty thousand miles, emitting a dim glimmer a billion times fainter than its parent star. It took dozens of years of hard work to detect planets beyond the Solar System.

ASTROBIOLOGY GROWS UP

All of the major ingredients for the emergence of astrobiology were in place by the 1950s.[27] James Watson and Francis Crick had shown how genetic information could be stored and transmitted in the double-helix structure of DNA. In 1924, Russian biochemist Aleksandr Oparin wrote a highly influential pamphlet on the origin of life, and British biochemist John Haldane came up with a similar idea independently. They were biochemists and also Marxists; their ideology may have allowed them to jump in where Darwin feared to tread. In any case, they provided plausible speculations on how life might have evolved from simple chemical ingredients. Stanley Miller and Harold Urey put these ideas to the test by synthesizing amino acids in the lab under presumed primitive Earth conditions.

Some physicists and astronomers were convinced that the abundance of habitable zones in the Milky Way ensured the possibility of many biological experiments. If that was the case, it might be worth looking for advanced civilizations directly. In 1960, the young Frank Drake, a newly minted Harvard Ph.D., began a search for extraterrestrial intelligence using radio technology.

The National Aeronautics and Space Administration (NASA) has been an effective steward and booster of astrobiology. During the great era of planetary exploration in the 1970s, the agency administrated a small grants program for research related to life in the universe. NASA also managed a SETI (Search for Extraterrestrial Intelligence) program until it was terminated by Congress in 1993. Since 1996, NASA has put increasing resources into a growing web of astrobiology institutes located at universities around the country, administered at NASA's Ames Research Center. Federal funding for this initiative exceeds one hundred million dollars per year. But progress isn't guaranteed. NASA cut astrobiology by 50 percent in 2006 to pay for the Moon, Mars, and Beyond initiative, and a cloud hangs over future promising missions. The United States

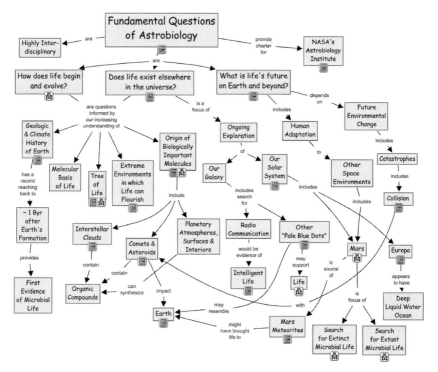

Figure 21. A concept map of major research questions in the field of astrobiology. NASA is the principal funding agency for the interdisciplinary community of researchers seeking answers to these questions.

doesn't have a lock on astrobiology; Europe, Japan, and Australia also have vigorous research programs (Fig. 21).

Astrobiology is compelling not just because it seeks to answer big questions but because it encourages scientists to think outside the box. Astronomers get to imagine all the places in the universe that might be habitable. Planetary scientists look at the ways life might alter a planetary surface or atmosphere. Biologists muse on the possibility of life without DNA or carbon. At biannual conferences hosted by the NASA Astrobiology Institute, cosmologists rub shoulders with geneticists and philosophers.

Interdisciplinary research goes against the grain of modern science, which has become increasingly specialized. Scientific knowledge from Giordano Bruno's time would have fit comfortably in diverse sections of a fat Sunday newspaper; polymaths like Leonardo da Vinci could work at the frontiers of several disciplines. Today's scientist has difficulty reading the research paper of a colleague in a different field. Often, he or she doesn't even try. Universities are organized into departmental "silos." Young scientists build reputations and are awarded tenure by becoming world experts in narrow subfields. The palpable excitement of astrobiology stems in part from scientists venturing out of the safe harbors of their own disciplines and creating new bodies of knowledge.

The continuing Copernican Revolution supports the expectation of life beyond Earth by showing that the basic ingredients—carbon, water, planets, and stars—are widespread in the universe. The "principle of mediocrity" asserts that conditions in our cosmic neighborhood are typical. However, we haven't yet detected Earth-mass planets, we know of no biology beyond Earth, and we have no idea if intelligence is a likely or inevitable consequence of biological evolution. As a prelude to speculating about the potential for life in the universe, we must learn as much as we can from the only known living planet.

2.

LIFE'S ORIGINS

> However, the macromolecule-to-cell transition is a jump of fantastic dimensions, which lies beyond the range of testable hypothesis. In this area, all is conjecture.
>
> —David Green and Robert Goldberger,
> *Molecular Insights into the Living Process*

The travelers look out onto the shore of a strange and hostile world. They don't leave the spacecraft; a probe shows that the atmosphere is mostly nitrogen, with smaller but lethal amounts of sulfur dioxide and methane and a hint of ammonia. Volcanoes belch gases into the sky, and the spacecraft trembles every few minutes from seismic activity. A young star, orange and bloated, perches on the horizon.

It's a young planet. The newly minted crust is still warm and plastic. Oceans have recently condensed from steam and are still kept warm and turgid by geothermal energy. Samples drilled from the crust show an age of two hundred million years—only 2 percent of cosmic time and the same fraction of the time the star will provide warmth to this planet.

Working swiftly, the visitors wrap up their experiments. It's not safe here. This soon after its formation, the planetary system is still strewn with debris. Every hour or so, the spacecraft shudders as a meteor slams into the ground nearby. There's a continuous light show overhead as smaller fragments burn up in the atmosphere. The large moon looming in the sky, which was splashed off an earlier impact, is a reminder of the potential for devastation.

Finally, results start coming in from a remote fleet of probes sent out a day earlier. Equipped with biosensors, they have fanned out across the landscape and the seascape. They found nothing larger than a sand grain, but the results are all consistent. There are microbes everywhere: at the edge of volcanic craters, near deep-sea fumaroles, floating on lapping lakes, buried in solid rock, even borne on currents of air. Life grips this young planet like a fever.

• • •

THE OLDEST THING IN THE WORLD isn't much to look at: a tiny speck of zircon crystal, not much bigger than the thickness of a human hair. It was found in Western Australia in 2000 in an outcropping of the rugged Jack Hills. Zircon is diamond's poor and plentiful cousin, familiar from cheesy cable-TV infomercials. But this crystal is special because it's 4.4 billion years old. Elements trapped inside it reveal what the Earth was like not long after it formed. It's remarkable that we can learn about our large planet from such a minuscule piece of it. To quote William Blake, we can "see a world in a grain of sand, and a heaven in a wild flower."

To understand how life is possible at all, we start by looking at the origin of its basic ingredients. As far as we know, the fundamental building blocks of life are carbon, nitrogen, oxygen, and hydrogen. The last two combine to make water, which is essential for all life on Earth.

If I took a representative sample of the "stuff" of your body (tiny enough that you'd never notice), chemical analysis would show a breakdown of about 55 percent oxygen, 24 percent carbon, 11 percent hydrogen, 4 percent nitrogen, 2 percent calcium, 1 percent phosphorus, and less than 1 percent of any other element. Most of the oxygen is bound up with hydrogen in the roughly 60 percent by weight of the human body that's made of water, and carbon is the major component of fats, sugars, proteins, nucleic acids, and the backbone of DNA itself.

If I took a representative sample of atoms from the Sun, I'd get a very different result: 73 percent hydrogen, 25 percent helium, 0.8 percent oxygen, 0.2 percent carbon, and 0.1 percent nitrogen atoms, plus tiny proportions of elements higher on the periodic table. Why is star stuff so different from life stuff? To answer this question, we must venture into stellar and cosmic cataclysms that took place long before the Sun and Earth formed.

Next, as a prelude to telling the story of the early Earth, we'll see how scientists keep track of time—not the familiar time of clocks and calendars but deep time, stretching back before recorded history, beyond even the fossil record. Scientists use the decay of the nuclei of atoms to measure the age of ancient things, so telling the story of life on Earth requires a detour into the physics of radioactivity. Only then can we begin to learn from the zircon.

Tracing the history of life becomes possible when we can date rocks. Life's story is coupled to the story of rocks. Working backward, we have a firm fossil trail for only half a billion years, little more than 10 percent of the age of the Earth. Before that, life-forms are microscopic, and the traces become indirect—scientists must look for evidence of an organism's activities. Sometimes all they can find are altered abundances of molecules and atoms, a mere whiff of life's existence. The age of the oldest organism is a matter of controversy, but it's not a lot younger than the Earth itself.

How did life begin? The trail can't be followed back that far, so scientists have worked forward from simple ingredients and an understanding of the conditions on the primeval Earth. They still have no reliable answer to a central question: how did simple molecules assemble themselves into long replicating chains and then cells? They also know little about the early experimentation that left all current living organisms with a single genetic code based on the DNA molecule.

Regardless of how and when life started, natural selection has sculpted it into an amazing variety of forms. Geology and biology on our planet are coupled in a profound and complex way, as we see when we consider the Gaia hypothesis.

Who are the travelers in the opening vignette? They might be our future selves, once we have mastered space technology and begin to search the nearest star systems for inhabited planets. But the same scene would have greeted aliens of unfamiliar function and form who ventured into our Solar System when it was young. They would have found the Earth to be a watery world rich in organic material, with primitive life full of promise and potential.

COSMIC CHEMISTRY

WHERE DID LIFE COME FROM? The universe is mostly dead and inert. It's made of 99.9 percent hydrogen and helium atoms, the two simplest elements, and in chemistry hydrogen and helium combine to make . . . nothing. On the other hand, a typical living organism is 40 percent carbon, nitrogen, and oxygen. Those three elements plus about a dozen trace elements on which life depends combine to give the richness of organic chemistry. The number of different molecules that can be made using carbon is essentially infinite. The origin of life begins with the birth of its chemical ingredients.

HELIUM AND THE BIG BANG

The atoms in our bodies share a strange and wild history. The universe today is old and cold, with its stars and galaxies spread across billions of light-years of almost perfectly empty space. But long ago, your atoms and my atoms and the atoms of all the creatures on Earth were joined in a titanic event of unimaginable power called the big bang. All life in the universe shares the kinship of a birth 13.7 billion years ago.

Very early in the expansion, when the universe was ten seconds old and a billion times smaller than it is now, collisions between protons were violent enough that some of them stuck together. In a large-scale version of the same process that causes the Sun to shine, hydrogen was converted into helium. After three minutes, all the fireworks were over—the expanding universe became too cool for fusion to work. One-quarter of the mass of the universe had been converted into helium.[1]

If nothing more had happened and the expansion had continued smoothly, this would be a singularly dull universe. Helium is inert, and hydrogen can combine with itself to form a molecule only when it's cool. It's hard to imagine biology in a universe with no chemistry.

THE STUFF OF THE UNIVERSE

Luckily for us, gravity created stars in the expanding soup of galaxies, and stars picked up where the universe left off, first fusing hydrogen into helium and then moving on to create even heavier elements, as we'll soon see. If we ask what the universe is made of, the answer is shown in Figure 22. The plot shows the cosmic abundance of elements across the periodic table, from hydrogen to uranium. The logarithmic scale is a bit misleading; if this data were plotted on a linear scale, you'd see only hydrogen and helium because all other elements are so rare. There are ten times as many hydrogen atoms as helium atoms. Hydrogen is thousands of times more common than the life elements (C, N, O), millions of times more common than aluminum or copper, and billions of times more common than gold or silver. Apart from hydrogen, everything else is a trace element.

Just how rare? Suppose a deck of cards represented randomly selected atoms in the universe. In one deck of cards, the aces and jokers would be helium atoms and the other forty-eight would be hydrogen atoms. You'd need thirty decks of cards before you'd expect to find one carbon atom. In the thirty decks of cards, there'd be a couple of oxygen atoms, too, but all the other cards would be hydrogen or helium. You'd need to search 3600 decks to find a single iron atom. Now imagine a seventy-five-foot cube (the volume of a small office building) completely packed with decks of cards—a total of two billion cards. If those cards were in direct proportion to the elements of the universe, there would be only one gold atom in the entire cube.[2]

How do we know what the universe is made of? Astronomers use remote sensing by spectroscopy to measure the composition of star stuff.[3] Each element has a unique set of sharp spectral features that acts like a fingerprint, so by identifying that fingerprint in starlight, astronomers can measure contributions of different elements. There's no exotic chemistry in space—all the elements seen

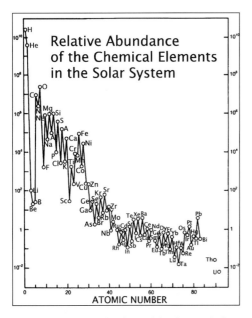

Figure 22. The cosmic abundance of the elements in the periodic table. The vertical scale is logarithmic, which allows heavier elements to be visible; they are all incredibly rare. The life elements—carbon, nitrogen, and oxygen—have concentrations of a few parts in ten thousand relative to hydrogen, and elements heavier than zirconium have concentrations less than one part in a billion.

in stars have counterparts on Earth, although at different relative abundances. Spectroscopy is exquisitely sensitive; for example, the heavy element thorium is readily detected in stars even though it's a trillion times less common than hydrogen.

CAULDRON OF THE ELEMENTS

Alchemy was the ancient art of transformation, a precursor to chemistry. At its heart was the dream of turning a base metal like lead into a precious metal like gold. It doesn't seem outrageous; both are dense, malleable metals of dull appearance in their natural state. Alchemy was protoscience rather than crank science (Newton was a famous practitioner), and its imagery has permeated popular culture, most recently in the blockbuster Harry Potter series of books and films. Alchemy features in the titles of two of the books, the symbolism in the naming of many of the characters, and in the person of Hogwarts headmaster Albus Dumbledore, who is stated to be an alchemist of renown, partner of Nicolas Flamel, a real-life alchemist from the fourteenth century. In literature, alchemy is used as a metaphor for the redemptive power of transformation and purification.

In real life, however, alchemy was doomed to failure. The essence of an element lies in its atomic nucleus, and this fortress cannot be touched by chemical means, which operate only on the outer shell of electrons. Alchemists need only have looked to the stars to find their philosopher's stone.

Every star is involved in the transmutation of elements. Because stars are large gas balls held together by gravity, the density and temperature rise smoothly as you move toward their centers. In the interior of the Sun, where the temperature exceeds ten million degrees, hydrogen is converted into helium by the process of fusion, which is the merging of atomic nuclei to form heavier elements. The protons in atomic nuclei have a positive electric charge, so they resist one another like tiny magnets. It requires a phenomenal temperature to force them to fuse.

Radiation from fusion is what keeps a star "puffed up." At every point in a star, the inward force of gravity is balanced by the outward force of radiation. Stars aren't expanding or out of control like bombs; their fusion is steady and measured. The energy released from the nuclear reactions reaches us as sunlight.[4] The best technology on Earth can keep a fusion reaction going for only a fraction of a second; the Sun does it endlessly and effortlessly. Each second, it converts a mass of hydrogen equivalent to twenty cruise ships into helium, and it will continue to do so for another five billion years.

A typical star like the Sun spends most of its life fusing hydrogen into helium. After the hydrogen is exhausted, the star loses its pressure support, and its core collapses. That compression continues until the ignition of a new nuclear fuel creates a balance at a new, higher temperature. Fusion is an unnatural act: it forces atomic nuclei to merge. The helium nucleus has two protons, so its positive electrical charge is larger than that of hydrogen, which has only one. The electrical repulsion is four times larger and it takes a temperature of one hundred million degrees to make helium fuse.

The next step is the key to life. It requires luck and a juggler's skill. Two helium nuclei fuse to make a beryllium nucleus, but beryllium is unstable, and it decays in a tiny fraction of a second. It's as if someone has a building block in his hands, but it crumbles to dust before he or she can add another piece. Occasionally, some beryllium survives, and if the energy levels are just right a third helium nucleus is added. Carbon is born. This two-step fusion is so tricky that it causes a bottleneck. As a result, the universe has three hundred times less carbon than helium.

Now it gets interesting. Add a hydrogen nucleus, and the atomic number increases by one. Add a helium nucleus, and it increases by two. In stars like the Sun, reactions peter out at carbon because small stars can't get hot enough for carbon nuclei to fuse. But in higher-mass stars, when an extra proton is added to carbon, it becomes nitrogen. One more and oxygen is formed.

Suitably warmed up, the cosmic element maker picks up the pace. Helium fused to oxygen makes neon. Helium fused to neon makes magnesium. Finally, at the awesome temperature of three billion degrees, two silicon nuclei fuse to create a nickel nucleus. Nickel quickly decays into iron, which is the most stable element and the end of the line.

In massive stars, creation of heavy elements is a crescendo. It takes ten million years to turn hydrogen into helium, half a million years to make carbon, and only six hundred years to make neon. Silicon fuses to make iron in less than a day. Imagine the frenzy of a roller-coaster ride that goes faster and turns in an ever-tighter spiral until it dumps you breathless and dizzy at the iron slag heap at the bottom.

The creation of elements up to iron releases nuclear energy because it results in a more tightly bound nucleus. Heavier elements are so unstable that they decay spontaneously in a process called fission. Since it costs energy to make elements beyond iron, stars sit tight. Low-mass stars end up with a seething core of carbon, with some nitrogen and oxygen. High-mass stars have cores of iron, not the solid iron of a wrecking ball but a bizarre, dense billion-degree gas.

CYCLES OF LIFE AND DEATH

If the story ended here, there would be no life. In fact, since most stars are less massive than the Sun, they trap their heavy elements and take them to the grave. They become cooling embers called white dwarfs, with cores so dense that a teaspoonful brought to Earth would weigh as much as a mountain. Their carbon-rich material is a crystalline form of carbon; Pink Floyd gave a nod to white dwarfs in their 1975 song "Shine On You Crazy Diamond."[5] High-mass stars entomb heavy elements like carbon, neon, oxygen, and silicon in layers as in an onion, with iron at the center.

Luckily, stars do share their bounty with the rest of the universe. As in the line from a song in an even earlier album by Joni Mitchell, "We are stardust, we are golden. We are billion year old carbon." After ten billion years in a stable state, the Sun, like all other stars, will eventually exhaust its nuclear fuels. It will become unstable and shuck off a layer of gas containing helium and carbon, blowing a huge smoke bubble into space. Heavier-mass stars lose even more mass—they drop off heavy sweaters of gas into the interstellar medium.

The most spectacular loss of mass occurs when the biggest stars die. Massive stars digest their elements slowly, and the steady absorption of neutrons ratchets slowly up the periodic table. At the end of these stars' lives, titanic explosions hold the key to the rest of the story of the elements.

With no more energy to be gained from nuclear reactions, the balance between gravity pulling and fusion pushing is disrupted for the last time. The floor drops out, and massive stars collapse in a spherical free fall. This implosion creates fantastic density and leaves behind a core of dense-packed neutrons (a neutron star is a gigantic atomic nucleus) if the star is fairly massive or a black hole if the star is extremely massive. But during the collapse, too much material rushes into a confined space, so most of it bounces back out in a spherical blast wave. This gas meets gas still falling in, and the collision creates temperatures of billions of degrees, enough to break through the iron "wall." Supernova!

Literally within seconds, explosive fusion creates the elements heavier than iron, all the way up to plutonium. These elements surf the blast wave and mingle with the thin gruel of gas between stars. For a few days, as witnessed by Brahe and Kepler and civilizations throughout history, a supernova outshines an entire galaxy. It's dazzling enough to see in broad daylight.

If you pause to reflect on your origin, it's as amazing as any tall tale told around a campfire. Our bodies—the iron in our blood, the oxygen in our lungs, the calcium in our bones, and the carbon and nitrogen that composes our genetic material—are formed from the detritus of generation upon generation of stars that lived and died before the Earth was born. Many of the atoms in the jewelry that we adorn ourselves with—gold and silver and platinum—are glittering relics of long-gone supernovas.

The Sun is a relative latecomer in the story of star birth and death. When the first stars formed, a few hundred million years after the big bang, they were pristine balls of hydrogen and helium. They had no planets because there wasn't anything to make them from. Over billions of years, stars forged heavy elements in their cores and ejected them into space, where they could become part of the gas clouds that would go on to form new stars. Generations of stars have handed down gas, steadily enriching it with the silicon and oxygen to make rocks and the carbon, nitrogen, and oxygen to make life.

As carbon-based life-forms, we borrow our carbon from the world around us as we grow in the womb—carbon that we'll return when we die. That same carbon is part of a much larger cosmic pageant of stellar life and death (Fig. 23). Life is, indeed, a strange and wonderful journey.

DEEP TIME

WE CAN'T TALK ABOUT ORIGINS without talking about time. Time is so trivial yet so profound that a definition is elusive. Newton imagined time as an intrinsic property of the universe, flowing linearly and immutably forward. Einstein blew this simple idea out of the water with his theory of relativity, showing that Newton's theoretical construct—a grid of clocks distributed through space, keeping perfectly synchronized time—could not work. Rather, time is supple and depends on the amount of motion and the local force of gravity. Atoms know nothing of time. Reactions flow back and forth, and time's "arrow" seems to be an emergent property of large collections of atoms.[6]

Human perception of time opens up another can of worms. It seems obvi-

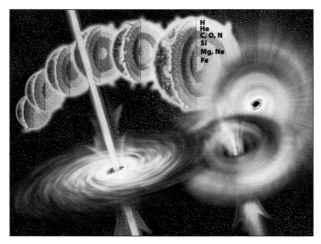

Figure 23. The existence of biology depends on a continuing process of star birth and death. For billions of years, stars have been creating heavy elements by fusion and ejecting them into space, where they become the raw material for future generations of stars and planets. The essential life elements have built up over time in this way.

ous: we feel time passing. Yet there's no actual sense that perceives time. We experience events in time and then assign them an order as they pass from the instantaneous "now" into our memories. The same series of events might be a blur to one person and a glacial progression to another. Timekeeping needs an objective way to keep track of events.

EVERYDAY TIME

The tyranny of time is a modern phenomenon, so pervasive that we rarely think about it. We keep time with our phones, on our walls, and on our desktops. At the obsessive end of the spectrum, you can buy a watch that connects wirelessly to the master atomic clock at the National Bureau of Standards, so you can plan your day to the nanosecond. Watches decorate the wrists of everyone from presidents to drug lords. Inside, a tiny crystal oscillates so precisely in response to an electrical impulse that it neither gains nor loses more than a few seconds per year. Watches are an overlooked miracle of technology; what other device is 99.99999 percent accurate yet often so cheap that it's disposable?

The history of the watch began with mechanical timekeeping devices perfected during Elizabethan times for the purpose of navigation on arduous ocean voyages.[7] Before that, Galileo had found that his pulse was an unreliable way to time experiments. Then he recalled a time in church when out of boredom he noticed a swinging altar lamp seemed to keep good time. This led to the pendulum clock. Throughout the Middle Ages and the Renaissance, clockmakers were elite artisans, and towns around Europe vied for their talents as they sought to acquire the status symbol of a clock tower. The oldest working clock is in Salisbury Cathedral in England. At the grand old age of six hundred years, it has ticked more than five hundred million times.

Before mechanical technology, all timekeeping was based on the cycles of the sky. Many divisions of time are based on astronomical phenomena: the day for the Earth spinning once, the month for a cycle of lunar phases, and the year for one Earth orbit of the Sun. Timekeeping had the practical purpose of regulating agriculture. Some ancient civilizations tracked time very carefully—four thousand years ago, the Babylonians kept a solar calendar accurate to five minutes, and Stonehenge used alignments of huge stones to predict solstices, eclipses, and other celestial phenomena.

The urge to keep time predates civilization. When humans were nomadic, they had to track the seasons and know when herds would migrate or berries would appear. The oldest human artifacts are calendar sticks from thirty to forty thousand years ago.

It's ironic that we keep time with greater and greater precision yet pay almost no attention to the vast span of time that stretches ahead of us. Danny Hillis wants to make a clock that will keep perfect time in ten thousand years. His "Clock of the Long Now" is a counter to the lack of long-term thinking in our culture. Hillis combines practical mastery with brilliant ideas. As an MIT undergrad he built a computer out of ten thousand Tinkertoy pieces. Ten years later, he built a computer called the Connection Machine that mimicked the human brain, and he served as the vice president of R&D for Disney's imagineering unit. He's said that the clock of his dreams "ticks once a year, bongs once a century, and the cuckoo comes out once a millennium." He's building it, with a perfect reconciliation of digital and analog technology that will persist and keep near-perfect time long after we're dead. The Clock of the Long Now excites people, but they find it strange—that's how hard it is for us to think of deep time.

TIME USING GRAVITY

How can we trace time in prehistory, when there were humans to bear witness but no written records? Everyday timekeeping uses a second as the short unit and a year as the long unit. For measuring geological time, a year is the short unit. The tilt of the Earth's axis causes seasonal climate changes that leave their imprint on the Earth in a number of ways.

One familiar tracer is tree rings. The rate of tree growth varies on a seasonal cycle, giving rise to a new layer or ring each year. But the oldest bristlecone pines are only five thousand years old, so this method doesn't reach into prehistory. To do better, we turn to places where erosion is absent and geological processes are dormant. On the Antarctic plateau, seasonal snows produce a new layer each year. A European team has drilled a deep core through the ice pack and derived a continuous record of climate dating back 750,000 years, although compression of the ice means that individual layers can't be distinguished in the early portion of the record. Similarly, layers of windblown dust on the loess plateau in China and sedimentation layers at the bottom of Lake Baikal in Siberia can be traced back several million years. These methods work for measuring relative ages—young layers are higher up, and old layers are lower down—but they're less reliable for measuring true ages because climate change could have led to times when ice or dust were not deposited.

Deeper time can be measured with a set of astronomical cycles that imprint periods much longer than one year. They're all caused by the interplay of gravity between the Sun, Earth, and Moon, and it's the predictability of the cycles that allows us to use them to track time. First, the shape of the Earth's orbit of the Sun varies from circular to 6 percent elliptical with a period of one hundred thousand years. Second, the tilt of the Earth's spin axis, which causes the seasons, smoothly varies from 21.5 to 24.5 degrees on a forty-one-thousand-year timescale. Lastly, the spinning and tilted Earth wobbles like a top, and the period of this precession is twenty-three thousand years.

The Serbian astronomer Milutin Milankovitch realized in the 1920s that these cycles must affect climate, since they dictate the amount of solar radiation received by different parts of the Earth's surface. Measuring these cycles in the rock record is very challenging, because the three cycles interact, and the effect of solar radiation on climate is not simple or linear. However, the hundred-thousand-year cycle has been convincingly detected in deep-sea core samples and in the advances and retreats of the ice pack. This takes us back ten million years or so. On longer timescales, the restless Earth jumbles the layers and alters most rocks so thoroughly that scientists need a different type of clock.

ATOMS AND TIME

Most things have resonant frequencies—rates of vibration or oscillation they favor. In general, smaller objects vibrate at a higher frequency. The biggest pipe of a church organ vibrates seventeen times per second, a frequency easier to feel than to hear. A standard tuning fork vibrates 440 times per second. Most modern watches contain tiny quartz tuning forks that vibrate 32,768 times per second.

Vibration is also a fundamental property of atoms, and since they're very small their resonant frequencies are very high or rapid. In the 1950s, scientists built a clock of unrivaled accuracy based on the natural frequency of the cesium atom. In 1967, a second was officially defined as 9,192,631,700 oscillations of a cesium atom, replacing the old second, which has been defined as 1/31,556,925.9747 of the Earth's orbit of the Sun in 1900. Atomic physics replaced astronomy for timekeeping.[8]

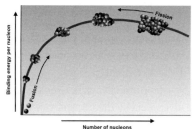

Figure 24. A graph of the stability or binding energy of different elements. Iron is the most stable nuclear configuration. All elements lighter than iron emit energy and become more stable by the fusion process, and all elements heavier than iron emit energy and get more stable by the fission process (radioactivity).

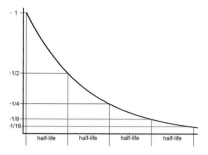

Figure 25. Radioactivity is a random process for an individual atom, but a large collection of atoms decays at a well-defined rate. In each successive half-life, half of the parent atoms decay into daughter atoms. Assuming the number of daughter atoms at the start of the process is known, the ratio of parent to daughter atoms gives the age of the rock containing the radioactive material.

Harnessing the vibrations of atoms to keep time is one thing, but it doesn't allow us to measure deep time. For that, physicists looked to another property of certain types of atoms: the nuclear process called radioactivity. All very heavy elements and some light ones spontaneously decay into lighter elements. For example, uranium decays slowly into lead. Radioactivity occurs as atomic nuclei try to find their most stable or tightly bound configuration. Up to iron in the periodic table, nuclei continue to get more stable, but elements heavier than iron show an increasing tendency to emit particles or energy and turn into lighter elements (Fig. 24). Sometimes there is a series of decays before a stable element is reached. In a radioactive process, the starting atom is called the parent atom (or isotope), and the end result is called the daughter atom.

A single radioactive atom makes a lousy clock. That's because of a weird aspect of the quantum world: individual events are not deterministic. A uranium atom has a 50 percent probability of decaying in 4.5 billion years, and that's all we can say. A uranium atom in front of you might decay tomorrow!

However, if you start with a lot of parent atoms, they'll steadily and predictably decay into daughter atoms. The time taken for half of the parents to decay into daughters is the half-life (Fig. 25). After 4.5 billion years, half of a sample of uranium atoms will have decayed into lead atoms, and after another 4.5 billion years half of the remaining uranium atoms will have decayed, leaving only one-fourth in the original sample.

How can something that is individually unpredictable produce a reliable outcome? It's precisely the tactic that gamblers make use of when they play the odds. Dumb gamblers play roulette and constantly shift their bets between red and black, trying to guess the next outcome. That's as pointless as trying to guess when a radioactive atom will decay. They'd get the same result—50 percent success—if they just kept the bet on black or red all the time. (Though roulette is always a fool's game since the house cut ensures an average success rate of less than 50 percent.)

A better analogy is a pan of popcorn on the stove. If you stared at one kernel, you would not be able to say when it would pop, nor would you be able to say which kernel would pop next. Like single atoms, single kernels are unpredictable. But by cooking a particular type of popcorn, you could observe how long it took for half of the kernels to pop. If you cooked more batches of the same popcorn under the same conditions, your measurements would home in on a smaller window of time in which half the kernels would pop. That's the popcorn half-life.

THE CLOCKS IN THE ROCKS

Now we can see how scientists measure the ages of ancient rocks. Radioactive decay can be used as a clock because the half-life is a constant of nature for a particular element and decay process. The nucleus is a tiny fortress at the heart of each atom. Its decay is governed by incredibly strong "glue" that holds protons and neutrons together. Chemical reactions, by contrast, involve arrangement of atoms and the exchange of electrons. The force that governs chemistry is a trillion times weaker than the nuclear force, and it operates on a scale thousands of times larger than the nucleus.

Take a radioactive sample and boil it, freeze it, or drop it in acid; its half-life will be the same. Subject it to magnetic fields or variations in gravity—no change. The half-life of uranium is the same whether the atoms are in the laboratory, deep within the Earth, or in a comet at the edge of the Solar System.

To measure the age of rocks, geologists need a sample in which the material was fixed (or solidified) at a particular time and in which the initial presence of daughter atoms is either very low or known. Radioactivity will steadily cause parent atoms to turn into daughter atoms, so the ratio of daughter to parent atoms gives the age of the sample (see Fig. 25 again).

The most famous form of radioactive dating uses carbon-14. The vast majority of carbon atoms are stable carbon-12 (a nucleus of six protons and six neutrons), but one in eight hundred billion is carbon-14 (with two extra neutrons), which decays with a half-life of 5,730 years. Any living thing takes in all forms of carbon from the atmosphere, but when it dies the tiny amount of carbon-14 within it begins to decay without being replenished. As a result, the ratio of carbon-12 to carbon-14 in formerly living things steadily increases. The time since the living thing died is measured by comparing the fraction of carbon-14 in it to the fraction of carbon-14 in the atmosphere. This technique has been used to show that the Shroud of Turin was a fake, and it's commonly used to date human artifacts such as fragments of cloth, bone, or wood.

Rocks have never been alive, so carbon doesn't work for measuring their ages. Also, after many half-lives the proportion of parent atoms gets so low that it's hard to measure. A clock with a really long half-life is needed to measure the age of the oldest rocks. Uranium-238, with a half-life of 4.5 billion years, is perfect.

Perfect, with one proviso: dating using the "clocks in the rocks" is reliable only when used on a rock that has not had its ingredients rearranged since it formed. In other words, since the technique is based on the decay of parent into daughter atoms, there has to be some way of knowing both how many daughter atoms were present when the rock first solidified and that the parent and daughter atoms have been trapped there ever since. Uranium-238 decays into lead-206 with a half-life of 4.5 billion years, and as a bonus it is found alongside uranium-235, which decays into lead-207 with a half-life of seven hundred million years. Comparing ages from two or more different decay processes provides a cross-check. In practice, geologists use a variety of methods and many rock samples to get reliable ages.

Which brings us back to zircon. This mineral is brilliant for radioactive dating because it incorporates uranium into its lattice in place of zirconium, but it rejects lead. (Lead is the daughter atom, so that ensures there was no lead in the zircon to start with.) It's chemically inert and tougher than nails, so it's immune from weathering and plate tectonics. These little crystals are found in the oldest granite formations, and they yield some of the best evidence that the Earth is billions of years old.[9]

FIRST TRACES OF LIFE

To DISCOVER ANCIENT LIFE, you have to find an ancient rock, and that's a challenge. Roger Buick notes that if you pick up a rock, regardless of where you live, it's not likely to be more than a few hundred million years old. That's less than 10 percent of the age of the Earth. In human terms, it's like searching high and low for a hundred-year-old, when all you see around you are kids, a few teenagers, and the odd adult. Think of the pockmarked lunar surface—that heavily impacted terrain mirrors our own collision history. But on Earth those scars have been erased by wind and water, and the surface has been re-built many times by the surge of magma from the interior.

Imagine you buy a house and find the fridge in the kitchen is full. You know the age of the fridge, but none of the contents has a label. How would you esti-mate ages? By deduction—by comparing the thickness of the rind on the cheeses, by seeing which tub of yogurt had the most mold, by finding out which item smelled the worst. Geologists can also tell a lot about rocks by the way they look, feel, and smell, but they tether their inferences to the iron-clad physics of radioactive decay.

THE ART OF READING ROCKS

Buick is animated as he names the places on Earth where you can find rocks more than three billion years old: Greenland, South Africa, Western Aus-tralia (Fig. 26). The last is his stomping ground, in the country that gives his accent a distinctive twang. Buick has shoulder-length hair and fiery, intense

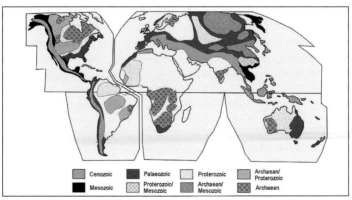

Figure 26. The continental masses of the Earth, showing the typical ages of the rocks. The very oldest rocks, those more than four billion years old, are extremely rare. The Cenozoic is the most recent era; the Archaean runs from 3800–2500 my ago.

eyes. His post-1960s attire and friendly demeanor belie intense intelligence. Much as it disappoints those who like to think that academia is a meritocracy, subtle advantages are still conferred by a tweedy look and an Oxbridge accent.

Each year, Buick sets off from his office at the University of Washington in Seattle and heads for places with no coffee shops or bookstores. He spends long enough in the field to sprout a wild man's beard and toss off concerns of personal hygiene. He prefers to work alone. The bush is harsh and austere. He's had a vehicle blow up on him, and he's had to walk thirty-five miles with no water. It's 120 degrees in the summer and humid. Australia hosts many of the world's most venomous snakes and spiders—this is not work for the faint of heart.

Years of scrambling through wilderness and smashing rocks with his hammer have given Buick an uncanny ability to discern the oldest formations. He is a field geologist—an unabashed rock hound. The ability to "read" rocks is a particular kind of skill. Astronomers look at smudges of light on a digital image and infer the history of galaxies billions of light-years away. Chemists take one whiff from a test tube and know its contents. People like Buick stare at a rock and infer the tumult of tectonic forces acting over geologic time.

Like other branches of science, geology has gone high tech. Many geologists work from satellite maps and use remote-sensing techniques. Not Buick. He goes into the field with a topographical map and a well-trained eye and then scrambles into gullies and up hillsides. In Australia's western outback, the rocks are red and weathered, and Buick is in his element. But the first time he went to Greenland, he was lost because the rocks were grainy and black, shattered by freezing, and sculpted by glaciations. His perception was like that of a newborn child. Sometimes it's the similarities in different terrains that are noteworthy. On one of his trips, Buick had a toy Mars rover with him. He put it among the ochre rocks of the arid Australian desert and was delighted to see it make a perfect tableau of the "Martian" landscape.

FINDING THE OLDEST ROCKS

The oldest piece of the Earth isn't a rock at all. It's a zircon crystal from the Jack Hills of Western Australia (Fig. 27). Roger Buick wasn't in on the discovery, but he knows the region well. The zircon, and others like it of similar or younger age, formed in molten granite ten miles down and eventually made their ways to the surface by erosion of overlying material, where they were embedded in younger, sedimentary rocks.

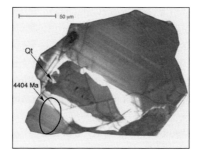

Figure 27. The oldest known object on Earth, a tiny sliver of zircon crystal. It's smaller than the thickness of a human hair and 4.4 billion years old, found in the Jack Hills region of Western Australia. This image was made by a technique that shows the interior structures. The relatively cool formation temperature means that the Earth could have had liquid water and perhaps been hospitable for life within 150 million years of its formation.

Since the Earth is so heavily altered, geologists reach with difficulty back to the beginning. Rocks at least three billion years old are found on all continents. The oldest specimens are four-billion-year-old rocks from Acasta in northwestern Canada and 3.8-billion-year-old rocks from Isua in western Greenland. These rocks are dated by a variety of radioactive techniques, so we know their ages to a precision of a few percent.

The backstop to the age of the Earth is set by pristine material that reaches us from the outer Solar System. Radioactive decay methods applied to meteorites give an age of 4.54 billion years with an uncertainty of less than 1%. With a time frame for Earth history established, let's see how far back life can be traced. Paleontologists and geologists work by establishing reliable landmarks before they can set new ones that reach even farther back.

A TIME CAPSULE OF LIFE

Most life leaves no trace. Soft parts decay and decompose; hard parts are eroded by abrasion and weathering. The organic material is recycled into the biosphere, blown on the four winds. Ashes to ashes, dust to dust. Tracing life's history is detective work based on the rare situations when life is preserved at all. Life is entombed intact for long periods of time only in rocks, so the stories of life and rocks are intertwined.

Well-preserved humans have been found from five thousand years ago, like the tattooed warrior on the Russian steppes who was found buried with his weapons and his hash pipe. The oldest well-preserved human is a ten-thousand-year-old mummy found in a cave in Nevada. These burial sites are dioramas of the lives of our ancestors. Freezing is better for preserving tissue than desiccation. An intact mammoth, twenty-one thousand years old, was raised from the permafrost in Siberia in 1999, its flesh fresh enough that dogs in the discovery team tried to eat it. Bacteria from even earlier in the Pleistocene era, thirty-two thousand years ago, were found in a frozen pond in Alaska and brought back to life.

Amber! Romans valued a bit of it as much as they did a slave. Ancient Greeks thought it was tears from Apollo's daughters as they mourned their dead brother Phaëthon. Artisans have carved and traded it for ten thousand years. It is sap that flowed out of plants millions of years ago and polymerized and hardened, essentially turning into plastic (Fig. 28). Since the resin can entomb flowers and ants and bees before it dries and hardens, it provides a window on several hundred ancient species.

While amber can preserve the form of insects and microbes almost perfectly, the premise of *Jurassic Park* is unlikely to ever be realized. The oldest intact DNA comes from a plant trapped for four hundred thousand years in the Siberian ice. Only tiny fragments of DNA have ever been recovered from insects trapped in amber.[10] These include pieces from a seventeen-million-year-old magnolia leaf, a thirty-million-year-old termite, and a 130-million-year-old weevil.

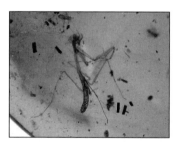

Figure 28. When plant or tree sap fossilizes, it turns into hardened resin similar to plastic. Spores, insects, and even small animals can be trapped within the resin in a good state of preservation for up to 150 million years. This praying mantis is twenty million years old. Despite appearances, the DNA within the creature has mostly broken down, so reconstruction of its entire genome is unlikely.

BODY FOSSILS

The story of life must be read through fossils. Fossilization is the slow process of preserving biological remains in rocks. It usually takes thousands of years, though under special circumstances it can occur as quickly as a few decades. Fossils are generally preserved by sedimentation, so we know the most about life-forms that lived in coastal regions, ponds, and shallow seas. Igneous and metamorphic rocks have usually been too brutalized for large fossils to survive in them. Soil is also bad news because it's often acidic.

Fossilization turns bone into stone. A fossil can also be an imprint or a cast of life rather than the life-form itself. Less than

1 percent of the Earth's creatures have been preserved as fossils, so it's tricky to infer the full biological diversity from such a small sample. An organism with no hard body parts is usually out of luck. Dead bodies are picked clean by scavengers or quickly rot, and bones are ground into dust or dissolve over time. It takes a quick death, under special conditions, to leave a corpse that will still look good millions of years later (Fig. 29).

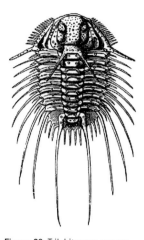

How does it work? Sometimes the carbon from the living form is replaced or transformed into a different form of carbon after death. Plants can be reduced to a black carbon film when some of the organic compounds vaporize and disperse, leaving the detailed structure of cell walls visible. In a tar pit, oily chemicals leach into bones and slowly entomb them in asphalt. More often, carbon that forms the "backbone" of life is replaced with silicon and calcium. When fallen trees are covered with water, the silica dissolved in the water can replace the cellulose in the wood, molecule by molecule. The wood is perfectly preserved or petrified, stained pretty colors by different minerals.

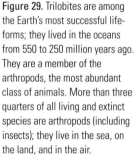

Figure 29. Trilobites are among the Earth's most successful life-forms; they lived in the oceans from 550 to 250 million years ago. They are a member of the arthropods, the most abundant class of animals. More than three quarters of all living and extinct species are arthropods (including insects); they live in the sea, on the land, and in the air.

In most fossils, the organic material has been totally destroyed, and the skeleton may be replaced by calcite or quartz or pyrite; the result is a faithful reproduction of the shape of the creature, but all details of the interior tissues and organs are lost.

Long-dead creatures can leave more indirect evidence of their bodies or body parts. These include footprints, tail prints, burrows, and even feces. Fossilized dung is called a coprolite. Bits of fossilized crushed material in coprolites tell us what extinct animals ate. One beautiful specimen from Saskatchewan was a *T. rex* turd the size of a small pillow, containing fragments of the head frill of a triceratops.

About half a million years ago, looking back in time, the record of animal fossils peters out like a road winding into a thick fog.[11] Before the Cambrian era, from 543 to 490 million years ago, when life was mostly in the oceans, organisms had no hard bodies or skeletons to leave any evidence of their function or form. Imagine an ancient jellyfish: made of 99 percent water, it's like a ghost even before it dies.

During the vast time span from 2.5 billion to 543 million years ago—an interval called the Proterozoic—fossil remains are mostly bacterial. At the end of the Proterozoic, life was reeling under a series of bone-chilling glaciations (metaphorically, since life then had no bones), but well-preserved worms and feathery-looking creatures have been found in the Ediacara Hills of southern Australia. From before about seven hundred million years ago, the only traces of life are microfossils. It takes very special conditions to preserve tiny creatures, and the best microfossils are found in beds of chert, a very hard sedimentary rock in which silica has replaced living tissue. Our ancestors used cherts like flint to make tools and arrowheads. At Bitter Springs in central Australia, 850-million-year-old chert beds were formed by chemical precipitation, so even the 3-D structure is retained. The beds contain exquisite fossils of bacteria, algae, and fungi.

Over most of the Proterozoic, life on Earth was fairly simple and bacterial (Fig. 30). From three billion years ago until their gradual demise about 700 million years ago, the king of microfossils was the stromatolite.

TRACE FOSSILS

Roger Buick was a first-year grad student when he found a stromatolite that was more than three billion years old. Stromatolites look like cabbages, and they've been called the Rodney Dangerfield of fossils; they don't get much respect because they're not as pretty as trilobites and ammonites. A stromatolite is a trace fossil, something left behind as the result of an organism's activities rather than the organism itself. Stromatolites represent some of the Earth's earliest life-forms. Their descendants are found in tidal areas today—including Shark Bay in Western Australia, close to where Buick grew up. He chuckles when asked if humans might prove to be as durable as microbes.

Figure 30. In southern Ontario, the Gunflint chert is a banded rock formation with beautiful stripes of red jasper. In the 1950s, two-billion-year-old microfossils were found in the chert that are still among the oldest fossils known. This is a cyanobacteria fifty microns long, the remains of one of the organisms that produced the oxygen in the air we breathe.

Buick explains that stromatolites are formed by bacterial mats, built up layer by layer by microbes called cyanobacteria. These tiny creatures can't be seen themselves with the naked eye; what you can see is the remains of the "city" they built. Think of Pompeii—the ancient city in Italy, entombed and preserved by ash from a volcanic explo-

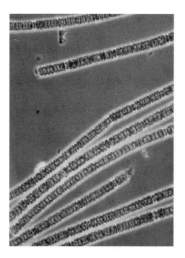

Figure 31. Earth's most successful life-forms. Cyanobacteria date back 3.5 billion years and their descendants are still found in many environments on land and in the water. These tiny photosynthetic organisms created almost all the oxygen in the atmosphere, and they led to the development of plants.

sion nearly two thousand years ago. Almost no human trace survived due to the intense heat, but it's easy to see the buildings and courtyards and infer from them the way the people lived. In the same way, stromatolites are trace fossils left by cyanobacteria. They deserve our respect, quite apart from the fact that they've persisted for 3.5 billion years (Fig. 31). Long ago, these tiny photosynthetic organisms created all the oxygen we breathe.

MOLECULAR FOSSILS

Long after the Cheshire cat has faded, we hope to be able to trace his chemical smile. From the first two billion years of Earth's history, life has left no cells or body parts. The only evidence of life are tiny imbalances in the abundance of individual atoms and molecules. Due to the subtlety of these tracers, a fierce debate rages over the age of the earliest life.

To graphically explain the idea of molecular fossils, Buick turns macabre. After killing me and burying me in sediment, he explains, he'll heat and squash my body until only organic ooze remains: oil. This is the cool part—he cheerfully points out that he could use the ooze to deduce that I had been an advanced organism that reproduced sexually and was made of complex cells with nuclei. He could even guess my cholesterol level; cholesterol is a molecule restricted to complex organisms, unique because it isn't found in bacteria and survives surprisingly well in geological environments.

Some molecules are reliable tracers of photosynthesis. Others trace more exotic metabolic processes. Buick has an ancient chunk of barite on his desk. He can scratch it and release hydrogen-sulfide gas—smelling distinctly of rotten eggs—from inclusions within the crystals. The toxic whiff points to bacteria that metabolized sulfur compounds in a warm pond 3.5 billion years ago.

The research is challenging because each of the chemical tracers can also occur naturally in situations where they are not tied to any biological activity. So the inference of life is based on localization within particular rocks, which makes it less likely to be contamination from a random source, and on a pattern of these biomarkers that doesn't depend on only one type of evidence.

LIFE'S EARLY FRONTIER

Stromatolites, molecular tracers, and even individual recognizable microfossils can be found all the way back to 3.5 billion years ago, so evidence that the Earth has been continuously inhabited since then is strong.

It's strong, but not totally compelling. This is a confusing crime scene. There are no living witnesses. Countless creatures have trampled across the delicate evidence, which has also been squashed and heated and scoured by various chemicals. Rivers of lava have flowed, and meteorites have rained down over the eons. Few geological formations exist. Despite this, scientists pluck up their courage and push back even earlier. They must rely on atomic tracers, and now the detective work gets really hard.

In addition to a short-lived radioactive isotope, carbon has an enduring isotope. About 99 percent of carbon in nature is familiar carbon-12, and just over 1 percent is carbon-13. Bill Schopf at UCLA, who has done important work on ancient microfossils, has been a proponent of carbon isotopic ratios as tracers of early life. In seawater, dissolved carbon dioxide and calcium combine to make calcium carbonate, which rains down on the ocean floor to form limestone. In living organisms such as plants, carbon dioxide is snared by an enzyme in the first step of photosynthesis, which later produces glucose. The two types of carbon atom bounce around inside a cell like Ping-Pong balls. The lighter carbon moves faster, so it hits the enzyme more often and is more readily absorbed. Limestone produced in oceans containing photosynthetic microbes would have five parts per thousand less of the light, more abundant form of carbon than rocks that never hosted life. This difference is easy to measure with modern techniques of mass spectrometry.

With carbon as a tracer, photosynthesis can be tracked through geologic time; it's found all the way back to 3.5 billion years ago. The same life tracer has been found more controversially in the oldest sedimentary formation ever studied—3.8-billion-year-old crystalline graphite from Isua in Greenland—by Manfred Schildlowski of the University of Mainz. But this last evidence is merely a hint of life.

Why should we care about the age of the oldest organism? The primeval Earth was peppered by debris left over from the formation of the planets; it was an inhospitable place of volcanoes, earthquakes, and toxic air. If life could exist there, we might think it could exist on countless planets beyond the Solar System. If life began quickly on Earth, life may be likely to form on any similar planet. Some time in the first half-billion years or 10 percent of the history of the Earth, the motor of life first turned over. But how?

LIFE IN A BOTTLE

CHEMISTS TELL A JOKE among themselves, when nobody else is around. Life is impossible, they say: we've put simple chemical ingredients in water, we've added energy in the form of heat or electricity, and all we ever get is an organic sludge. We never see replicating molecules. We never make a cell. The astronomer Fred Hoyle once said that the act of assembling the simplest living organism from simple molecular ingredients was as unlikely as a tornado whipping through a junkyard and assembling a jumbo jet. Yet somehow it happened. Was it blind luck? And if it somehow happened here, could it happen somewhere else?

THE MILLER-UREY EXPERIMENT

In 1953, Stanley Miller, a young University of Chicago graduate student, flipped a switch and sent an electric current through a glass flask containing water and a mixture of methane, ammonia, and hydrogen (Fig. 32). His advisor, Harold Urey, suggested the experiment but then tried to dissuade Miller from going ahead with it, thinking that it might take months or years to yield a useful result. Both men were amazed by what happened. Within hours, the liquid turned distinctly pink and then very red. Within a day, there was a brown film on the inside of the flask. Within a few weeks, 4 percent of the contents of the flask had been converted into thirteen of the

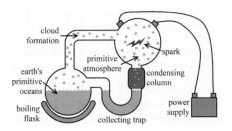

Figure 32. A schematic view of the apparatus used in the classic experiment by Miller and Urey in the 1950s. A mixture of methane, ammonia, hydrogen, and water vapor is introduced to a lower flask and boiled into an upper flask, where it has energy introduced in the form of an electrical discharge. After a week of continuous operation, the water in the trap is covered with a brown film containing amino acids and other hydrocarbons.

twenty amino acids used in life on Earth—amino acids are the building blocks of proteins, which combine to create the genetic code. Any high school chemistry lab can repeat the original result.

Miller presented his results in a seminar; their reception was predictably electric. When the noted physicist Enrico Fermi asked if this process could actually have taken place on the primitive Earth, Urey leaped up and said, "If God didn't do it this way, he missed a good bet." The scientific community was surprised and skeptical. When Miller and Urey's paper was submitted, one reviewer refused to believe the result and delayed publication. (By chance, it appeared within a month of Watson and Crick's description of the DNA molecule.) But there's a catch to this classic experiment.

Miller and Urey took their cue from planetary scientists, using a gas mixture with a composition comparable to that of the giant planets at the time of the Solar System's formation.[12] Gases rich in hydrogen are able to combine with carbon and build more complex molecules. However, on a terrestrial planet such as Earth, volcanoes produced an early atmosphere rich in carbon dioxide. In alternative Miller-Urey experiments where carbon dioxide replaces hydrogen-bearing molecules, there's a lower yield of organic compounds.

The original experiment also used an electrical discharge that was unrealistically strong—an attempt to simulate the effect of lightning. However, there were a number of other energy sources on the early Earth, such as ultraviolet radiation, volcanoes, seismic shocks, deep-sea vents, and even impacts from space. Regardless of the nature of the energy source, all "life in a bottle" experiments produce similar results if gases rich in hydrogen are used.

INGREDIENTS FROM SPACE

In addition to organic compounds that formed in the primordial soup of Earth's atmosphere and oceans, life probably had assistance from space. Raw organic materials are deposited on the Earth's surface at a rate of ten million kilograms per year. The rate was one hundred thousand times higher during the bombardment that preceded the earliest evidence of life, when debris left over from planet formation was abundant. Both comets and meteorites can transport complex molecules to Earth. More than seventy amino acids were found in the Murchison meteorite after it landed in 1969, eight of which are among the twenty amino acids used by terrestrial life. Surprisingly, when scientists simulate impacts, amino acids not only survive the crashes, many join to form polypeptides or miniproteins—another step along the road to life.

Meteorites also deliver a vital ingredient: phosphorus. This reactive element is the fifth most important biological element after carbon, hydrogen, oxygen, and nitrogen, but stars make very little of it (Fig. 22). Phosphorus is five times rarer in the cosmos and eight times rarer in ocean water than it is in bacteria. Phosphorus is critical to life because it forms the backbone of DNA, and it's also the major ingredient in ATP (adenosine triphosphate), life's fundamental fuel. The most common phosphorus-bearing mineral on Earth, apatite, does not give up its phosphorus easily. However, meteorites contain the metallic mineral schreibersite, which releases phosphorus-rich compounds into water.

Many variations of the Miller-Urey, or "life in a bottle," experiments have been carried out since 1953. None has produced any living organism or even a replicating molecule, but they've generated most of the simple building blocks of life.[13] In addition to producing amino acids, the experiments have yielded sugars, fatty acids, and all of the bases used by DNA and RNA. By adding carbonyl sulfide, which is found in volcanic gases and deep-sea vent emissions, one group at Scripps Institute saw amino acids combine four at a time into polypeptides, a level of complexity not observed by Miller and Urey.

Perhaps we shouldn't be surprised that a few weeks in the lab can't duplicate a process that probably took millions of years on Earth. Yet the results of the "life in a bottle" experiments seem a little disappointing. They take us from simple molecules with three to five atoms up to organic molecules with a few dozen atoms. Compare this to the simplest proteins, which have thousands of atoms, and DNA strands from the most primitive organisms, which have tens of millions of atoms.

THE NATURE OF LIFE

IT'S A LONG ROAD from organic mud to the exquisite biochemical machinery of a cell. If this is a journey to the top of a high mountain range, we are stuck in the lower foothills of the Hindu Kush while the upper peaks beckon in the misty distance. There may be many paths to the top, but they look awfully difficult (Fig. 33). Like all journeys in science, this one is accomplished by persisting, taking one step at a time, and recognizing that progress involves failed ascents and the occasional spill. But how will we know when we've arrived? What's the boundary between life and nonlife?

LIFE USES ENERGY

The definition of life is surprisingly difficult for biologists to agree on.[14] Everyone agrees on the attributes of growth and reproduction. But flames and crys-

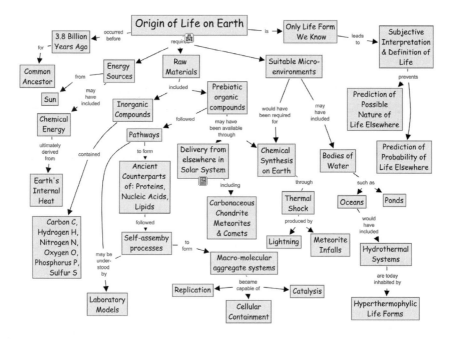

Figure 33. A flowchart that summarizes the complex processes that led to the formation of life on Earth. Since there is so little physical evidence from the first 20 percent of the planet's history, much of this progression is still speculative.

tals can grow and propagate. Viruses can reproduce, but they can't do so without a host cell. Most people also agree that life stores and organizes energy. But so does a star. A more sophisticated concept that's central to life is metabolism, which derives from the Greek word for change. Living organisms take in matter and energy from their environment and then release it in the form of waste and heat. The biochemical pathways are complex—some materials are broken down, and others are built up from smaller components—but there's always a process of flux in and out.

The idea of life as a flow of matter and energy makes perfect sense. We eat hamburgers, but we don't become the hamburger. Rather, we extract energy from the hamburger and rearrange its atoms to make more of us. Every five days, you get a new stomach lining. You get a new set of skin cells every six weeks and a complete replacement of your liver cells every two months. Each year, 98 percent of the atoms in your body are replaced. Constant replacement of components and energy flow within an unchanging structure is the hallmark of living organisms.

Important insights come from physics. In his 1945 book *What Is Life?* Erwin Schrödinger, one the pioneers of the quantum theory, discussed life as a thermodynamic process. Thermodynamics is the study of heat flow—heat being defined as microscopic motions of atoms and molecules, the most unstructured form of energy. Energy in the universe is conserved absolutely; it can't just appear or disappear. If energy seems to disappear, it's usually because it has been turned into heat. When a windup toy stops moving, it's because the energy in a coiled spring has turned into motion and then heat. When a rolling car comes to a halt, energy stored in the chemical bonds of fossil fuel turns into motion and then into heat in the brake pads and the tires.

Where does life's energy come from? Many living organisms such as plants gather light energy directly from the Sun. Others, higher up the food chain, obtain their energy from the stored energy in the chemical bonds of simpler creatures, but the original source is the same.[15] Life's energy use is not (nor does it need to be) perfectly efficient. When sunlight hits the leaf of a plant, about 30 percent triggers photosynthesis, while the rest of the energy is dispersed and heats up the leaf and the air around it. (If the light hit a sheet of paper instead, all of it would heat up the paper and the air around it.)

The suite of chemical reactions involved in a living organism is a metabolism. Some reactions break down ingredients to release energy for vital processes; others synthesize vital ingredients. The activity of even the simplest cell seems audaciously complex, but all of the arrangements and rearrangements of atoms and molecules work within the cast-iron bounds of the law of conservation of energy.

LIFE BATTLES DISORDER

A common misconception is that life violates the universal tendency toward disorder.[16] To see why this isn't true, we need to take a detour into physics. The Second Law of Thermodynamics can be expressed in a number of ways, but the one most useful to a discussion of life is the fact that in all the transactions of energy in the natural world, the proportion of disordered or heat energy tends to increase. The physicist's measure of disorder is called entropy.

Entropy is not the same as chaos. Everyday analogies of shuffled cards or lost socks create an appealing mental image but are incorrect parallels because they involve human arrangements of macroscopic objects. Entropy is defined in terms of microscopic states of atoms and molecules, and it's directly related to heat. When an ice cube melts in a glass of water or a noxious odor spreads across a room, the entropy increases because the structure in the ice crystals is lost or the concentrated state of the gas has been lost. In both situations, the result is a higher degree of uniformity or microscopic disorder and mixing.

Another way to think about entropy is in terms of the number of equivalent

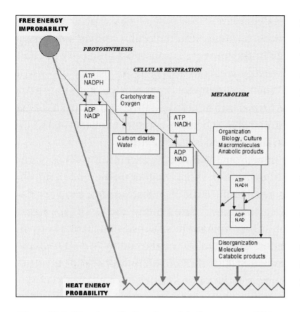

FREE ENERGY
IMPROBABILITY

PHOTOSYNTHESIS

CELLULAR RESPIRATION

ATP
NADPH

Carbohydrate
Oxygen

METABOLISM

ADP
NADP

ATP
NADH

Carbon dioxide
Water

Organization
Biology, Culture
Macromolecules
Anabolic products

ADP
NAD

ATP
NADH

ADP
NAD

Disorganization
Molecules
Catabolic products

HEAT ENERGY
PROBABILITY

Figure 34. In this schematic view, the metabolic processes of living organisms are intermediate stages in the conversion of solar radiation to heat, which increases entropy and so produces a more probable end state. ATP and ADP are the energy-storing molecules, and NAD and its cousins are enzymes that facilitate chemical reactions.

microscopic states. Ice crystals are very particular states of matter because the frozen water molecules have fixed orientations and can't go anywhere. When they melt, the water molecules disperse. The state of a water molecule is the same on average anywhere in the glass. When the toxic gas is concentrated in one small part of the room, it's a very particular arrangement. However, when the gas has fully mixed, all parts of the room are equivalent, equally contaminated. We see in this description the origin of our sense of the arrow of time. Ice cubes do not spontaneously reassemble, and farts do not magically reconverge on the offending emitter. Natural processes move toward the largest number of equivalent microscopic states—where everything is uniform, and everything is the same. Energy disperses. Entropy increases. Sigh.

Life, viewed in isolation from its surroundings, seems to show an amazing degree of order and organization. But neither cells nor complete organisms are closed systems—they take in matter and energy from the outside. In living tissue, the overall result of the metabolism is the increased amount of disordered energy, even though energy has been used to create new chemical structures. On the scale of the entire planet, the biosphere is an intermediary for the conversion of energy from its concentrated form in the sizzling Sun to its dispersal in the frigid depths of deep space (Fig. 34). The same will be true of life on planets around other stars as well.

LIFE STORES INFORMATION

It's only a small step to translate the idea of entropy to the idea of information. Life stores information—in the localization of its chemical ingredients, in the specific functions of the metabolism, and most obviously in the base-pair sequence of the DNA molecule. We tend to think of information as a fact or a piece of data, but there's a much more general way to define it.

If you have information, you know something. Looked at the other way around, gaining information means a loss of uncertainty. The more information you have, the less unsure you are. Sounds reasonable, doesn't it? Let's take the noxious gas from the earlier example and reduce it (thankfully) to a single molecule of hydrogen sulfide. If you know where the person who emitted it was standing, you know exactly where it started. You have all the information you need to stay well clear. But time goes by, entropy increases, and eventually that single molecule might be anywhere in the room. So the probability of it being in any particular part of the room is low, but it's the same everywhere. The story is the same for the entire fart. After time has passed and disorder has increased, there will be roughly equal numbers of hydrogen-sulfide molecules everywhere in the room. We can see that more information corresponds to reduced entropy or disorder.

How might you lower your uncertainty? Divide the room in half with a partition and ask if the molecule is on one side or the other. The answer to this question must be yes or no—one bit of information.[17] Then divide the half of the room with the molecule in half again and ask the question again. You have another bit of information, more loss of uncertainty. After following this procedure for a while, you can localize the molecule and express this as a certain amount of information.

Now we can pull these concepts together and see how they relate to life.[18] In his classic book *The Origins of Life*, Leslie Orgel said, "Living organisms are distinguished by their specified complexity." What did he mean by that? There are many things in the universe that aren't random, like planets and galaxies, but they are not specified. Life is not just complex; its function depends on information contained in the arrangement of matter. We can define information content as the minimum number of steps needed to specify a structure. If amino acids have combined randomly to make a polypeptide, we need to define only the proportions of amino acids within it. But to specify an enzyme, we must say which acid occupies each position in the enzyme's sequence. So the enzyme has higher information content.

Life harnesses preexisting matter and arranges and rearranges it. New structures and their specific arrangement correspond to an increase in information. But this is just a temporary diversion in the universal increase in entropy and dispersal of energy. Living creatures spend down the Sun's high-temperature solar radiation into lower-temperature heat radiation and store some of that energy in chemical bonds. An organism uses that energy to live, and then the energy in those same chemical bonds can be the fuel for a second organism, which eats the first one. Lunch isn't free, but it's tasty until the check comes due.

THE MECHANISM OF EVOLUTION

The attributes of life discussed so far omit a crucial element. Individual organisms utilize energy, organize matter, and store information, but they also reproduce. Something extra then comes into play: life evolves.

Evolution implies a deeper connection to the environment than just the use of it as a source of energy and a repository of heat. Life is sculpted by the environment. All organisms are different, even if their initial information content is identical. Biologists focus on the distinction between the genotype—the genetic information of a member of a species stored in its DNA—and the phenotype, a complete description of the appearance and physiology of the entire organism.

Suppose you had a bake-off party where everyone was given the same recipe for chocolate-chip cookies. The recipe is a genotype, an analogy for genetic instructions. If you gathered the cookies and laid them out on a table without any way to tell who had baked which one, you'd see obvious variations due to the different choices of time and temperature for baking. That's the phenotype, the influence of the environment, which includes the baker. It would also be hard, if not impossible, to group the cookies by the original baker. Even within the same batch, there will be variations because the effects of the environment can be both subtle and profound.

Knowing the exact DNA sequence of an organism does not allow us to predict success in the environment. If seedlings are cloned and placed in different soils and at different elevations, the outcomes are all different but not in a way that simply correlates with local conditions. Identical twins have identical DNA but different fingerprints and different susceptibilities to disease. Your immune system has been affected by a lifetime of chance exposures to different antigens. There are millions of VW Beetles out there, all built according to the same factory blueprint. But some are sitting in mint condition in garages, others have been driven so hard all the moving parts are worn out, and others have low mileage but have had their appearances altered by rust or accidents.

So it's simplistic to reduce life to the information content of DNA. Your three-billion-long base-pair sequence of DNA would fit on a DVD, but it doesn't say all there is to say about you. Each organism is a unique consequence of the interaction between the genome and the environment.

The environment imprints on an organism, and this affects its success, just as twins separated at birth may have very different paths through life. But if genetic material were unaltered from generation to generation, the range of outcomes would always be the same because there would be no way for individuals within a population to get bigger or faster or smarter or otherwise adapt to the environment. Evolution requires genetic variation.

Genetic variation occurs naturally during reproduction.[19] The mechanism for copying DNA within a cell is excellent but not perfect; a little information is "lost in translation." Small changes or mutations are also caused by external influences, such as radiation and chemicals. Mutation doesn't always affect reproductive success; most mutations are neutral and neither favor nor disfavor the organism. Evolution doesn't always move toward greater size or complexity. Some kinds of bacteria have endured for billions of years, and creatures can even get simpler during evolution, such as cave dwellers that lose the capability of vision.

Living creatures reproduce as slightly imperfect copies, and the variations are successful or unsuccessful based on their fitness in the environment. The accumulation of small changes can lead to substantial variations in function and form and, eventually, to new species. This is the mechanism of natural selection, which is the heart of Darwin's theory of evolution.

More than two thousand years ago, Aristotle wrote, "Nature proceeds little by little from things lifeless to animal life in such a way that it's impossible to determine the line of demarcation."[20] It is still infuriatingly difficult to define the simplest living organism, and since it probably arose in the first five hundred million years after the Earth formed we may never have direct evidence of its emergence. But scientists are now able to paint a very plausible picture of how it happened.

HOW DID LIFE START?

WHEN WE SPEAK OF Mary Shelley's gothic novel *Frankenstein*, Dr. Frankenstein is often confused with his monster. Dr. Frankenstein is a brilliant scientist who is obsessed with the nature of life. He painstakingly constructs a creature and breathes into it the spark of life. But the life-form turns malicious after suffering discrimination and rejection and turns on his creator. The confusion is natural—both Frankenstein and his creation are monsters. Both are also tragic figures; in Dr. Frankenstein's case, his good intentions are darkened by the hubris of stepping beyond the bounds of natural knowledge to forge a superhuman race and unlock the secrets of life itself. The story resonates deeply by drawing on our almost obsessive idea that science will give us the ultimate power to solve our problems while unleashing forces we can't control.

Jack Szostak, a molecular biologist, is no Dr. Frankenstein, and his creations are not monsters. With appointments at Harvard Medical School,

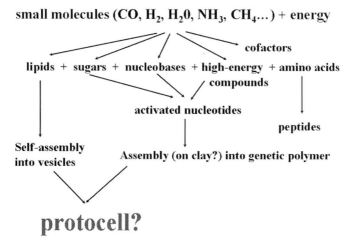

Chemical evolution....

small molecules (CO, H_2, H_2O, NH_3, CH_4...) + energy

cofactors

lipids + sugars + nucleobases + high-energy + amino acids

compounds

activated nucleotides

peptides

Self-assembly
into vesicles

Assembly (on clay?) into genetic polymer

protocell?

Figure 35. A possible path by which simple molecules might combine to make a primitive type of cell, with replication and genetic material coded in polymers that are a fraction of the size of the DNA in modern cells.

Howard Hughes Medical Institute, and Massachusetts General Hospital, he is widely recognized as a gifted researcher. Szostak is a mild-mannered man with a smooth, lineless face and a soft Canadian accent. His voice is thoughtful and measured, but he becomes animated when he talks about his research, which won him the Nobel Prize in Medicine in 2009.

Szostak is trying to reproduce in the lab events that occurred when the Earth was young. Without an exact model of how it happened originally and without any direct evidence to guide him, he's attempting to create a living cell from scratch (Fig. 35).

AN INTELLIGENT DESIGNER?

The daunting magnitude of the task becomes clear when we remember Fred Hoyle's comment about the junkyard and a tornado. Miller-Urey experiments take us only a tiny fraction of the way from simple ingredients to the genome of an advanced life-form.

To put it another way, how could random, unguided processes have generated something as complex as a cell? If a vast army of monkeys pounded away at typewriters, we would be surprised if they came up with a meaningful sentence. Yet arrangements of the four base-pair letters, ACTG, have somehow led to the intricate poetry of life.

Biblical literalists believe in a young Earth, meticulously traced back through the lineages of the begats in the Bible, and in the spontaneous creation of humans and all other species within seven days. Because science strongly supports the idea of an ancient Earth and a long time frame for the development of life, literalists must turn their backs on the entire edifice of science. Astronomy, biology, geology, and physics lose all traction if their interlocking web of evidence is denied.

A recent permutation of creationism acknowledges that the Earth is old and even accepts the mechanism of natural selection but argues that life is too complex to have arisen naturally without intervention from an "Intelligent Designer." Intelligent Design has roots two hundred years ago, when the English clergyman William Paley came up with the watchmaker analogy. If you found a watch in a field, he wrote, you'd infer that such a fine and intricate mechanism couldn't have been produced by unguided natural forces. It could have been made only by an intelligent being. The same reasoning, he said, can be applied to the level of molecules and cells. Surely the odds that tiny molecules would spontaneously combine into complex macromolecules are too low for life to have arisen by chance. The complex workings of a cell—interdependent components, none of which can work without the others—become evidence of an Intelligent Designer, essentially a euphemism for God.

By definition, Intelligent Design isn't a scientific theory: its central premise can't be confirmed by experiment, and it makes no new hypotheses to test. It attempts to poke holes in evolutionary theory, but the criticisms have been firmly rebutted. There is no place for it in a science classroom, where the ideas involve reproducible evidence and predictions that can be tested.

On the other hand, the amazement that underlies the theistic interpretations *does* deserve to be addressed. The "tornado in a junkyard" analogy for evolution is evocative, but our amazement ebbs when we see why the analogy is flawed.

In the junkyard, the 747—an intricate, highly ordered piece of machinery—is produced by random chance. But evolution is *not* pure chance. Evolution selects some assemblages over others, based on their fitness to the environment. A tornado slams parts together and rips them apart with no discrimination, so it totally fails to represent natural selection. Further, in the analogy, the tornado turns a pile of parts into a fully assembled airplane in one step. Evolution, by contrast, uses a huge number of incremental steps to go from something simple to something complex, with each step subject to natural selection. It's not so miraculous when we can "connect the dots" and see the intermediate steps on the path to complex organisms. Last, the goal in the analogy is to produce a jumbo jet. But evolution has no goal; it selects for what's useful now, not for what might become useful in the future.

In the sixth century B.C.E., the Chinese philosopher Lao-tzu said, "A journey of a thousand miles begins with a single step." We turn now to life's early steps.

FROM ATOMS TO MOLECULES

How did the primeval Earth march from simple atoms to complex molecules? Before the first cell existed, the process of synthesis was critical. Carbon is an atom that naturally and readily forms chains with itself and with other elements. At the level of the simplest ingredients, atoms found one another due to random collisions; but once the atoms stuck together, new collisions could continue the process. Molecules grow bit by bit.

As an analogy for small things bumping into one another and making something bigger, take all the hearts from a deck of cards. Start with the cards well mixed. Now suppose the molecule you want corresponds to all the hearts in order; how likely is that to occur by chance? Shuffle the cards repeatedly, and you may see the occasional run of three or even four cards in a row—but not seven or eight and never all thirteen. If it takes ten seconds to shuffle and inspect the cards, probability theory says you're unlikely to see all the cards in sequence until you've been shuffling for more than a year.

But now let's add the feature that atoms tend to stick together in particular ways. Take the same hearts and shuffle them until the two comes out on top. Set it aside and shuffle the rest until the three comes out on top. It attaches to the two, so set it aside. Keep going until the four is on top, and so on. How long does it take to get the full sequence? About eight minutes.[21] Building something bit by bit is far easier than throwing the bits together randomly.

"Life in a bottle" experiments don't ever make life, but they show that collisions and natural chemistry can easily go from molecules of a few atoms up to monomers of a few dozen. Monomers are simple units that can combine to form long, repeating chains called polymers. Monomers include amino acids, sugars, and fats or lipids—rings and small chains that are the basic building blocks of organic chemistry. However, they're too simple to carry out the high-level functions of a cell.

As another example of how easy it is to combine atoms to make the building blocks of life, more than 140 different molecules have been detected in interstellar space, where the reaction rates are millions of times lower than in a liquid like water. These include the amino acid glycine. In lab conditions that simulate interstellar space, four amino acids formed on the surface of dust grains, showing that the building blocks of proteins can occur in the deep freeze of space with no water present. Radio astronomers have also discovered sugar and alcohol in the vast gas clouds that form stars—they just need some citric acid to have the makings of a good margarita.

RNA IS SPECIAL

Strolling through foothills is easy, but soon the terrain is steep and treacherous. How did molecules ascend in complexity to the point where they could carry significant information and replicate?

All modern organisms carry their genetic information in nucleic acids—RNA and DNA. The instructions of the genetic code are specific sequences of nucleotides, which are the building blocks of nucleic acids. The code specifies the amino-acid sequences of all the proteins that an organism needs to live. But nucleotides are synthesized only with the help of proteins, and proteins are synthesized only according to the prescription of a nucleotide sequence. DNA embodies both information and action. It's a classic chicken-and-egg situation: these two complex systems work in concert, yet it seems very unlikely that they could have arisen independently and spontaneously.

The first glimpse of a way out of this paradox came in the 1960s, when Carl Woese, Francis Crick, and Leslie Orgel hypothesized a precursor biology called "the RNA World." This was a time when RNA could link amino acids into proteins and replicate without the help of those proteins. The speculation became plausible in the early 1980s when Nobel laureates Thomas Cech and Sidney Altman showed that RNA catalyzes many of the chemical reactions essential to life. Before this, enzyme activity had been known only in proteins. RNA is an active participant in the chemistry of life, not just a passive messenger. RNA enzymes are called ribozymes.

The extent of RNA activity didn't become clear until fairly recently because the functional units in the cell are fantastically tangled bundles of molecules, like a ball of yarn after a cat has played with it. One of the most important molecular "machines" in a cell is the ribosome. It decodes messenger RNA into synthesis proteins that do most of the heavy lifting in biology. The ribosome is a universal decoder. Think of a DVD player that can "translate" any DVD into a movie. In this analogy, the DVD is messenger RNA, the DVD player is the ribosome, and the movie is the protein product. Researchers were excited to find that RNA is the conductor of the protein symphony in the ribosome. RNA can even control the expression of genes, another function once thought to be exclusively the preserve of proteins. We've neatly sidestepped the chicken-and-egg problem—RNA embodies both information and action in a cell.

At some point in history, the RNA World was superseded by the duality of DNA and proteins. DNA's double helix is more stable and more suitable for the long-term storage of information. Although RNA continues to serve as a versatile helper in the modern cell, all organisms now use DNA for their genetic basis.

THE FIRST REPLICATOR

The biggest uncharted terrain now becomes the transition from a soup of simple molecules, particularly amino acids and nucleic acids, to the RNA World. As Leslie Orgel has said, "Anyone who thinks they know the solution to this problem is deluded." But, he added, "Anyone who thinks this is an insoluble problem is also deluded."

Jack Szostak is one of the pioneer cartographers for early life, and he knows the terrain as well as anyone. The problems are legion. Long strands of RNA tend to fall apart in water, and water even makes polymerization difficult; Szostak has called it a "horrible, corrosive, toxic substance." The best place to make proteins may have been shallow tide pools or warm environments where water could evaporate. RNA has a backbone of alternating sugar and phosphate units, with the bases A, C, U, and G attached. In lab experiments, it's difficult to form the C and U bases under plausible primitive Earth conditions, leading to speculation that the earliest life used a two-letter genetic code or even some other set of bases.[22] RNA was likely preceded by simpler replicating molecules that have left no trace in modern biochemistry.

The construction of very large molecules faces an extreme version of the card-shuffling problem mentioned earlier. A simple protein may have one hundred amino acids in a particular sequence, and there are 20^{100} or 10^{130} ways to combine twenty different amino acids, the vast majority of which have no potentially useful function. Similarly, the simplest gene has about 220 nucleotide bases, which can combine in 4^{220} or 10^{130} ways. These are prodigious numbers. If you imagine a grid of points with the spacing of protons in the atomic nucleus, it would take a grid the size of the universe to encompass all these possibilities!

However, as Szostak points out, there are trillions of places in the shallow oceans and ponds of the early Earth where such interactions could have occurred. Also, chemical reactions take a fraction of a second, so if primitive biology took as little as a few million years to develop there are many trillions of time steps for reactions. Even something very unlikely will occur somewhere.

Also, complex chemistry is not random. When chemicals have reactions in which one of the products catalyzes its own production, the system is called autocatalytic. Catalysts accelerate chemical reactions by factors of thousands without being consumed. (In living cells, the catalysts are a particular type of protein called an enzyme.) As a result, growth in complexity is rapid and non-linear.

Computer simulations show self-organization and rapid progress through the statistics of possible reactions. The physicist Freeman Dyson modeled these networks and showed that with ten different monomers as building blocks, it takes only ten billion steps and a short time to build polymers with ten thousand monomers. In terms of complexity, this is two-thirds of the way to a full-scale RNA molecule. Computers show only combinatory possibilities; they can't yet make realistic models of specific modes of catalysis.

Recent research on chemical networks shows that small organic molecules like amino acids can catalyze the formation of other small organic molecules like sugars and nucleic acids. The feedback loops in the network steadily build more complex reactions. Networks like this may be able to create all the active ingredients needed for RNA.

All of this takes place in the lab or a computer. Where in the real world did the first fragments of RNA or other early replicating molecules emerge? Clay surfaces are ideal because the lattice of microscopic particles in them forms a template to allow complex molecules to build one unit at a time and to replicate using the mineral layers. Amino acids on clays can be built to fifty-monomer lengths in two months, and fifty-monomer chunks of RNA can be built in only two weeks. Life may have been baked like crepes on a seashore rather than cooked in a watery soup.[23]

More extreme environments may also have played a role in the start of life. It sounds wild, but in the late 1980s Günther Wächtershäuser proposed a metabolic basis of life using fool's gold near hydrothermal vents. Pyrite and other combinations of iron, nickel, and sulfur could have catalyzed crucial biochemical reactions billions of years ago in superheated water near smoking volcanic chimneys on the seafloor. The temperature and chemical gradients near a vent greatly speed up reaction rates. Metallic mineral clusters still play a key role in modern bacteria.

Once RNA fragments exist, the path forward is clearer. In a modern cell, RNA has a specific and limited catalytic ability. But Jack Szostak and his group have expanded the envelope of life processes by creating new kinds of catalytic RNA or ribozymes in the lab. These ribozymes do many important tasks: join together two pieces of RNA, produce complementary strands, and copy their complements and duplicate themselves. All of the requirements for growth and replication are satisfied.

THE FIRST CELL

Life today does not appear as chemicals floating in a tide pool or as complex molecules clinging to a mineral surface. All forms of life, from the tiniest bacterium to the mightiest sequoia, are made of cells. Once a primitive cell is created, the path to the summit is clear. As microbiologist Lynn Margulis has said, "To go from a bacterium to people is less of a step than to go from a mixture of amino acids to a bacterium." So how did the first cells form?

As Jack Szostak notes, making a container for chemicals is no big deal. It's been known for decades that fatty acids, fairly abundant molecules on the prebiotic Earth, can naturally form tiny enclosures. They have one end that hates water and another end that loves water, so they self-organize into spheres— think of a drop of mayonnaise in water. This kind of membrane is critical for life because it isolates and concentrates the results of chemical reactions, while allowing new ingredients to flow in and waste to flow out. These primitive protocells are called vesicles (Fig. 36), and they can grow and divide in several interesting ways (Fig. 37).

Life's original container need not have been a lipid membrane. RNA falls apart in warm water, but researchers have found that ice contains tiny compartments that hold molecules in one place. Small RNA pieces trapped in freezing and thawing ice can self-assemble and grow. At the other temperature extreme are iron-sulfide chimneys on the seafloor. They are made of many cavities, and the microscopic chambers concentrate ingredients in the same way modern cells do, by allowing in small reactant molecules but trapping larger products. This would allow the formation of replicating molecules, and then the protoorganisms complete their evolution into cells by developing a flexible lipid membrane.

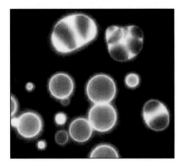

Figure 36. The progression from simple to complex proceeds too slowly without a container to concentrate chemical reactions. Fatty acids in solution can form vesicles that contain polymers yet are permeable. Vesicles grow and split, and RNA can react and grow within them.

THE BIRTH OF EVOLUTION

Jack Szostak has also shown how the mechanism of natural selection can emerge, even without cells. In one experiment, a new ribozyme with the ability to join two RNA chains was evolved through repeated selection and amplification. After eight successive rounds, the speed of the best catalyst had improved by a factor of three million. He thinks that self-replicating RNA could increase its number compared to the

Pathways for growth and division

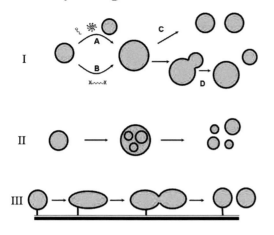

Figure 37. Vesicles or simple cell membranes can form easily from fatty acids in water solution. There are a number of ways they might replicate: by acquiring nutrients and growing until they split (I), by forming enclosures that grow within and then break out of a larger vesicle (II), or by stretching on a mineral substrate (III).

more inert forms around it. With minor mutations and successive rounds of copying, the original replicator would acquire new abilities. "Life starts simple," he says, "beginning with one gene, probably a replicase, and accretes additional functionality over time." A replicase is an enzyme that facilitates the replication of single-stranded RNA.

Szostak has also found that clay minerals again play an important role, because they accelerate the formation of vesicles, or primitive cells, by a factor of one hundred and bring RNA fragments into the vesicle. In a world of vesicles with different ingredients, only a small fraction will contain RNA sequences that can copy themselves. But vesicles containing those versions will swell and divide (see Fig. 37). Over time, they will come to dominate.

At this point, we have primitive precursors of the attributes of living organisms: containers that concentrate chemical activity, macromolecules that can carry information and replicate, and a mechanism for conveying selective advantage to the best replicators. The last is crucial because it means that Darwinian evolution could have operated in the dark ages before DNA and modern cells (Fig. 38).

The story just told is a patchwork quilt of lab experiments, computer simulations, and plausibility arguments based on physics and chemistry. The evidence to support it is indirect. Given planetary playgrounds for chemistry to act on, similar events could have occurred elsewhere in the universe. Was early life on Earth like a golem, formed in clay and mud, or was it like a robot built from metallic minerals? Was its first home Darwin's famous "warm pond," pockets of ice in a frozen wilderness, or fiery deep-sea vents? "These are big questions," Jack Szostak says. "Anybody who thinks has to be grabbed by these." Asked why he is attempting a project as difficult as creating a cell from simple ingredients, his measured scientific tone falls away to reveal the excitement of a teenager: "Because it would be *so* cool!"

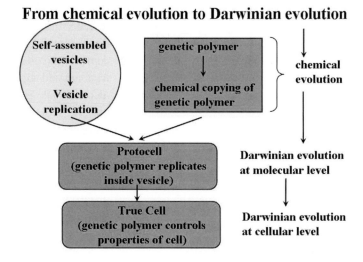

From chemical evolution to Darwinian evolution

Figure 38. The key transition in early life is replication with variation, where the success is determined by the environment, as in Darwinian evolution. Pure chemical evolution is not shaped by natural selection, but natural selection arose before the first true cell did.

ONE EARTH

FORTY YEARS AGO, James Lovelock was designing a NASA experiment to test for life on Mars. He realized that Mars and Venus, with atmospheres of carbon dioxide and not much else, were chemically dead. But Earth, with its oxygen and a relatively large amount of methane, was far from chemical equilibrium. Those gases are interactive, so they must be in constant circulation, and the pump for that circulation is life.

A realization came to him quite suddenly, while he was sitting in an office at the Jet Propulsion Lab in Pasadena. The Earth's atmosphere was unstable, yet it maintained a constant composition for long periods of time. What if life not only created atmospheric gases but also regulated them, keeping them at a level favorable for living organisms?

GAIA

With Lovelock's realization, the Gaia hypothesis was born. He has continued to develop the theory with microbiologist Lynn Margulis, popularizing it in several books. Gaia in its mildest form is uncontroversial, saying that organisms in the biosphere interact with the environment in a complex series of feedback loops. Gaia in its strongest form has made Lovelock a lightning rod for strident criticism, for it says that the Earth is a single giant organism. But the criticism is somewhat unfair. Lovelock never said that the self-regulation of the planet is purposeful or that the Earth as a whole has genetic material and is subject to natural selection.[24] Rather, he pointed out many ways in which living organisms participate in large-scale processes, with the result that conditions remain favorable for life.

We've looked at life's early stages from the perspective of microbial organisms, and all living creatures are obviously dwarfed by the planet they inhabit. But life isn't a layer painted onto the surface or a special extra ingredient. The Gaia hypothesis asks us to look at life as a vast interconnected system.

Biology has profoundly altered our planet. The biosphere extends from miles within the crust to the base of the stratosphere. Earth has been continuously inhabited for at least 3.5 billion years. Human civilization still depends on underground reservoirs of ancient fossilized creatures. The global energy usage of our technological society is twenty times smaller than the energy use of the microbial biomass.

AN INTRICATE DANCE

Over the Earth's history, geology, biology, and the atmosphere have been locked in an intricate dance. Carbonates have been steadily laid down on the ocean floor, and the emergence of photosynthesis caused the oxidation of the atmosphere and the oceans. The "pollution" of Earth's atmosphere by oxygen might have caused the destruction of many organisms, but other species were able to adapt and the survivors evolved to become plants and animals. Twenty percent of the molecules in the air were produced by the respiration of tiny microbes that evolved several billion years ago.

One example of this interplay is the constant temperature of the Earth, despite the Sun increasing its heating of the Earth by 25 percent since it formed. Another is the salinity of the oceans. The oceans maintain a salt concentration of 3.4 percent, which living creatures match. If the ocean salinity rose above 5 percent, it would be catastrophic for basic cell functions, such as preserving the membrane. Yet oceans are only at one-tenth of their salt saturation levels despite the continual introduction of new salt by weathering of rocks, and they have maintained this level through events that should abruptly change salinity, such as meteorite impacts and periods of glaciation. There's still no good model for how ocean salinity is regulated. However, it must be relevant that bacteria make up one-third of the ocean's biomass and 80 percent of its biologically active surface area due to their high surface-area-to-volume ratio. Plus, they pump salt.

The classic example of Gaia in action is the carbon cycle. Carbon dioxide is a trace component of the Earth's atmosphere, but it's biologically crucial as a reactant in photosynthesis. Under normal circumstances, life processes don't affect the net amount of the gas, but volcanoes emit lots of it. Left unchecked, the extra carbon dioxide would warm the atmosphere, as it's a greenhouse gas. One way carbon dioxide is removed from the atmosphere is by rock weathering, where rainwater and the gas combine with rock particles to make carbonate. Carbonate deposits sink to the ocean floor and eventually into the mantle, where some of the carbon dioxide that is trapped in limestone will be recycled into the atmosphere by volcanoes. Since soil bacteria are more active at high temperatures and assist in weathering, they act to cool the planet. The entire cycle is a massive feedback loop.

Back in the 1960s, Gaia combined with images of the Earth from Apollo 8 to give us a sense of the fragility of the ecosystem and the interdependence of all its parts. We're currently witnessing a rise in global temperature and carbon dioxide, with a strong likelihood that it's caused by human industrial activity. Perhaps there's a feedback loop that will nudge the planet back to cooler temperatures. But we may have shifted it so far from equilibrium that we ignore the consequences at our peril.

The evidence of the Earth is that life started quickly and radiated into an amazing variety of ecological niches. The chemistry that led to the first replicating molecule and the first cell is universal. Gaia is good news for astrobiology because it implies that life will have profound effects on its environment, making it easier to detect life on planets orbiting distant stars.

3.

EXTREME LIFE

> Bacteria on Earth can live five kilometers below the surface. They can live on nothing but rock and water, extracting energy from chemical reactions rather than sunlight. Life on Earth, and perhaps Mars and other planetary bodies, may have originated in such strange environs, and if so, the subsurface of water-rich planets, asteroids, and satellites might be home to a rich diversity of microorganisms.
>
> —**Jeffrey Taylor,** Hawaii Institute of Geophysics and Planetology

In the dream, you are in an ice cave. It is starkly beautiful, suffused in blue light from an outside source. There's nothing to eat, no sustenance, just the angular planes of ice crystals. It is stunningly cold, well below freezing. Your breath billows in front of you; perspiration forms a frozen rind on your neck. You can't stay here long. Then you notice creatures working industriously along the far wall of the cave. They're oblivious to the intense cold. From the strange smell, you guess that they have antifreeze running through their veins. This place is clearly their home.

Then you awake—not to your bed but to another strange world. You are on the shores of a river, with canyon walls that rise up and disappear in the gloom. The river is acrid and filled with the worst kind of industrial effluent. The water is so acidic that it sizzles as it passes over the rocks, which are themselves discolored by chemical residue. The smell is foul and metallic, and it almost makes you gag. As your eyes get used to the twilight, you see shadowy figures in the water. Amazingly, they are unperturbed by the toxic environment. Some of them are splashing and playing, some are drinking the water, and others are gathering lumps of metal from the sediment on the riverbank. The scene would be idyllic if it were not so bizarre.

You wake again, with a start. But you are still not in your room. You're encased in a metal shell, something like a submersible. A porthole in front of you is made of glass several inches thick; you sense the phenomenal pressure of water beyond. By your hand there's a switch. Flicking it illuminates a fantastic scene beyond the porthole: smoky fumaroles emerging from fissures where the magma glows dull red, as well as rocks crusted with colorful minerals and crystals. The water shimmers with intense heat, and you can feel it leaching into the submersible; this is another place you cannot stay long. Wonder and claustrophobia are warring within you. Then you notice graceful creatures gliding through the gloom. They are translucent in this place, where sunlight never reaches. They graze at the edge of the deep-sea vent, just yards from a seam that reaches down miles into the crust. You sense that they have lived here for eons.

You wake once more. This time it is to the familiar landscape of your bedroom. You marvel at the lucidity of the dream; the real world seems a bit disappointing by comparison. Another realization hits you. In your dream you had been miraculously shrunk to microscopic size. The tableaus you explored would pass unnoticed in the everyday world.

Then you awaken.

•   •   •

FOR BETTER AND OFTEN FOR WORSE, humans have left their marks all over the Earth. We're proud of our role as nature's generalist—not as swift as the gazelle or as strong as the gorilla or as agile as the goat, but pretty darn good at everything. (This pride should be slightly dimmed by the knowledge that most people would be unable to run down a wildebeest or escape from a leopard.) We take collective credit for the cars and planes and computers that our ingenuity has wrought. Alone among the species, we have gained dominion over the planet through technology. Humans are endlessly plucky and adaptable; it seems we can do anything.

Yet in truth, we're frail. From the safety of our living rooms, we admire the achievements of people who conquer Everest or cross great deserts. But without the assistance of technology, we couldn't live beyond Earth's temperate zones. We cannot survive for extended periods outside a forty-degree temperature window. We can survive underwater only as long as we can hold our breath.[1] Without water to drink, we'd die in three days. In any of the situations described in the opening vignette of this chapter—being subject to extreme cold, immersed in toxic waste, or near a deep-sea vent—we would swiftly and surely perish.

Microbes as a group, on the other hand, are hardy. Microbes are resourceful. With no other purpose than to reproduce, microorganisms on Earth have adapted to a stunning range of environments. When we talk about microbes, we're using a single term to span a dizzying array of microscopic organisms, the genetic and metabolic diversity of which is not yet well explored by scientists. There are microbes in all of the three main branches of the tree of life and millions of different species, so in a general sense Earth is a microbial planet. Their evolutionary success should make us blush at the hubris of thinking that the Earth was made for us.

Microbes are ubiquitous. There are more bacteria in your gut than there are humans who have ever lived. Think of them as doughty little superheroes, doing heroic jobs under nearly impossible conditions. (If it helps, you can imagine them clad in tiny masks and capes and boots, though in practice they are nearly spherical and unexceptional in form.) Their names are designed to convey their particular superpower, so they're not quite as catchy as the names

of comic-book superheroes. There's Thermophile, who emerges from the in-
ferno unscathed. There's Psychrophile, who shrugs off extreme cold. There's
Endolith, who does his best work encased in rock. There's Acidophile, who en-
ergizes by bathing in battery acid. And there's Barophile, who withstands pressure that would bring a lesser superhero to its knees.

These superheroes are the main characters in this chapter. Collectively,
they're called extremophiles, a term coined thirty years ago. Extremophiles can
be found at temperatures above the boiling point and below the freezing point
of water, in high salinity, and at pH values ranging from pure acid to pure base.
There are microbes that can live deep inside rock and others that go into a
freeze-dried "wait state" for tens of thousands of years, only to be reanimated
by water. In addition to familiar metabolisms such as forms of photosynthesis,
microbes possess metabolisms based on methane, sulfur, and even iron. Ancestors of modern extremophiles were among the earliest living things on Earth.

The term "extremophile" is unfortunately anthropocentric. These microbial
environments seem extreme only to us; to an extremophile, they're normal.
Microbes are linked individually and collectively to their environment, which
gives rise to the concept of a biosphere. Seeing the weirdness of life on Earth expands the definition of a biosphere and makes it more likely that there will be
habitable places elsewhere in the Solar System. Most cosmic environments are
inhospitable, but "inhospitable" is life's middle name because primitive organisms thrive in such an amazing range of physical conditions.

The sheer diversity of terrestrial organisms is important in framing the
prospects of biology beyond Earth. Moreover, the seemingly established statement that all life on Earth depends on the Sun might be wrong. The existence of
extremophiles on Earth makes it plausible that life at the microbial level exists in
the Martian permafrost or in the frigid oceans of Europa or in a half dozen other
harsh environments in the outer Solar System. Advanced forms of life like
mammals may be unusual, because they are able to thrive in such a narrow
range of physical conditions. Across the cosmos, planets covered by a web of extremophiles may be the norm. It's difficult to feel kinship with *Bacillus infernus,*
and even scientists must guard against anthropocentric thinking. We say that
life is hardy, that life is resourceful. Perhaps—on millions of planets beyond the
Solar System and with no deeper meaning than rocks or clouds—life just *is.*

THE TREE OF LIFE

FOR A CENTURY after Darwin published his theory of natural selection, biology
textbooks presented evolution in terms of a "tree of life." Historically, there
were two main branches to the tree of life: plants and animals. The tree had hu-

mans at the apex, lesser plants and animals as lower branches, and microbes at the root (Fig. 39). It conveyed not only our centrality in the scheme of creation but also a clear sense of inevitability in the progression from bacteria to us.

APPEARANCES CAN BE DECEIVING

It turns out that both aspects of the traditional metaphor of evolution are wrong. There wasn't a stately or inevitable progression from pond scum to people. Evolution is capricious, opportunistic, and contingent. As Stephen Jay Gould once noted, if one hundred identical Earths evolved from a primeval state under the warming rays of one hundred Suns, it's unlikely many of them would harbor mammals after 4.5 billion years, let alone *Homo sapiens*.[2] Recent evidence shows that life is a sprawling bush—not a neat tree. The entire animal kingdom, from ants to elephants, is concentrated in a single peripheral twig.

The study of the evolutionary history of life is called phylogeny. For centuries, scientists deduced evolutionary connections from the physical appearances of organisms. Unfortunately, a similarity of function and form is a crude and sometimes misleading way of measuring evolutionary connection. For example, sharks and dolphins share the superficial similarities of side fins, a dorsal fin, and a torpedo-shaped body. However, sharks are cold-blooded fish that have lived in the ocean for 450 million years—one of the most successful creatures on Earth. Dolphins, on the other hand, split off from sharks in the evolutionary tree and moved onto land three hundred million years ago. These vertebrates evolved into warm-blooded mammals and returned to the ocean about fifty million years ago. Dolphins are much more closely related to us than they are to sharks!

Why is appearance such a poor guide to evolutionary similarity? Similarity may only be skin-deep; often, the expression of one or two genes leads to a creature's shape, coloring, or surface pattern. Dolphins and sharks both evolved for fast swimming to catch prey. Their forms are sculpted by the environment rather than reflecting deeper genetic similarity. More important, creatures with external or internal skeletons have lived for only the last five hundred million years. Most organisms are microscopic prokaryotes and eukaryotes, and for most of the Earth's history they left no fossils and few traces of any kind.

Since we can't reconstruct the tree of life from the sizes and shapes of current species, we have to turn to the blueprint itself.

PHYLOGENETIC TREES

Modern phylogenetic trees are based on sequences of genes. Genes represent the smallest unit of the genetic code in an organism that expresses function

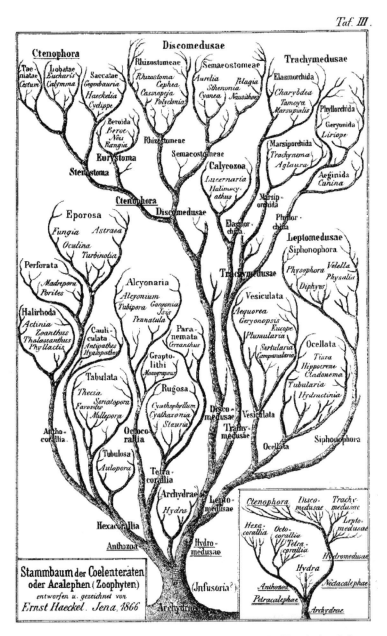

Figure 39. This tree of life was published not long after the death of Darwin, but similar versions were used in textbooks until about thirty years ago. Labeled in Latin, the hominid line is near the apex of the tree of evolution, with natural selection pruning away the branches of lower organisms. The metaphor conveys a sense of inevitability about our emergence after billions of years of evolution—a metaphor that has since become controversial.

and form. To trace as far back in time as possible, RNA is preferred over DNA, and the tool of choice is a small chunk of the RNA that comprises the ribosomes of all types of cell. Ribosomal RNA is the part of the machinery that translates genetic information into working parts. Expression of genes occurs in two steps. The first is the transcription of the information encoded in DNA into a molecule of RNA. The next is the translation of the information encoded in the nucleotides of RNA into a defined sequence of amino acids in a protein. This combined process is so important that it's referred to, with ironic self-awareness, as the "central dogma" of biology.

Mutations create variations in the base-pair sequence of DNA (which are then transcribed into RNA sequences). Variations accumulate over time. Every new species has a slightly different DNA sequence from its predecessor. Two species with similar DNA probably diverged recently in evolutionary history, while two species with very different DNA probably diverged a long time ago.[3] Our genetic code is a "living fossil" that can be used to trace our origins.

The method works because DNA and RNA proved to be so successful at storing and transmitting genetic information. Every form of life on Earth uses the same genetic code. Our common ancestor had a robust architecture. It has suffered damage from the elements over time, and different occupants have added layers of paint and porticos and even extensions, but the basic floor plan remains, and it's still visible.

A useful analogy is found in the study of languages. Spanish and Portuguese emerged in neighboring countries. The languages diverged hundreds of years ago but continue to share many common characteristics. Both of these languages are part of a larger group of Romance languages, which emerged from provincial dialects of Latin (called "vulgar Latin") fifteen hundred years ago. English is a hybrid language, in which vulgar Latin mixed with Norse dialects that evolved into modern German. Romance and Germanic languages share even earlier roots with countries across Asia in what is called the Indo-European group. These in turn emerged from a protolanguage that spanned much of Europe and Asia five thousand years ago. Reaching back before recorded history, we imagine that all human communication began with simple utterances—language fragments with a hint of grammar—tens of thousands of years ago.

Language trees suffer from some of the same problems as genetic trees. Just as most species that have ever lived are now extinct, most languages that have even been spoken are now dead. With migration and conquest, linguistic "DNA" has been shuffled, so forming a unique set of evolutionary linkages is difficult. Words called cognates can reach back across the centuries to a time of common language. For example, the English "star" is "estrella" in Spanish, "aster" in Greek, "stjerne" in Norwegian, "estêre" in Kurdish, "str" in Sanskrit,

Phylogenetic Tree of Life

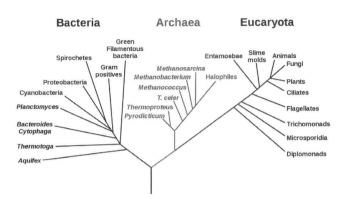

Figure 40. The modern tree of life based on the analysis of differences in the base-pair sequences of mitochondrial DNA, from research by Norman Pace and others. Distance along the lines of each of the three major families of organisms gives a sense of genetic diversity. The *Homo* line would be a tiny twig emerging from the minor branch called Animals, seen to the right along the lineage of the eukarya (or eucarya), but there are many simple organisms along the other two lineages that are equally displaced from the last common ancestor. The vertical arrangement of organisms corresponds roughly to increasing time.

and "setare" in Persian. This is a pleasing reminder that we were once all united in wonder as we looked at the night sky.

Figure 40 shows the modern tree of life based primarily on pioneering work by Carl Woese of the University of Illinois and Norman Pace of the University of Colorado. Evolution flows upward in this diagram. The distance between any two points represents the degree of genetic difference between organisms. Since genetic differences accumulate, distances in the tree can be converted into evolutionary times, but molecular "clocks" are not that reliable, so it's hard to label the tree with times. In this version, it's presumed that there's one organism and one genetic sequence from which all subsequent life descended, termed the last common ancestor.

There are three major domains of life: eukarya, bacteria, and archaea. Bacteria and archaea are entirely microbial (single-celled organisms), and eukaryotes are also mostly microbial protists and fungi but also include all plants and animals. Archaea are a primitive branch of microbes that was unknown before the 1970s. Major categories of organisms are twigs emerging from each of these three main branches. A species is too small a subdivision to be visible. Now imagine zooming in on the part of the tree labeled "Animals." You'd see a dense forest of fifty million tiny twiglets—everything from sponges and squids to worms and woodpeckers to bees and buffalo. One of the twiglets is us.

INTIMATE STRANGERS

The phylogenetic tree is a striking realignment of our view of life. We think of the Earth as being dominated by creatures like us, but in terms of biodiversity the lineage of mammals is a small tributary. Roughly 5,400 mammal species are outnumbered by 8,200 reptiles, 10,000 birds, and 29,000 fishes, while the sum of all 58,000 vertebrate species is dwarfed by 290,000 species of plants and 1,200,000 species of invertebrates, 950,000 of which are insects! (Three hundred thousand of those are beetles, which led the biologist J. B. S. Haldane to comment that God has "an inordinate fondness for beetles.") Counting species is not an exact science. Evolutionary biologists argue vehemently about taxonomy, and they tend to divide into "lumpers," who try to group organisms economically, and "splitters," who see differences wherever they look.[4]

The concept of species doesn't help us make sense of microbial diversity. If we could get over the fact that microbes are too small to have legs or wings or eyes, we would be amazed at their pervasiveness and their capabilities. Microbes probably make up 90 percent of the biomass of the planet. They're found in every environment humans have ever studied. They process and recycle not only carbon but also the other elements critical for life: nitrogen, sulfur, iron, and phosphorus. Their metabolic sophistication rivals that of any larger creature. And remember, they were doing all this three billion years before multicelled organisms arrived on the scene.

James Staley, a microbiologist at the University of Washington, calls microbes "intimate strangers." They inhabit our mouths and our skin and our guts, yet we know very little about them. Part of the reason is that very few can be cultured in the lab. No natural microbial community has ever been fully characterized, and many environments haven't been studied at all. The microbial universe is largely unexplored.

Given our overall state of ignorance, what's been learned? Figure 41 shows the forty kingdoms of bacteria known as of 1999. (A kingdom is a large grouping; animals fall in one kingdom.) Molecular techniques have been essential in this process, because few of these bacterial lines can be studied "in captivity." GenBank is the public database of DNA sequences funded by the National Institutes of Health. It holds full sequences for fifteen thousand bacterial species, but they come mostly from only four bacterial kingdoms; the archaea and viruses are largely unexplored. Recent work has taken the number of bacterial kingdoms up to one hundred, of which only fifteen have been cultured in the lab. One Norwegian team found five thousand different bacterial species in a gram of soil, and a ton may contain four million. The ocean is nearly as fecund. A teaspoon of seawater has more DNA than the human genome, and the oceans may support two million different kinds of bacteria.

Figure 41. A phylogenetic tree for bacteria, showing the forty major kingdoms known in 1999. This tree is unrooted, which means that the times of deviation from the ancestral line are not represented. The horizontal bar shows ten changes in the nucleotide sequence, which is the measure of evolutionary separation. Recent work takes the number of kingdoms to more than one hundred (for comparison, all animals on Earth form one kingdom).

THE UNITY OF LIFE

Emergence from a single genetic strain is evidence for a unified origin of life on Earth. We like to dwell on differences, but under the surface all organisms share the same organizational backbone of DNA.

Take humans, for example. The phylogenetic research that traces human origins back to a common ancestor in Africa also shows the degree of genetic variation in people from different races and cultures. Imagine comparing the DNA of two unrelated people living in the same village anywhere in the world. The typical genetic difference between them is five times greater than that between people plucked from any two places in the world. Attributes such as skin color and facial features are dictated by the expression of a handful of genes. Genetically, a typical American may have more in common with a Bantu herder or an Inuit fisherman than with their next-door neighbors.

There is 99 percent overlap between our DNA and that of monkeys and

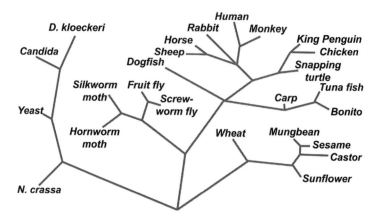

Figure 42. The commonality of genetic material as measured by deviations in the amino-acid sequence of the protein Cytochrome c. Compared to humans, there are accumulated differences of less than 1 percent for rhesus monkeys, 12 percent for dogs, 16 percent for chickens, 18 percent for rattlesnakes, 19 percent for tuna fish, 32 percent for moths, 39 percent for wheat, and 40 percent for yeast. These differences originate from the time of divergence of the species in the tree of life.

apes; we have 85 percent in common with dogs and two-thirds (67 percent) in common with moths (Fig. 42). Take an extra moment to glance at a banana the next time you eat one: you share half of your DNA with it. We also share half our DNA with simple yeast. And while we might not feel any kinship with unicellular fungi, the elegant spiral staircase within every cell bears testament to our common origins.

The new tree of life is exciting for the way it changes our perception of evolution, but it's still a work in progress. Tracing evolution back to the earliest branching points is difficult; even the simplest living organisms are a lot more complex than the last common ancestor. Consider this: the simple bacterium *E. coli* has 4,500 genes—only six times fewer than we do! Moreover, it's difficult to measure time accurately in the tree, since mutation rates were presumably much higher before cells developed the machinery to correct DNA transcription errors.[5]

Details of the branching pattern should not be taken too literally. Mutations are largely random, so differences between the same portions of genetic code for a group of organisms do not uniquely define their phylogenetic relationship. Making a phylogenetic tree is like using a construction toy with rods and connectors. The length of the rod gives the genetic "distance" between two organisms. There are many ways to hook up a large number of rods while preserving the same overall orientation. So the same phylogenetic data are used to infer different branching patterns. None of these caveats affects the broad conclusion that three major domains of life parted company about three billion years ago.

A GENETIC BAZAAR

However, there's another biological mechanism that does cast a shadow over the tree of life. Mutation is not the only mechanism that can alter genetic information. In the process of lateral gene transfer, organisms can leapfrog the slow and steady process of mutation and adaptation by grabbing a bundle of genetic material with a function that has been preprogrammed. Gene transfer is rare in eukarya but common in bacteria. It can occur between bacterial species by the action of viruses. For example, gene transfer is the way bacteria gain resistance to antibiotics.

Evidence that gene transfer complicates the tree of life comes from the fact that many eukaryotes use enzymes that are of bacterial origin, not archaeal, despite the fact that eukarya and archaea are closer in the phylogenetic tree.[6] By some estimates, the majority of the genetic material in microbes may have participated in the genetic bazaar of the early Earth, where symbiotic relationships were common, and cell functions were freely exchanged. In fact, in 2007 biologists found evidence for DNA from the bacterial parasite Wolbachia in 70 percent of invertebrates they studied—but in many cases it was the entire genome of the parasite that had been incorporated into the host.

If gene transfer was the dominant form of experimentation early in life's history, the tree of life may have to be redrawn (Fig. 43). The effect of gene swapping or genes jumping ship would be to make the overall organism less important. Organisms might not have persisted, and evolutionary relationships between them were probably very complex. The story of life would be a

Figure 43.
A schematic variation of the tree of life that shows the effect of lateral gene transfer. Bundles of DNA can move between different organisms, so the early branching points of the phylogenetic tree are very uncertain. The entire idea of a last common ancestor may not make sense. Most scientists now believe that life emerged chaotically after millions of years of biochemical experimentation.

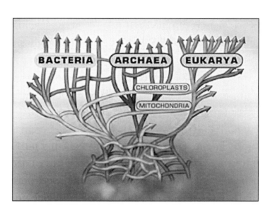

story of genes. Where does this leave the idea of a single origin to life? Rather than a single ancestor, there may have been a set of related organisms that contributed to the modern genome. Before that, there was a maze of chemical experiments and forms of life not here today. The tree analogy still works, but it's a banyan tree, with a huge and complex root system sitting under the spreading branches.

However messy its origin, nature tries to be efficient and tidy. Our genetic strain outcompeted all rivals and spread rapidly across the planet. Every living creature, from the smallest virus to a blue whale, uses DNA for its genetic code. All other forms of life became food for the victors.

OUR DISTANT ANCESTORS

WE'RE ALL DESCENDED from extremophiles. There are many ambiguities in the construction or interpretation of the tree of life, but researchers agree that organisms closest to the root operated under environmental conditions that would be intolerable to us. The earliest common ancestor of all life on Earth was probably thermophilic, or heat loving. When proteins from ancestral bacteria were resurrected for study in the lab, they performed best at a sizzling 150°F (66°C). The descendants of these organisms still live in hot springs like those in Yellowstone or close to deep-sea vents. When life started, Earth was nearly as inhospitable as Dante's version of hell. Magma, lightning, steam, and sulfur don't bring to mind Darwin's "warm little pond."

PROTISTS ARE NEAT

Asked if extremophile researchers tend to look like their objects of study, just as pet owners resemble their pets, Lynn Rothschild laughs warmly. She's a senior scientist at the NASA Ames Research Center in northern California. A biologist by training, she was sure from the age of eight that she wanted to study the wild and invisible microcosm of microorganisms. Every day she rubs shoulders with geologists, astronomers, and space scientists, and the interdisciplinary nature of the work is both a challenge and a source of excitement. True to her passion, Rothschild has written that "normal is passé; extreme is chic." In the popular culture, everyone's into extreme makeovers and extreme sports. Extreme life takes this idea to the limit: physical extremity as the essence of being.

Rothschild's voice is as enthusiastic as a third grader's when she says she works on protists because they're "neat!" Protists consist of a broad group of protozoa, algae, and some kinds of fungi. All eukaryotes—including us—are either protists or descendants of protists. They are the base of the food chain in the world's oceans and the source of 70 percent of the oxygen we breathe. They're the most abundant form of life in polar regions. They have a lot to tell us about the limits of life because they thrived billions of years ago, when all life was microscopic.

Three billion years ago, the Earth didn't have an ozone shield, and solar UV radiation poured through the atmosphere, causing DNA damage to any organism near the surface. Life evolved strategies to deal with the radiation, such as repairing DNA rapidly or making pigments such as melanin to absorb and distribute it. Some microbes even learned how to live in tiny enclaves inside rocks.

Rothschild's husband, Rocco Mancinelli, is also a biologist. He works at the nearby SETI Institute. His research focuses on halophiles, microbes that love salt; in fact, they can't survive unless their environment is at least 25 percent salt. There's a good-natured rivalry in play. He works on bacteria, but she thinks protists have more personality and that their flamboyant world is better than anything Steven Spielberg could dream up. It's easy to imagine dinner at Rothschild and Mancinelli's house as an adventure, caught between their conversation about food irradiation and heavy doses of salt.

THE FATHER OF EXTREME LIFE

The three hundred researchers at the annual International Conference on Extremophiles in Baltimore also look reassuringly normal. Karl Stetter gives the keynote address. A slim man in his mid-sixties, Stetter has a weathered face and wears a leather cowboy hat at a jaunty angle. His perfect English still carries the clipped tones that betray his German origins. He loves microbes of all kinds, whether they're in his beer or in deep-sea hydrothermal vents.

Stetter is the father of the field of extremophiles. Recently retired from the University of Regensburg, he spends every spare moment at volcanoes in Iceland or hot springs in Siberia. He knew he had found his calling when he realized that he loved the smells of sulfur and ammonia. Stetter has a lab that cultures extremophiles and supplies scientists around the world with samples for culturing. In 2002, he discovered a tiny creature in a hydrothermal vent off the coast of Iceland. Named *Nanoarchaeum equitans,* or "the ancient dwarf who rides the fireball" after its habit of hitchhiking on a larger heat-loving microbe (Fig. 44), it has the smallest genetic code of any known living thing.[7] A biotech company called Diversa has already locked up exclusive rights to potential commercial applications based on this tiny organism.

Extremophiles were central to the early development of life, but they

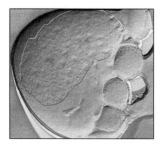

Figure 44. This electron micrograph shows the creature with the smallest genetic code known. Four cells of *Nanoarchaeum equitans,* each about half a micron across, are attached to a larger microbe, with which it has a parasitic relationship.
Nanoarchaeum equitans was found near a deep-sea vent. Similar organisms were among the first forms of life on Earth.

shouldn't be thought of as an evolutionary backwater. They occupy a substantial fraction of the real estate on the tree of life, and their descendants thrive today. Extremophile talents—like the ability to create antifreeze proteins—have evolved multiple times, implying a degree of evolutionary convergence or necessity. In other words, natural selection often comes up with similar solutions to the problems of physical duress. Extremophiles have shed light on protein folding, and they recently led to the discovery of the twenty-second amino acid, pyrrolysine. Their future, like their past, is full of exceptional promise.

Dozens of different metabolic designs branch from the base of the tree of life. The full range of microorganisms and their functions are still only vaguely understood, in part because although they can be isolated, few can be cultured in the lab. If global warming upsets the food chain in the oceans, or if land animals are threatened by a future meteor impact, we should be confident that microbes will soldier on, flying the flag for life on Earth, adapting and surviving as they have for four billion years.

CHAMPION EXTREMOPHILES

IT'S TIME TO MEET our tiny superheroes. You won't find them in the Earth's rolling grasslands or verdant forests. Some would even die at room temperature. In the "city" of life, they hold the jobs nobody else wants—toiling in the boiler room, rescuing people from nosebleed altitudes, roaming the perimeter fence in the depths of winter. The Greek root "phile" means "lover," but peering out from our frail bodies, it's difficult to imagine that extremophiles actually enjoy their work. Are they merely masochistic? Some can't continuously withstand a harsh environment; they go into a state of suspended animation until conditions improve. But others truly thrive while living at the edge.

Taken as a whole, extremophiles challenge fundamental assumptions about our notion of what life is and what normal is. Life on Earth began in extreme conditions, probably near high-temperature, toxic hydrothermal vents. We simply have a different sweet spot than extremophiles. In the spectrum of living organisms, it may be our fragility that's unusual. And while most extremophiles are microscopic, many are animals with physiologies not totally dissimilar from our own.

ALMOST INVULNERABLE

Meet *Bacillus infernus*, the "bacillus from hell," which withstands a combination of great heat, pressure, and acidity. This hardy creature was discovered by microbiologist David Boone in a deep drilling project in Virginia. *Bacillus infernus* lives several miles underground, where the pressure is hundreds of times greater than on the Earth's surface. That far below the surface, an organism is detached from the conventional biosphere. It exists independent of the Sun's rays, doesn't use photosynthesis, and doesn't consume the organic material from other formerly living organisms. Rather than breathing oxygen, it breathes iron and manganese dioxide.[8] Life is hard when you live in a rock, so *Bacillus infernus* only divides about once every thousand years.

Now meet *Deinococcus radiodurans*, "Conan the Bacterium," which can tolerate radiation thousands of times more intense than a dose that would kill a human (Fig. 45). This tough customer was originally found in a can of meat that had been sterilized by radiation but had spoiled nonetheless. The feisty *D. radiodurans* has the amazing ability to repair damage to its chromosomal DNA—usually within twenty-four hours. It does this by keeping five stacked copies of its genome ready to be transcribed. It also protects itself by forming a tough outer layer of lipids that can survive both the vacuum of space and punishing UV radiation. Microbes like this might have hitchhiked on meteoritic debris between the planets and moons of the Solar System and even beyond.

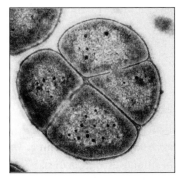

CUTE BUT TOUGH

Not all hyperadaptive organisms are microscopic. First identified by J. A. E. Goeze in 1773, the remarkable tardigrade is no bigger than the dot above this "i," but it's a multicellular creature with its own phylum (Fig. 46). Often called "water bears," tardigrades have five body segments, four pairs of clawed legs, and a single gonad. They also have a multilobed brain, digestive and nervous systems,

Figure 45. *Deinococcus radiodurans* is a remarkable organism with the ability to withstand vacuum and extreme doses of radiation. It can repair its DNA much faster than more complex organisms can, and it is an indicator of how robust life might be elsewhere in the universe, greatly expanding the number of potential habitats for life.

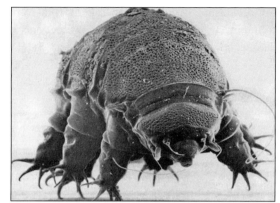

Figure 46. The hardy tardigrade, or "water bear." Tardigrades are about two hundred microns across and live in all climatic zones on Earth. They are light enough to travel on wind currents. More than one hundred species are known, but there may be hundreds more yet to be discovered. They have been found trapped in hundred-million-year-old amber, so they can be thought of as one of the Earth's oldest and most successful life-forms.

and separate sexes. More than 1000 distinct species of tardigrade have been discovered. Perhaps only a mother could love this somewhat intimidating arthropod, but they're worthy of our respect because of their extraordinary adaptability.

Tardigrades live in all climate zones from arctic to rainforest and can handle temperatures from −300 °F to 300 °F (−185 °C to 150 °C) and pressures from vacuum to one thousand atmospheres. This is the full range of temperature and pressure found anywhere on Earth, spanning the coldest winter day at the North Pole, the hottest summer day in Death Valley, the top of the highest mountain, and the deep ocean floor. And move over, Conan: tardigrades can also handle one thousand times the radiation dose that would kill you or me.

Tardigrades survive extreme conditions by going into a freeze-dried suspended animation called cryptobiosis.[9] The tardigrade forms a hard, waxy exterior called a tun, which renders it impervious to the elements. (Tuns should be standard issue for all superheroes.) The tardigrade survives this way for decades and then reanimates. Light, desiccated tuns disperse on the wind and are easily carried long distances by animals.

Technically, tardigrades aren't considered extremophiles because they don't naturally thrive in outrageous conditions; they're simply able to survive in them for a while. A tardigrade is always waiting for something better. Scientists are studying the mechanism of cryptobiosis because it may lead to strategies for humans to cheat death. Suspended animation is certainly a requirement if we're ever to travel to the stars.

REDEFINING NORMAL

If extremophiles are not just exceptions to the rule, then their range of adaptation mechanisms is telling us something important about life in the universe.

We still haven't hit the limit of the range of conditions in which life can be found on Earth; new creatures are found flourishing in *Guinness Book of World Records* habitats every year. The conditions that all large mammals, including humans, need to survive—moderate temperature, continuous access to liquid water, no intense radiation or extreme chemical conditions—are found only one place in the Solar System: Earth.

On the other hand, conditions hospitable to known microbes are found elsewhere. Low temperature and low moisture? That's like the surface or subsurface of Mars. High temperature and high pressure? Sounds like a description of the upper regions of Venus's atmosphere. Frigid water? The subsurface oceans of Jupiter's moons Europa and Callisto and the geysers of Saturn's moon Enceladus come to mind. Strange forms of biochemistry not strictly based on carbon? Saturn's moon Titan and Jupiter's moon Io might be relevant. We quickly reach a tally of half a dozen moons and planets, and that's based only on the *known* range of terrestrial extremophiles. Life elsewhere would form and adapt to be sculpted by local conditions, so it's hard to speculate how wide the entire range might be. This consideration increases the likely number of habitable worlds elsewhere by a substantial factor. Extremophiles redefine normal. How do they do it?

SOME LIKE IT HOT

Extremophiles were first discovered in the hot springs of Yellowstone National Park just over forty years ago. Yellowstone is still the prototypical site for the study of microbes with resistance to high temperature and acidity (Fig. 47). As the temperature of water rises, most life is challenged. At 104 °F (40 °C), rivaling a very hot summer day, oxygen doesn't dissolve well in water, so organisms like fish will die. Above 167 °F (75 °C), chlorophyll degrades, and photosynthesis is impossible. Above 212 °F (100 °C), the boiling point of water at sea level, cells shouldn't be able to control the flow of molecules entering and leaving. Worse, the thread of life should unravel. At this temperature, DNA and proteins denature, so they can no longer retain the complex shapes that dictate their functions. (If all the letters in this sentence turned rubbery and lost their shape, the sentence would lose its meaning.) Apparently, this doesn't always happen.

The champion thermophile, the enigmatically named Strain 121, can grow when it's as hot as 250 °F (121 °C) (Fig. 48). More controversially, extremophiles near deep-sea hydrothermal vents might even live in superheated water over 300 °F (150 °C). These organisms survive by reaching deep into the chemical toolbox for adaptive strategies. For example, they use only the sturdiest enzymes,

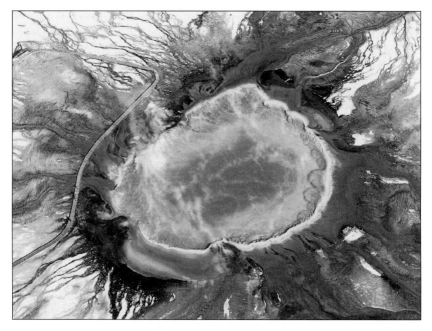

Figure 47. The Grand Prismatic Spring at Yellowstone Park gains much of its scientific interest from thermophilic (heat-loving) bacteria that live in and around the boiling water. Iron and hydrogen sulfide in the water power these microbial metabolisms. The scale is given by the road that crosses the image. One type of thermophile first recovered from Yellowstone is the basis for the multibillion-dollar DNA-copying industry.

ingest salts that shield the DNA double helix, and incorporate saturated fats to bolster their cell membranes. Heat tolerance is a legacy of the conditions on the early Earth. These deep-sea survivors could ride out the heavy bombardment that wreaked havoc with surface dwellers early in the Earth's history.

LIVING AT THE EDGE

At low temperatures, chemical reactions slow down, and life gets sluggish. When water freezes, it expands by 10 percent and causes cells to rupture. Yet there's a nematode that can survive all its water being frozen. Organisms ranging from microbial colonies to the tiny insect called a Himalayan midge remain active down to 0 °F (−18 °C). Some do it by insulating themselves from the external environment with hard lipids. Others take in salts that act like antifreeze. Many cell lines remain viable but inactive in liquid nitrogen, in its frigid basement at −321 °F (−196 °C). Biodiversity in the coldest parts of the world is amazing. At the base of the food chain below whales and penguins there's so much algae that the ice is often discolored and there's enough krill to easily outweigh all humans on the planet.

Water is thought to be essential for life, yet there are organisms that survive with surprisingly little of it. The Atacama Desert in northern Chile and the high valleys in Antarctica are among the most desolate places on Earth, so dry they're perfect labs for the scientists who test-drive equipment and experiments intended for Mars. A small variety of lichen and insects are found there, even though it may not rain for decades at a time. Making do with little is one thing, but there's another strategy when the desiccation is total. Rather than just giving up, organisms maintain a breezy optimism by going into hibernation. Some spores and cysts appear to be able to survive for a million years in suspended animation, and bacteria manage the same trick when they're trapped in salt crystals. To reanimate, just add water!

Pressure is also no problem at all for creatures that have adapted to it. Life is found on the deepest seafloor of the Marianas Trench, seven miles down, where the water pressure is 1,100 times the air pressure at sea level—such an environment would instantly turn a school bus into a crushed lump of yellow metal the size of an armchair. Single-celled organisms called foraminifera are found in these sediments. Their fossil record goes back more than 550 million years. They feed on sunken organic matter, but it's a truly challenging environment because high pressure is usually accompanied by extreme cold or heat, and there's no light for photosynthetic organisms to thrive on.

LOST WORLDS

Deep-sea life is central to thinking about extremophiles beyond Earth because its discovery was so unanticipated. For this, we can thank a stubby submersible called *Alvin*. *Alvin* is the size of a small bus and cruises at a plodding one knot, but its titanium hull can withstand the mountainous weight of an entire ocean. In 1977, *Alvin*'s crew discovered hydrothermal vents off the Galapagos Islands. The crew was amazed by a thriving ecosystem that lived without solar energy (Fig. 49). Three hundred species congregated near superheated water from the "black smoker" vent, ranging from bacteria and iridescent shrimp to giant clams and red-tipped worms that look like ten-foot lipsticks. Twenty times fewer people have visited this exotic world than have stood on top of Everest.

More recent discoveries had added the important information that heat from an active vent is not required for life in the deep sea. The "Lost City" is an extensive hydrothermal field on the Atlantic seafloor that has endured for at least thirty thousand years. Water there is warmed by chemical reactions, not

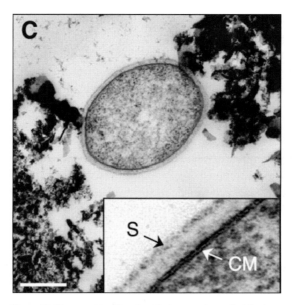

Figure 48. The record-breaking, heat-loving organism called Strain 121 respirates using iron the way aerobic animals like humans use oxygen. The white bar is a millionth of a meter, so Strain 121 is microscopic. Its cell has a single-layer envelope (S) and a cytoplasmic membrane (CM) controlling the flow of nutrients.

volcanic activity. At a dozen other sites around the world, including under the Antarctic ice shelf, ecosystems thrive without sunlight. At the frigid bottom of the Gulf of Mexico, scientists have found crabs, limpets, and bivalves. Colonies of pink worms burrow into mounds of methane ice and live off hydrocarbons that bubble through the crust.[10]

MEET THE TITANS OF TOXICITY

At this point, it will be no surprise to hear that microbes can handle all kinds of chemical extremes. If they could think, they'd think that humans are pathetically frail—then they'd roll their eyes, if they had eyes.

Take salt, for example. If you found yourself adrift at sea in a small boat with no supplies, you'd die of thirst long before you died of hunger. Tortured by the endless expanse of blue liquid, you might get desperate enough to drink seawater. It would be lethal. With a salt concentration of just 3.5 percent on average, seawater causes dehydration by increasing the osmotic pressure inside cells. Water is drawn out through cell membranes, and the DNA inside breaks down.

Figure 49. Black smokers are hydrothermal vents that are found near mid-ocean ridges where geothermal energy penetrates the thin crust. There, sulfur-bearing minerals dissolved in superheated water crystallize when they meet cold water near the ocean floor, building a chimney and precipitating to turn the water black. Entire ecosystems thrive near the chimneys.

But that's a scenario limited to human experience—champion salt-loving microbes make their homes in salt flats, inland seas, and briny pools on the seafloor. They use ions and simple sugars to protect their cell function. Green algae called *Dunaliella salina* can handle salinity ten times that of seawater, at which point the salt will actually precipitate out.

The pH scale describes the concentration of hydrogen ions in a solution. This is important for biological processes because hydrogen ions use their electrical charge to cajole, nudge, and generally facilitate chemical reactions. A substance with low pH is called an acid; a substance with high pH is called a base. The scale is logarithmic—a difference of one unit corresponds to a factor of ten difference in the concentration of hydrogen ions. The pH of pure water is 7, and our cells can't stray far from this without serious dysfunction.

The Iron Mountain mine is in the shadow of Mount Shasta in northern California. The site was first mined in the 1860s gold rush and has since become one of the Environmental Protection Agency's worst toxic sites. Exposed pyrite in the tailings contains sulfuric acid, and the acid runoff has in turn leached out heavy metals like arsenic, cadmium, and zinc. But *Ferroplasma acidarmanus* likes it there just fine. This microbe is an archaean, a descendant of Earth's earliest forms of life. It grows best at a pH of 1 and a temperature of 240 °F (115 °C).

To see how impressive this is, consider that the lower acidity of lime juice is able to unravel or "cook" the protein in fish, allowing for the prudent pleasures of ceviche (fish or other seafood marinated in citrus). Lime juice has a pH of 2, similar to vinegar. The hydrochloric acid secreted by our stomach linings has a pH of 1. But *F. acidarmanus* can handle a pH of zero, which corresponds to battery acid. It does this by using protons in a careful electrical balancing act that fends off the worst effects of acidity within the cell. Sulfides from the metal ore are converted into sulfuric acid, and its cell membranes contain tiny pumps to drive out the heavy metals. Ironically, this tiny superhero also works for the dark side; *F. acidarmanus* is a major cause of acid mine runoff in the United States.

The opposite extreme of a very high pH environment is also a challenge to cell chemistry. Extremophiles can be found at a pH of 9, corresponding to baking soda, and a few types can handle pH of 11, equivalent to ammonia. Both types of adaptation probably date back several billion years.

There are organisms that metabolize sulfur or iron or potassium or methane and others that breathe pure carbon dioxide or live in organic solvents. Visitors to high deserts will be familiar with desert varnish, a thin, colorful patina that coats sun-baked rocks. The patches of red and black are bacterial colonies. Unlike our cells, which gain energy from glucose, these bacteria grab trace amounts of manganese and iron from the air and get energy by creating oxides. Their surface layer includes cemented clay particles that protect them from desiccation, extreme heat, and solar radiation. They've essentially made their own adobe dwellings!

STRATEGIES FOR SURVIVAL

One microbe's toxin is another microbe's tonic. Every breath we take is laced with the oxygen that keeps us alive. Our very distant bacterial ancestors painstakingly developed this efficient aerobic metabolism, but oxygen is a volatile and reactive gas that binds to many of the cell's crucial components and enhances radiation damage to DNA. (Oxidation accelerates aging and cancer, and a multibillion-dollar sector of the vitamin industry exists solely to combat the fact that we are aerobic creatures.) Aerobic microbes exuded a gas that was toxic to many of the other life-forms on the planet. They had to either adapt to the newly flammable atmosphere or find new environments.

Mutation is a key ingredient of natural selection, but it's also a survival strategy. Without mutation, we might never have developed hands or eyes or a brain. But with too much mutation, a species is in constant flux and can't gain traction in its environment. So extremophiles have had to learn to adapt to radiation. In humans, even mild doses create reactive oxygen products that attack cell function and cause cataracts, sterility, and cancer. Rapid DNA repair allows *D. radiodurans* to withstand the radiation dose that would be felt one hundred yards from an atom bomb. In an environment of low temperature and low water content, fewer damaging free radicals are created, so radiation resistance probably developed as a by-product of adaptation to desiccation.

Survival often involves keeping a low profile. Bigger isn't better in the world of microbes. The surface area presented by DNA declines rapidly as the size of the genome decreases, so small microbes are more easily able to protect their DNA. Such robustness was crucial early in the Earth's history, when the atmosphere did not have enough ozone to be an effective shield, and it's relevant to astrobiology since many space environments have high radiation levels.

Another survival strategy is to find a stable environment. As implausible as it seems to live in a rock, at least it's pleasantly dull. Rock such as granite is not impenetrable; there are many tiny air pockets, cracks, and fissures. Water seeps in and brings nutrients. If food is really scarce, organisms can slow down their life cycle to reproduce once every hundred years—several million times slower than the half-hour division rate of bacteria in the lab. The luckiest microbes in a rock get a penthouse view from under surface layers of crystals in their own little solarium.

Stability is also found at great depths under the Earth's crust. Several kilometers down, the temperature is more than 200 °C, and atmospheric pressure is thousands of times greater than at sea level. Granite is rich in heavy radioactive elements, so microbes are forced to deal with the punishing radiation, too. But the real problem is finding something to eat so far from the surface food chain. There's evidence that microbes metabolize hydrogen seeping up from the interior. A thousand feet under the seafloor, an even more bizarre type of microbe eats volcanic glass and leaves behind tiny acid trails. Scientists think that the total deep-rock biomass may far exceed the mass of all other life on the Earth, and its metabolic diversity is essentially unexplored.

If life on Earth can exist without the benefit of sunlight, then we have to relax the definition of a habitable zone. Geothermal energy will be available on any large planet or any moon gravitationally flexed by the planet it orbits. If life on Earth can dwell in rock and use hydrogen as a nutrient, then the surfaces of many terrestrial planets are fair game. In the movie *Fantasy Worlds,* which has been touring planetariums since 2005, biochemist George Fox collaborated with animators to imagine an alien planet suitable for each type of extremophile, where they would thrive rather than just survive. Figure 50 shows examples of microbes that can grow at temperatures below the freezing point of water, and in places without any significant oxygen. The discovery of life on Mars might not raise the eyebrow of an extremophile on Earth.

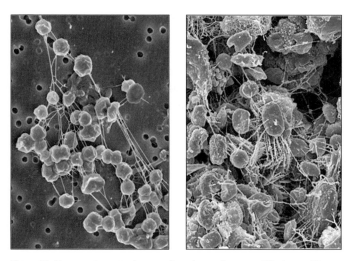

Figure 50. Electron microscope images of two forms of extremophile that could potentially survive under the surface of Mars. The left side shows one-celled organisms called methanogens, which get energy from hydrogen and carbon dioxide and turn their waste into methane. The right side shows one-celled organisms called halophiles, which flourish in very salty water and which can handle high doses of radiation. Both types of bacteria are found in the Antarctic.

EXTREME INDUSTRIES

LET'S RETURN TO the "intimate strangers" living inside your gut. Humans live in a symbiotic relationship with many kinds of bacteria, and, contrary to popular perception, most are beneficial to us. Of the million or so types of bacteria that are thought to exist, fewer than 1 percent have been fully described, in part because they are difficult to isolate and culture in the lab.

MINIATURE INDUSTRIALISTS

Extremophiles are just a subset of a large class of poorly understood microbes in the bacterial and archaean domains of life. Interest in them is not purely academic; the biotech industry has taken note of their remarkable abilities. As a consumer, you'll find them in laundry detergent, makeup, and diet pills. They put bounce in your bread, white in your paper, and color in your food.

The central use of extremophiles in biotechnology is in the creation of enzymes. An enzyme is a protein that catalyzes, or speeds up, a chemical reaction, often by a factor of thousands or even millions, while remaining unchanged itself. Enzymes are essential to life because most chemical reactions in a cell would occur too slowly without them.

Enzymes that work well under extreme conditions are at the heart of a wide range of industrial processes. Enzymes from extremophiles that tolerate heat are used to bleach paper, treat human waste, and control baking and brewing. Extremophiles that love the cold are used to make ice cream and artificial snow and to tenderize meat. Extremophiles that can handle salty environments are used to deliver drugs into the body and to modify food textures and flavors. Extremophiles that like alkali (high pH) conditions propel a multibillion-dollar annual market for laundry detergent; thanks to them, we're all wearing whiter whites. That's what the package means when it says the contents are "biological."

These hardy microbes also act as the cleanup crew when industry runs amok. They break up oil spills, scrub the sulfur out of coal and gas emissions, and neutralize nasty bleaches. If you have a really toxic site to clean up, naturally you call on Conan the Bacterium. *Deinococcus radiodurans* makes solvents like toluene and heavy metals like mercury harmless while at the same time shrugging off intense radiation. The U.S. Department of Energy is using Conan to make headway at three thousand Superfund sites around the country.

We finally have the set piece we've been waiting for: good against evil. On one side, *D. radiodurans*, Conan, the microbe that would laugh at kryptonite if it were real. On the other side, *F. acidarmanus*, let's call her Toxica, the microbe that emits effluent. We sit in our (tiny) ringside seats and watch the battle. Toxica attacks first, spitting acid and arsenic. Conan swallows the poison and quickly repairs the acid damage to his DNA. The battle continues. The ground shakes. All the action happens on a stage no larger than the head of a pin.

THE MIDAS TOUCH

Investing in companies that use extremophiles is one way to make money, but there's an even more direct approach: harvesting bacteria that make gold. When John Watterson of the U.S. Geological Survey panned for gold in Alaska, instead of melting down his yield for hard cash he looked at samples under an electron microscope. To his surprise, instead of solid nuggets of gold he saw a lacy pattern of tiny cylinders and rods. The cylinders looked just like *Pedomicrobium* bacteria. Normally, *Pedomicrobium* lives in water that's rich in dissolved minerals and builds layers of iron and manganese oxide around itself, like a shell. In this case, it used the most readily available metal: gold. Similar deposits of "gold-plated bacteria" have since been discovered in Venezuela, China, and South Africa.

The villain of the James Bond movie *Goldfinger* liked to asphyxiate his victims by painting them with gold. Normally, the outcome would be the same for bacteria. Gold would cause suffocation by stopping up the tiny holes in the cell walls through which food comes in and waste goes out. If that's the usual case, how does *Pedomicrobium* manage it? Instead of reproducing by splitting into two cells, it sends out a stalk above its gilded cage. A new cell then grows from the end of the stalk—imagine blowing a cell "bubble." New bacteria are deposited at the end of the colony, and as growth continues the result is an expanding sphere of golden sarcophagi.

Nobody knows how these bacteria distill the normally inert metal to twenty-four-carat purity. Unfortunately, there's no prospect here for a get-rich-quick scheme yet; it takes a year to grow a layer of gold the thickness of a human hair. Perhaps genetic engineering can speed this process up and turn some biotech entrepreneur into a modern-day Midas.

PHOTOCOPYING DNA

The most important use of extremophiles is to copy DNA. Bioengineering and related fields rely on the ability to quickly amplify tiny amounts of DNA—a technique called PCR, or polymerase chain reaction. Normal enzymes can make DNA only at low or moderate temperatures, when the DNA is coiled. Scientists needed an enzyme that was viable at a temperature where DNA unravels, or denatures, because only then is the master plan exposed.

The first step of PCR separates the twin strands of DNA by heating them to 194 °F (90 °C). Next, short sequences of nucleotides called primers get the process started by making a copy of the first few nucleotides. To do this, the solution must first be cooled to 131 °F (55 °C). Last, a type of enzyme called a polymerase is used to make a complete copy of the template DNA. The three steps all occur in the same vial, and the complete cycle takes about two minutes. But a temperature sufficient to unzip the DNA ladder breaks down normal polymerase, so it would have to be topped up after each cycle.

Enter *Thermus aquaticus.* These heat-loving bacteria were first discovered in Yellowstone National Park in 1969. They give their name to Taq polymerase, which works comfortably at 194 °F (90 °C). As long as the vial is stocked with primer and polymerase, the cycle can be repeated, and each newly synthesized piece of DNA will act as a new template. After thirty cycles, a billion copies of a single DNA strand may have been made.

This turbocharged process won scientist and surfer Kary Mullis a Nobel Prize in 1993. He thought up PCR while cruising in his Honda Civic to work at the Cetus Corporation in the northern San Francisco Bay area. Cetus gave him a bonus of ten thousand dollars for the idea and then sold the rights to LaRoche Pharmaceuticals for three hundred million dollars. Mullis, meanwhile, has always resisted working in the biotech industry or academia. He has a surfer's tan and bleached hair and lives across the street from a surfing spot made famous in Tom Wolfe's *The Pump House Gang*. This counterculture figure has spawned a rapidly growing billion-dollar-a-year industry.

TINY SUPERHEROES TO THE RESCUE

In the evolutionary arms race between men and microbes, the microbes might be winning. As new diseases emerge, infectious agents are rapidly developing resistance to our best antibiotics. According to the National Institutes of Health, two million people get bacterial infections in U.S. hospitals every year, and more than one hundred thousand die from the infection, a number that has gone up 700 percent in the past fifteen years. Most of the infections are resistant to the drugs used to treat them, and strains of meningitis, tuberculosis, and pneumonia are mutating to the point where they will be resistant to *all* current therapies.

The problem is genetic diversity. Nearly all our microbes, and so nearly all of our antibiotics, come from one branch of the tree of life. Almost all current antibiotics come from a single family of soil-based germs called *Actinomycetales*. Reliance on this one strain, combined with overprescription of antibiotics and their use in soaps and meat, has worked against us. The infectious agents easily swap DNA with other species and acquire resistance. Soon, we will be shooting blanks.

The pharmaceutical industry is barely able to stay one step ahead of these newly resistant organisms. A brute-force approach is used to sift through many thousands of ingredients, seeing how they work in combination and hoping for a few that are effective in battling human disease. Nearly half of all drugs on the market are derived by natural product screening, looking for beneficial activity in the vast number of molecules and compounds found in living organisms. This work has been extended to extremophiles, and several dozen new antibiotics are under development. By reaching across the tree of life to the archaean, the issue of resistance and genetic diversity can be addressed (for a while). With hope, the best benefits to us from life's wild frontier are still to come.

A CONSPIRACY OF GERMS

When we use the unusual capabilities of microbes to do our bidding, it's a triumph of modern technology. Our sense of accomplishment is dimmed by the fact that we're engaged in a winner-take-all evolutionary arms race with these tiny organisms, and so far the germs have fought the trillion-dollar-a-year biotech industry to a standoff. It's just as well the little buggers can't talk to one another, right?

Well, they can. For decades, scientists thought that bacteria were little more than single-minded opportunists, efficient machines for self-replication. The geneticist François Jacob has written that the sole ambition of a bacterium is to produce two bacteria. But how to explain the behavior of the Hawaiian bobtail squid? This squid lives in knee-deep coastal waters, and it hunts after dark. On moonlit nights, when its shadow on the sand would make it visible to predators, it turns on a light "organ" that emits a blue glow, which perfectly matches the light shining down through the water. The agent of this cleverness is a community of luminescent bacteria called *Vibrio fischeri,* which the squid takes in from seawater and regulates in a hollow chamber of its body.

Microbiologist Woody Hastings noticed that a population of *V. fischeri* in the lab doubled every twenty minutes, but the amount of the cell's light-producing enzyme stayed the same for four or five hours, spread among an increasing number of cells. Then, when the population increased vastly, it would begin to glow. From the perspective of a single *V. fischeri* cell, this makes perfect sense. Emitting light is very expensive for an organism metabolically, and a lone glowing cell could never be seen in the vast ocean. But how did the cell know when the community had reached a critical mass? Hastings's student Ken Nealson speculated that they secreted an unknown molecule he called an "autoinducer," which accumulates in the environment until it reaches a critical concentration. In effect, the bacteria can keep track of their numbers and vote, like a group of legislators, only when there are enough members present—it's called quorum sensing.

Quorum sensing is now a mainstream idea. Bacterial communication gives them distinct advantages in a battle with an apparently superior foe. Small numbers of bacteria can remain inactive and so avoid the immune-system response of a larger host, sending a signal to leap into action only after they have accumulated a greater number. This strategy of stealth and communication may explain how the bubonic plague spread and killed millions in the Middle Ages. Communication allows bacteria to act in concert, like a multicelled organism, thereby removing one of the major distinctions between bacteria and eukarya.

Microbial life is much richer than we ever might have imagined. It's intricately networked and highly social. With cell-to-cell networking, microbes can track changes in their environment, conspire with other members of their species, communicate with their hosts, and form mutually beneficial alliances with members of other species. This is all behavior once thought to be only in the domain of advanced organisms such as ants, bees, and people. We might cavil that this collective strategizing isn't accompanied by central processing, as in a brain. But in an astrobiology context, the richness of communal microbe behavior is fascinating. Quite possibly microbes on other worlds have evolved behaviors that match all our criteria for intelligence.

COSMIC HITCHHIKERS

THE HARDINESS OF EXTREMOPHILES is spurring new interest in panspermia, the idea that life was seeded on Earth from an extraterrestrial source. The idea originated with Anaxagoras more than two thousand years ago and was proposed in its modern form by the physicist Hermann von Helmholtz in the late nineteenth century. Panspermia doesn't address the issue of how life started on Earth; instead, sidestepping it by saying that it started somewhere else. Even if we are confident that life on Earth needed no external assistance, panspermia is a useful idea because it provides a mechanism for the spread of life or life's raw materials among planets and solar systems in the universe.

RIDING THE INTERPLANETARY SHUTTLE

The idea that rocks can move around our Solar System is uncontroversial. Three dozen Martian meteorites are known, and their origin has been confirmed beyond doubt. Apart from geological forms that are peculiar to the red planet, they contain tiny bubbles of trapped gas, the chemical composition of which exactly matches the atmosphere of Mars as measured by Viking and other landers. In fact, a ton of Martian rocks fall on Earth every year, though most of them get buried in sand or ice, disappear into the oceans, are too small to recognize, or lie unnoticed among terrestrial rocks.

The Allan Hills meteorite ALH 84001 sparked a raging debate and introduced the possibility that we might be descendants of Martians. This meteorite is an ancient Mars rock, dating back to soon after the formation of the Solar System. So if life formed early on Mars and was transported to Earth after an impact, that life could be the basis for all our genetic information (Fig. 51).

Even if the evidence for fossil life-forms in ALH 84001 is inconclusive, the meteorite got researchers to seriously consider the possibility that life might survive a perilous interplanetary journey. When any rock is ejected by an impact, it's subject to an acceleration of tens of thousands of g's, which would be fatal to us but is no problem for a microbe. Rocks that are near the impact site but not actually pulverized by the impactor get bounced into space. (Think of

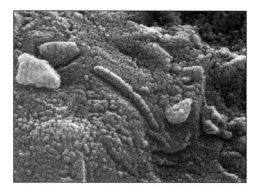

Figure 51. The electron micrograph of an interior section of ALH 84001, the notorious meteorite found in Antarctica after its long journey from Mars. Gases trapped in tiny pores in the meteorite exactly match the composition of the atmosphere sampled by Mars landers. Some have claimed the tiny elongated forms are microbial fossils, but this is hotly disputed.

sitting on a trampoline when a huge person lands near you.) In deep space, the dangers include vacuum, temperature extremes, cosmic rays, and ultraviolet radiation. But an inch of rock provides protection from all these hazards.

The final threat to life comes from reentry. When the surface of a meteorite melts as it travels through the atmosphere, magnetic crystals in the rock take up the orientation of the Earth's magnetic field. But deep inside the Allan Hills meteorite, the crystals had random orientations, showing that the interior was never hotter than a midsummer day. Rock is a poor conductor of heat; a minute of careening through the atmosphere is not enough time for the fiery surface to transmit its heat to the center. Any meteorite larger than a potato will keep its cool as it plunges to Earth.[11]

A THIMBLEFUL OF STARDUST

In the early spring of 2006, on a twentieth consecutive day of rain, Don Brownlee left his sturdy wooden houseboat in Seattle and traveled to the bone-dry Mohave Desert, where a small payload had just drifted to the ground under a canopy emblazoned with the NASA logo. Inside a sealed capsule was the first sample of the distant Solar System ever brought back to Earth. Using a gossamer-thin aerogel—think of smoke turned into a solid—held out like a tennis racket to the side of the spacecraft, Brownlee's Stardust mission captured material from the tail of Comet Wild-2. The probe survived a buffeting while it passed within 150 miles of the comet—jets of gas, dust particles, and chunks of rock fizzed from the nucleus. The aerogel gathered about fifty million particles, none larger than the head of a pin. This tiny sample—less than a gram—is stardust.

Comets contain pristine material left over from the formation of the planets, so we expect their composition to tell us about the ingredients in distant solar systems. Wild-2 spent 4.5 billion years in deep space until a chance encounter with Jupiter swung it close enough to the Earth to reach with a spacecraft. As Brownlee points out, comets are half water-ice and 10 to 20 percent organic material, two crucial components of life. They travel on looping orbits through the Solar System, spending most of their time hundreds of times farther from the Sun than Pluto. When they pass through the inner Solar System, a small fraction of them hit terrestrial planets. This is probably the mechanism by which Earth got some of its water and most of its organic material.

Stardust's comet rendezvous in early 2004 was followed by a more spectacular encounter on the Fourth of July 2005, when NASA's Deep Impact shot a meter-long copper bullet into Comet Tempel-1 as it flew by. The eight-hundred-pound projectile hit at 23,000 miles per hour, and the parent spacecraft made observations of the plume of material that was ejected. Sometimes rocket science really is kid's stuff. As Jet Propulsion Lab scientist Don Yeomans said, "I can't believe they're paying us to have this much fun!"

These missions have confirmed that comets are rich in organic materials. The Stardust mission found molecules with nitrogen-oxygen bonds, an essential part of the architecture of DNA, proteins, and enzymes. No claim of life in a comet is yet plausible because our techniques are still too crude to look for replicating molecules or evidence of metabolism.[12] Nonetheless, comets are perfect for giving life a jump start in our Solar System or others.

LIVING AMONG THE STARS

Moving material from one terrestrial planet to another in one solar system is one thing, but the distance to the nearest stars is hundreds of thousands of times greater than the distance to Mars. Most debris that travels among the planets never leaves the Solar System. The small fraction that does typically waits tens of millions of years to be ejected into interplanetary space. For example, ALH 84001 traveled a long, meandering journey to the Earth, taking sixteen million years to get here.

However, the speed required for ejection is such that a rock may take only about one hundred thousand years to reach a nearby star once it has been ejected. Surprisingly, the transit between solar systems is the quickest part of the process.

Though it remains questionable whether life can survive long enough to make the trip, scientists are cautiously optimistic that it could. Russian scientists have been able to revive and culture bacteria, yeast, fungi, and other microbes from Siberian permafrost that's more than two hundred thousand years old. One- or two-million-year-old bacteria have been revived from deep within the arctic ice pack.[13] More exciting, but subject to some controversy in the research community, Raul Cano at Cal State Polytechnic extracted spores from the guts of bees that had been entombed in amber for twenty million years. He added nutrients and within weeks was able to grow *Bacillus sphericus* bacteria from the ancient spores. Lab tests confirmed that the bacteria were not modern contaminants. More controversial evidence supports the reanimation of 250-million-year-old bacteria within a salt crystal from a mine in New Mexico. Even setting aside this last claim, microbes seem hardy enough to make the long trip to a nearby star.

Unfortunately, showing that something *could* happen isn't the same as showing that it *has* happened. What fraction of rocks kicked up by an impact contain life? How many life-harboring rocks does a planet eject? How many of those are captured by another star system? What fraction land on a potentially habitable planet? Without answers to these questions, panspermia is pure speculation.

Several researchers have attempted the calculation, with intriguing results. Jay Melosh at the University of Arizona has estimated that about fifteen rocks per year can be ejected from the surface of terrestrial planets due to impacts. This adds up to sixty billion over the time that life has existed on Earth. However, the numbers that are life bearing will be much smaller. In his model, the rocks spread out through interstellar space, and a small fraction is captured by other star systems. Of these, another tiny fraction actually lands on a terrestrial planet in a different system.

Overall, panspermia is a very inefficient process. If life arises spontaneously on the surface of a terrestrial planet, there is no need for panspermia to explain its presence, even if the mechanism does operate. According to the models of Melosh, the Earth has probably seeded one other stellar system with life over its history, although we cannot know if it landed on a planet hospitable to life. In star groups or clusters and in binary star systems, the situation is much more favorable. Where star densities are high, life-transfer events might take only a few million years. We should expect long-lived star clusters to be richly cross-pollinated with life.

HOW STRANGE CAN LIFE BE?

WHEN NEWTON FORMULATED his law of gravity, it was radical and bold. He based it on the orbits of planets and moons in the Solar System, but in calling it a "universal" law he used the inductive reach of the scientific method to apply it far beyond familiar shores. Its first big prediction—the return of the comet that would bear Edmund Halley's name—was not confirmed until 1758, long after Newton's and Halley's deaths. Newton's law has since been applied to black holes and binary pulsars and galaxies, which are types of objects that were unknown to Newton. His theory has proved to have brilliant explanatory power, which is why it's referred to as a law of nature, but it couldn't have been used to predict the existence of these objects.

Biology has no equivalent of a law of gravity. The chemical ingredients of life are universal because they're made in stars, but the biological mechanisms inside a cell were developed in the specific environment of the Earth. With one example of life to study, we've no idea if life elsewhere is just like terrestrial life, utterly different, or doesn't occur at all. However, we can gingerly apply induction and address the potential for life elsewhere. We're really asking this question: how strange can life be?

CHANCE AND NECESSITY

There's a concept in natural philosophy called the argument from design. It's attractive but logically dangerous. As stated by noted twentieth-century philosopher Bertrand Russell in a 1927 article titled "Why I Am Not a Christian," "everything in the world was made just so that we can live in the world, and if the world were slightly different, we could not live in it." This argument is used by theists to argue for an intelligent creator—essentially stating that humans are too special to have arisen as a random outcome from a history of mutation and evolution. The argument from design was first critiqued, devastatingly, by the philosopher David Hume in 1779 and subsequently by other academics such as Russell and Jacques Monod.

The argument from design is made irrelevant by Darwin's theory of natural selection. Modern biology and paleontology describe convincingly how complexity grows with time and how the environment shapes function and form.[14] Evolution isn't perfect, and the environment is in a continual state of flux, so species show important telltales like vestigial organs and relics, but the mechanism of natural selection is seen operating on life in all its guises.

The first notion to rebut is the idea that life is so improbable that it must be a fluke. A New Jersey election commissioner named Nick Caputo was charged with fraud because in forty of the forty-one elections that he oversaw, Democrats appeared at the top of the ballot. The odds of this occurring in the random drawings that Caputo claimed to have used were one in fifty billion. However, the New Jersey Supreme Court refused to convict the "man with the golden arm" because even very unlikely events can occur by chance. (They did, however, force him to change his method of drawing names, arguing wryly that they wanted to avoid "further loss of public confidence in the electoral process.") Clearly, this is an example of design, not chance; Caputo was a Democrat, so he used a method that favored his party.

To put it another way, these two sequences of flipping a coin have exactly the same probability of occurrence: HHTTHTTHHTHT and TTTTTTTTTTTT. But you'd be right in guessing that the first arose by chance while the second resulted from me leaning on the "T" on my keyboard. As we saw in the last chapter, life on Earth arose out of a huge range of combinatorial possibilities, but it rapidly converged on narrower chemical and biological pathways. According to Monod, if a phenomenon occurs with low probability and also conforms to a prespecified pattern, the interpretation could be Intelligent Design (or a form of human intervention, as in the Caputo example) or necessity. The crucial idea in modern biology is that evolution results from the random process of mutation combined with the deterministic process of natural selection.

This isn't the same as saying that advanced forms of life *have* to be the way we see them now. As we have seen, Stephen Jay Gould argued strongly that contingency affects evolution. His speculation leads us to wonder about the role of chance or contingency in biology itself.

Rather than claim serendipity, it's more accurate to say that life on Earth has been very selective. It depends on only two dozen of the elements in the periodic table, works with only one of the two possible orientations ("handedness") of building-block molecules, and uses a single type of molecule to code genetic information. It employs only twenty amino acids from among thousands available and ten thousand proteins from among an essentially infinite number that are possible. Are these selections inevitable? Could they have been made differently? How do we begin to examine the sufficient and necessary conditions for life?

LIFE 1.1

One way to answer these questions is through the study of synthetic biology. Steve Benner, a biochemist at the University of Florida, has been thinking about "weird" life for fifteen years. He exudes boyish charm, has a shock of thick hair tinged with gray, and speaks with a slight southern lilt. In 1988, as a young researcher, he organized a conference in Switzerland called "Redesigning Life." Senior scientists objected, convinced that the title would lead to riots over scientists tampering with nature.

The storm abated, and the conference proceeded; currently, synthetic biology is a rapidly advancing field. Benner has made a career of asking deep questions about the nature of life on Earth. He wants to know what aspects of biology are optimal solutions within the constraints of physics and chemistry. He wants to know if our biochemistry contains any relics of experiments much earlier in the history of life. And he wants to know which of life's features are accidents, where the initial conditions might easily have led to a different outcome.

Scientists such as Benner have developed a growing toolbox that allows them to reengineer microbes. The year after the Swiss conference, he persuaded cellular enzymes to accept an unnatural base pair into their DNA. More recently, Peter Schultz at the Scripps Institute created a molecule called 3-fluorobenzene, which forms a base pair with itself rather than a

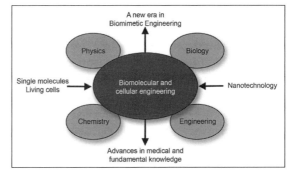

Figure 52. The potential of biomimetic engineering—imitating, copying, and learning from biological systems—has just begun to be explored. This research spans many disciplines, from fundamental physics to robotics. For astrobiology, we hope to learn about the degree of historical contingency in the evolution of life on Earth.

partner. Snuck like a Trojan rung into the ladder of DNA, it was readily replicated by polymerases in the cell. These experiments add eight new "letters" to the alphabet of life. For the first time in four billion years, the syntax of life, which consists of A-T and C-G pairings, has new linguistic possibilities.[15]

Tinkering with mechanisms inside a cell is a wide-open project with unknown and untamed possibilities. Schultz has figured out how to add nearly one hundred unconventional amino acids to the proteins in bacteria. Proteins are the workhorses inside a cell, and new proteins will have different functions. A protein is expressed when an enzyme reads the DNA base sequence and transcribes it into RNA. Protein specificity comes from the fact that the transfer RNA recognizes codons (the sixty-four possible sequences of three base pairs), and the codons map to specific amino acids. Each time Schultz inserted a new amino acid, the protein that was expressed behaved differently.[16]

This research has enormous practical importance (Fig. 52). Bacteria can be tweaked to sniff out explosives or neutralize nerve gas. They can be modified to make insulin or the antimalaria drug artemisinin, so rare in nature that it's very expensive to produce. As cells turn into tiny drug factories, it may become possible to treat diseases by fixing defective cell functions or promoting the growth of cells that attack intruders.

LIFE 2.0

Tweaking fundamental biochemistry that has been in place for four billion years is radical. An even more audacious approach involves inventing entirely novel tools for the toolbox, or building life from the ground up. In 2000, scientists inserted two devices into the simple bacterium E. *coli,* which lives in the human gut. A group at Princeton put together three interacting genes in a way that made the bacteria emit light and blink regularly, like Christmas-tree lights. Meanwhile, a group in Boston set up two genes to interfere with each other's function. In doing so, they created the equivalent of a toggle switch and endowed E. *coli* with a rudimentary digital memory. Essentially, they controlled a set of on-off states by biochemical means instead of with a semiconductor, in which the same thing is done with a flow of electrons. It took years to develop these tricks, but MIT biologist Drew Endy foresees a time when man-made biological mechanisms may far outnumber the products of eons of natural selection.

Endy is the inventor of BioBricks. They don't look that impressive—the dozens of vials on Endy's desk seem to contain only clear, viscous liquid—but they represent a revolution in the making. Each BioBrick is a chunk of DNA that, when inserted into a cell, causes a protein to do something useful. They're standardized so that they send and receive the same biochemical signals and interact well with one another. One BioBrick sends a high signal when its input signal is low and vice versa. In other words, this is a *not* operator. Another emits a signal only when it receives signals on both inputs: the logical *and* function. With enough *not* and *and* functions, it's possible to do any computation.

Endy and his colleague Thomas Knight have created a registry of over 3000 different BioBricks, and they make them freely available to other researchers. Making components for "squishy" computers that work millions of times slower than the silicon kind doesn't seem very promising, but the eventual goal is to engineer functions into living organisms. The researchers encounter the problem of persistence. Their tiny devices have to work in the busy and messy world of a cell, not in a sterile vial on a lab bench. In the world of the cell, they tend to mutate and break.[17]

How far can we take the idea of building life? Eckard Wimmer stunned the world in 2002 when he announced that he and his team had built a live poliovirus from mail-order segments of DNA and a genome map that's freely available on the Internet. The implications for bioterrorism are obvious—what if someone could synthesize Ebola, smallpox, or anthrax? Even worse, what if synthesized germs could be endowed with resistance to antibiotics? The rate of progress is dizzying. More recently, genome icon Craig Venter put together a virus that infects bacteria. It took him only three weeks to do what had taken Wimmer three years. In 2010, his team synthesized an entire bacterial genome from scratch.[18]

Bacteria are much more complex than viruses. The simplest bacteria have just over five hundred genes; their DNA sequences have been mapped, even though all the functions aren't understood (Fig. 53). It's simply a matter of time before we see bacteria synthesized in the lab.

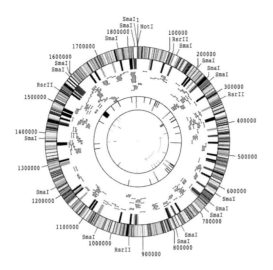

Figure 53. In 1995, *Haemophilus influenzae* was the first organism to have its genome completely sequenced. The numbers indicate base pairs; each little bar is a gene, though not all their functions are known. The map is drawn as a circle to represent the actual shape of the genome, which is made of a single strand of DNA fused together at the ends.

EXPLORING THE BIOLOGICAL LANDSCAPE

This experimentation casts light on the original question: how strange could life be? Though there are many definitions of life, let's say for the moment that life is a self-sustaining chemical system capable of adaptive evolution. Researchers don't all agree on the universality of our biochemistry. Norman Pace at the University of Colorado, who has done pioneering work on phylogeny, believes that a core set of biochemical processes will power life in all cosmic settings. Steve Benner also knows that cells are exquisitely designed biochemical machines. But he thinks the rich possibilities of chemistry allow for widely differing possibilities for the function and form of living organisms.

Let's also accept the primacy of carbon and water in building life: two cosmically abundant ingredients with special properties that facilitate interactions and complexity. Life on Earth is constructed from two types of biopolymers: nucleic acids and proteins. They represent yin and yang, or information and action. However, other genetic codes are also feasible—perhaps they would use an expanded lexicon of amino acids. It's also easy to imagine life using a single biopolymer—instead of our double combination—that can both replicate and evolve, since this probably happened on early Earth. At the level of metabolism, Earth life probably doesn't exploit the full range of potential ways to extract energy from an environment. A cell is an efficient way to concentrate chemical reactions, but life might exist without any "container." In essence, the variations of hypothetical biology are mostly limited by the imagination.

Benner describes the situation in terms of a landscape. Our biology is like a pleasant valley that supports a rich biota. We can see how life developed in this valley from the simpler and hardier organisms that live on the high plateaus and rocky peaks. But how do we know this is the best or the only valley? There may be places beyond the horizon that are even more verdant or "lost worlds" with unfamiliar creatures. Similarly, our biology may be one of many possible "solutions" to the evolution of complexity. In different physical settings, the other solutions may be preferred. Given the limitations of lab biochemistry, the answer will come only from astrobiology. Countless realizations of life may already exist in deep space.

ARTIFICIAL LIFE

VIEWED FROM ABOVE, the creatures display a rudimentary intelligence. They eat, move, reproduce, and fight. They have a primitive form of vision. Food resources are finite, so evolution is shaped by the hand of natural selection. Closer inspection shows that each creature's brain operates as a neural net, like ours. Genetic code determines brain architecture: the number, size, and connectivity of neural clusters. The creatures learn during their short lifetimes, and, since their physiology is encoded genetically, all their characteristics evolve over multiple generations. They're fascinating to watch. Given time, different species emerge. They display a surprising array of individual and group survival strategies, including swarming, foraging, and attack avoidance.

LIFE IN A COMPUTER

Polyworld isn't real; it exists only in a computer. This digital ecology was invented by Larry Yaeger, professor of informatics at Indiana University and former chief scientific officer at Apple Computer. In Polyworld, the user is God. He or she can alter the genetic code, modify the environment, or switch sex on and off. Hundreds of generations can be run in a single session. Evolution unfolds in real time. Polyworld is one of a fascinating array of simulations that fall under the rubric of artificial life. Many of these simulations look like computer games, but they're far from trivial. They're as complex as life itself.

Larry Yaeger's creatures aren't much to look at: simple colored polygons that wouldn't retain the interest of a typical computer gamer. But clever algorithms and raw computer power give them interesting behaviors. They develop new behaviors over their short lifetimes and characteristics that evolve over multiple generations. It's fascinating to watch as they crawl over their digital landscape and play out miniature set pieces of life and death.

The idea of artificial life dates back to the middle of the twentieth century, with the computing pioneers Alan Turing and John von Neumann. Back then, the state of the art in computers was a behemoth called ENIAC. It was the size of a small house, weighing thirty tons, consuming two hundred kilowatts, and needing three full-time technicians to swap burned-out valves and resistors. But Turing and von Neumann imagined a time when computing would be cheap, easy, and portable. Turing was a troubled genius who killed himself at the age of forty-one. He invented the idea of an algorithm—a logical step-by-step procedure for solving a math problem in a finite number of operations—and the Turing test: in which a computer is judged "intelligent" on the basis of its ability to mimic human responses. Von Neumann was a giant of mathematics and science who invented the processing architecture that's still used in most modern computers.

Starting in the 1970s, researchers in the new field of computer science explored another von Neumann creation: cellular automata. Imagine a single square that's colored black or white, with each color representing on or off. Simple rules describe whether neighboring squares or cells turn black or white: any live (or black) cell with fewer than two neighbors dies (turns white) of loneliness; any live cell with three or more neighbors dies of crowding; any dead cell with three neighbors comes to life; and any live cell with two or three neighbors lives on unchanged.

It sounds too primitive to be interesting, yet John Conway at Cambridge University found surprising depth in this black-and-white, pixellated world (Fig. 54). Conway took the essential idea of cellular automata, most easily displayed as a one-dimensional horizontal sequence in which evolution with time is shown vertically, and he animated it in two di-

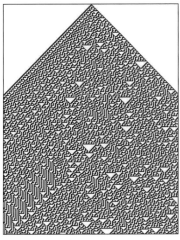

Figure 54. Cellular automata are computational systems that display an amazing degree of complexity. Depending on the state of a cell (black or white, on or off), the cells around it become black or white, and the pattern grows and propagates downward. In this example, a very simple rule leads to a mixture of order and chaos, where randomness contains islands of pattern and regularity. Cellular automata display some of the fundamental attributes of biological life.

mensions. His "Game of Life" features patterns that evolve and die, some growing like cancers and others creating endlessly changing patterns. It looks like a hybrid of a laser light show and animated fractals. Conway saw beautiful "gliders" and "guns" that generate an endless stream of new patterns. Guns act like wires that transmit information, so guns can combine to form *not* and *and* logic gates, the basis of all computation. It's been proven that the Game of Life is as capable as any computer with unlimited memory.[19]

Like Polyworld, the Game of Life isn't visually interesting enough to divert a young teenager for more than a few minutes, but scientists became very excited about the patterns generated in the software. (In any video game, everything is preprogrammed; nothing happens that the programmers didn't anticipate.) First, the patterns demonstrate emergence—complexity that derives from simple rules and a primitive starting point. Second, the patterns are self-organizing. Both attributes are central to biology.

Stephen Wolfram, the math prodigy who created the Mathematica software package, has taken the interpretation of cellular automata even farther. He's shown that cellular automata can be used to calculate transcendental or prime numbers, find quick solutions to differential equations, and generate behavior that is random on small scales yet predictable on large scales. This last attribute is striking because it mirrors the way the unpredictable quantum world maps into the well-behaved world of macroscopic objects. This world of tiny squares and simple rules can be used to show that there are axiom systems beyond those of traditional mathematics; in other words, the sum of the knowledge in all math textbooks is just a subset of all possible mathematics.

If anyone imagines that these conclusions are caused by the particular properties of two-dimensional arrays of cells, Wolfram has an answer for them, too. He shows that the same conclusions hold if there's no grid (i.e., if a network of connections is used rather than fixed cells), if there are more than two dimensions, and if there are constraints instead of rules. The formalism has almost unlimited scope.

A-LIFE

Chris Langton is the pioneer who organized the first international conference on artificial life at Los Alamos in 1987. He's one of only five resident faculty members at the Santa Fe Institute, a small think tank where physicists and computer scientists and economists rub shoulders. It was there that Langton articulated the reasons we should take the study and implications of artificial life seriously. He claims that it's artificial only in terms of components, not emergent processes. In other words, he argues that if these artificial components are properly implemented, the processes they support are every bit as genuine as the natural processes they imitate.

This is a big claim. Langton takes James Watson's famous statement that "life is digital information" literally. He's saying that if computational elements carry out the same functional role as biomolecules in natural living systems, they will be "alive" in the same way that natural organisms are alive. In other words, it's the process that counts, not the substrate.

One way to think about artificial life is that it's just made of different stuff from the life that evolved on Earth. Simplistically, the human genome is three billion base pairs written in a four-letter alphabet—information equivalent to an encyclopedia that can fit on a CD (although this is just the tip of an iceberg of information coded in the myriad ways that proteins interact with one another and the environment). By analogy, the brain is just an electrical network of one hundred billion neurons and one thousand trillion connections that works at a few kilohertz. If Moore's Law—the doubling of computer power and data density every eighteen months—continues unabated, computers will have this capability in 2020.[20]

At this point in the argument, many people call a time-out. They argue that while Polyworld and its cousins are impressive, they're still just simulations (Fig. 55). Pull the plug, and the lights go out. After all, only someone with no social life would confuse *The Sims* with real life, right? Similarly, most people weren't worried when world chess champion Garry Kasparov was beaten by the IBM computer Deep Blue in 1997. Deep Blue was a massively parallel machine capable of evaluating two hundred million moves per second, and it had been built expressly for the purpose of playing chess. If Deep Blue had also written poetry or sung the blues, people might have been more upset. Perhaps we feel threatened only when computers do something intrinsically "human." So is artificial life just a parlor trick or a profound insight into the essence of life?

Figure 55. Invented by Thomas Ray, Tierra is a simulation like Polyworld, with computational organisms subject to natural selection in a digital environment. The goal of the simulation is intended to evolve superior software. The most powerful medium for this experiment will be computers linked over the Internet. This metaphorical representation shows programs occupying computer RAM; mutations (lightning) cause random variations in the code while death (skull) culls the inefficient or defective programs.

The answer isn't yet clear. Some researchers take a pragmatic view. They hope that computer modeling will yield a deeper understanding of nature. To them, symbiosis is an emergent property in biology, and something like it can be found in the behavior of genetic algorithms. Others use complexity as a general framework for understanding life. But while the complexity of a neural net or a program is easy to measure, what's the relevant measure of complexity in a biological system? The number of DNA base pairs? The number of genes? The number of metabolic pathways? Or the number of species?

The computation approach has already yielded interesting insights. It's been shown, for example, that adding feedback sharply increases complexity in computational experiments. It's also been found that sudden changes in the environment help to advance complexity. The easy life turns out to be bad for evolution because it leads to stagnant genetic material. Extremophiles show that duress is a big part of biological success.

LIFE BEYOND BIOLOGY

Computers are useful in expanding the definition and understanding of life. Can we jump out of the box entirely? Does life really need carbon or water? Could we have all of the functional processes of biology without organic chemistry? At the most general level, life requires only thermal disequilibrium and an energy source. Life is characterized by islands of information that persist, reproduce, and adapt; computer algorithms can readily mimic this.

About fifty years ago, the astronomer Fred Hoyle wrote a science-fiction book called *The Black Cloud,* in which a huge interstellar cloud becomes an intelligent life-form. With gravity as the container, interacting networks of organic molecules form the genetic material of the entity. Unfortunately, in real life the density of interstellar gas is so low that interactions would take place hundreds or thousands of times slower than in a liquid medium on Earth. Hoyle's idea is implausible, but it's hard to rule out.

Other science-fiction ideas include self-organizing electric and magnetic fields. The possibilities may simply be limited by our imaginations. Even if we avoid flights of fancy, a more likely form of artificial life may be emerging under our noses. What if the genetic material (or DNA) for a life-form is a computer program and the container (or cell) is a machine?

If we take a snapshot of technology in the early twenty-first century, there are two distinct avenues to the development of machines. One involves robots. Currently, there are more than one million robots toiling in factories around the world, doing jobs too difficult or too tedious for humans. They might eclipse us in brawn, but their microprocessor brains are still puny. This will soon change.[21]

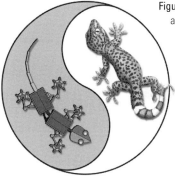

Figure 56. Live and robot geckos as symbols of natural and artificial life. There is a synergy between biology and engineering; biologists follow the top-down approach of understanding living organisms by studying their component parts, while engineers use a bottom-up approach of building complexity from parts that mimic life functions. Advances in miniaturization and computing mean that the natural and the artificial might one day seamlessly merge.

The threat and potential of robots is a central theme in science fiction, from Isaac Asimov's three laws of robotics, in the book and movie *I, Robot*, to the dystopian vision of Philip K. Dick's story "Do Androids Dream of Electric Sheep?" the basis for the classic movie *Blade Runner*. In these works of fiction, the risk of having robots do our bidding lies in their ever-increasing capabilities. Eventually, as robots approach self-awareness, humans have to face the moral implications of their invention.

We create robots, so in a sense they are our evolutionary progeny. It's up to us to decide whether we get the dream or the nightmare. (Or worse, it may be up to them.) The end result of this technology is difficult to predict.[22] But the stark dichotomy often presented in the popular culture—wars of man versus machine, robots eclipsing flesh and blood—could turn out to be misleading (Fig. 56). Our symbiotic relationship with robots may grow so stealthily that we fail to see the crucial moment when we need them but they no longer need us.

The second avenue of machine development, nanotechnology, is more subtle because it involves technology too small to see. Computers can now be miniaturized to the scale of molecules. Micromachines will eventually be able to swim through the bloodstream, drill through clogged arteries, and make detailed measurements of the body without invasive surgery. A major goal of nanotechnology is to fight disease from the inside. Genetically engineered microbes will be able to do hand-to-hand combat with germs and cancer-causing agents. Instead of replacing a defective or worn-out body part, we will have the capability to regenerate the organism from within.

WELCOME TO THE SINGULARITY

Life implies death. This implacable truth puts a boundary on our personal hopes and dreams. Every organism that ever lived has died, and every species that ever lived has gone extinct. Ray Kurzweil, however, is taking no chances. He doesn't tailgate. He gave up smoking and drinking long ago. He takes 250 diet supplements daily. He plans to live long enough to see mankind achieve immortality, an event he calls the Singularity.

Kurzweil is a brilliant inventor and futurist who started young. At age eight, he built a miniature theater in which a robotic device moved the scenery. By the age of sixteen, he'd built his own computer and programmed it to compose melodies. He invented the first optical character recognition, the first reading machine for the blind, and the first speech-synthesis and speech-recognition technologies. He's won the half-million-dollar Lemelson-MIT Prize for inventors and was inducted into the National Inventor Hall of Fame in 2002. The comparisons with Thomas Edison are both inevitable and apt.

Kurzweil wants to live long enough to see the Singularity, an event he thinks will take place around 2045. He started to pay close attention to his health after his father and grandfather both died from heart disease. He tracks up to fifty fitness indicators to fine-tune his programming. "What, then, is the Singularity?" He asks and answers the question in his 2005 book *The Singularity Is Near: When Humans Transcend Biology*: "It's a future period during which the pace of technological change will be so rapid, its impact so deep, that human life will be irreversibly transformed. . . . [T]his epoch will transform the concepts that we rely on to give meaning to our lives, from our business models to the cycle of human life, including death itself."

This is heady stuff. Rather than life-sized robot helpers, Kurzweil sees nanobots as the catalyst to transforming the human condition. Nanobots will keep us young forever by swarming through the body, repairing bones and organs, rejuvenating brain cells, and improving our genetic code by downloading new instructions from the Internet. He sees death as a "tragedy," not a natural or inevitable process.

Futurists such as Kurzweil are doing no more than extrapolating the exponential advance of technology that has fueled innovations in computer science and biology for the last two decades. Projecting forward, it seems inevitable that the nonbiological portion of our intelligence will be vastly more powerful than unaided human intelligence. "There will be no distinction . . . ," according to Kurzweil in his 2005 book, "between human and machine or between physical and virtual reality." This prospect might horrify many people, but Kurzweil notes that these advances could end hunger and poverty, and he insists his prognostications are value-neutral.

Technological change may point to a postbiological future, where machinery merges with the organism until we become a new entity. The term for this is "cyborg," a contraction of "cybernetic organism." Science-fiction writers who first wrote about cyborgs fifty years ago could barely have imagined that we would be on the threshold so soon. If genetic engineering is directed by a high-powered computer, evolution might occur more rapidly because it will be decoupled from the inexorable but inefficient molding of the natural environment. Such a computer might use a toolkit like BioBricks to build a better, bolder biology.

The implications for astrobiology are profound. Perhaps, on Earth and on other planets scattered through the cosmos, biological evolution is simply a phase of evolution that's succeeded by computational organisms built from the material that nature provides. Remember, there are planets that may have been hosting biology for five billion years before the Earth formed; they could be as advanced compared to us as we are to bacteria. Species that pass through the Singularity become immortal and join an increasing cohort of hyperadvanced citizens of the galaxy. If we overcome our troubled adolescence as a species, that might be the beacon that draws us forward.

4.

SHAPING EVOLUTION

> We came this close, thousands and thousands of times, to erasure by the veering of history down another sensible channel. Replay the tape a million times and I doubt that anything like *Homo sapiens* would ever evolve again. It is, indeed, a wonderful life.
>
> —Stephen Jay Gould, *Wonderful Life* (1989)

The asteroid streaks across the sky like a fiery messenger from the gods. As the sonic boom hits, small lizards scurry for cover, but some of the larger dinosaurs barely look up from their grazing. Seconds later, the asteroid heads out of the atmosphere into deep space, like a stone skipping off the surface of a pond.

It's been a close call. The incoming rock was the size of a large mountain. If it had been traveling a bit faster, or if the Earth had traveled slightly less far in its orbit, the asteroid would have hit head-on, causing utter devastation. Large creatures would have been killed almost instantly, and debris flung up into the atmosphere would have dimmed the Sun and disrupted the food chain, killing many more species. Instead, the Earth and its inhabitants shrug and continue their day.

Sixty-five million years pass. The dinosaurs diversify and continue their eons of dominance. Some learn to hunt cooperatively and invent simple social structures. Others develop metabolisms to deal with extremes of climate, and they expand their range on the planet. Mammals find successful evolutionary niches in the rain forest, but they have to survive by stealth and speed, so they don't evolve past shrews. Primates never emerge. As a result, Earth has no apes and no humans.

Instead, the most advanced animals live in the oceans, where they are immune from most fluctuations in the climate. Trapped in coastal regions by the success of the reptiles, some mammal lines gradually return to the water where they originated. New species emerge in the oceans. Driven by population pressure, they develop increasingly sophisticated adaptive strategies. The most successful of these creatures eclipse even whales and dolphins.

Earth's alpha species is descended from the giant octopus and is almost perfectly adapted to an aquatic life. It has no natural enemies, uses communal sensing, and lives in fluid social groups several thousand strong. Individual members of the species have cognitive skills, they experience a rich range of emotions, and they're aware of their mortality. The cohort transmits knowledge from generation to generation. The alpha species has appendages that can manipulate tools but has no need to venture far from the water. As a result, these animals will never build telescopes and wonder about the vast universe beyond the ocean.

•   •   •

LIFE'S DRAMA HAS MANY ACTS. For four billion years, players have entered and exited the stage, oblivious to the story unfolding around them. We're different. Self-awareness makes us the ham actors in this pageant. We relish our time in the spotlight; we strut and preen. It's hard not to feel the stage was made for us. But our time, too, will pass. We may not like it, but there's no script—evolution is pure improv.

Not only is there no script, but the stage is changing all the time. The players have their costumes and their protection from the elements: fur and feathers. The props are in place for life to unfold: carbon, water, and nutrients. A single warming spotlight hangs above. However, Earth is a restless planet; geological change gives some players more time and extra lines and writes others out of the script entirely. With little or no warning, life can be disrupted by external forces. A supernova explodes stage left. A star passes nearby stage right. A killer asteroid approaches. Fade to black.

Life on Earth isn't isolated from the larger environment like life in a petri dish. It's subject to hazards and influences from space. These range from torrents of cosmic rays as the Sun changes its spots to rocks the size of mountains hurtling in at almost unimaginable speed. When the Solar System formed, there was leftover debris, and some ended up on Earth. Even though most impact timings are impossible to predict, we know enough about space junk to say that the dangerous stuff gets here about once every hundred million years.

The unpredictability of life-altering impacts and the complexity of the interaction between life and its environment have led to a heated debate on the nature of evolution. Some scientists argue that contingency rules: random influences make it impossible to predict the outcome. Others look at the same data and conclude that evolution follows convergent paths: eyes and wings and brains are the inevitable solutions to life's problems.

Astronauts often comment on the fragile appearance of our planet. The full extent of the biosphere, from vents on the seafloor to spores floating near the stratosphere, is a small fraction of the size of the planet. On the other hand, while individual organisms and species come and go, biology is amazingly robust. There's complex interplay among air, earth, water, and living organisms. Our biosphere has been in a four-billion-year barroom brawl and has survived, bruised and battered but ready for more action. Life is the ultimate gamer.

BIRTH OF THE EARTH

LIFE NEEDS PLANETS. If we accept this premise, then the events that took place 4.5 billion years ago are of great interest. A cloud of gas and dust somehow turned into a star and a set of rocky bodies. How did it happen? As with the history of life on Earth, this detective story has evidence that isn't always easy to interpret. The crucial events took place long ago, with nobody there as a witness.

CLUES FROM THE CRIME SCENE

First, we note that the planets are tiny compared to the Sun. The mass of the Sun is five hundred times more than that of all the planets combined, and the sum of all comets, meteors, and asteroids is one hundred times smaller still. Rather than the main event, the planets are like a residue, scraps left over from the solar building site.

There are many important clues. Most of the planets orbit in a plane. That's why when you see planets in the night sky, they always appear in the strip traversed by the Sun—constellations along the strip form the zodiac, or circle of animals. Almost all of the planets rotate in the same direction they orbit the Sun. Their orbits are nearly circular.

Here's another curious fact: the planets' distances from the Sun increase in a roughly geometric progression, doubling with each successive one. In the 1760s, the German astronomer Johann Titius noticed this fact, although it was his colleague Johann Bode who first published the result. When Uranus was discovered in 1781, it fit the pattern. This focused attention on the gap between Mars and Jupiter, where a planet might be expected. Ceres was discovered in 1801; it's puny for a planet but turned out to be the largest chunk in a ring of rubble called the Asteroid Belt. Neptune didn't fit the pattern, although little Pluto did, as does Sedna, a minor outer planet discovered in 2004.

Any planetary detective encounters an immediate problem: how many planets are there? The comfort of a childhood mnemonic—My Very Excellent Mother Just Sent Us Nine Pizzas—has been disrupted as international astronomers have formally demoted Pluto to a mere "dwarf planet." The problem lies in the Kuiper Belt, a ring of icy bodies orbiting the Sun beyond Neptune, some of which may be as large as or larger than Pluto. Most astronomers would like to hold the line at eight planets and omit puny objects. This leads to controversy over the status of Pluto, Sedna, and recently discovered Eris (which had been provisionally named Xena, with her own small moon nicknamed, inevitably, Gabrielle, after the televised warrior princess's sidekick).

The Titius-Bode "law," as it's called, raises an interesting question in science. When is an apparent numerical pattern just a coincidence and when does it point to a deeper physical meaning? The relative planet distances are related to Pascal's number triangle, an interesting construction in pure mathematics in which each number is formed from the sum of the two numbers above it. Variations of this geometric spacing rule apply in miniature to the satellite systems of Jupiter, Saturn, and Neptune.

Is this numerology, or should a detective take it seriously? It usually depends on the length of the pattern and the plausibility of potential explanations. U.S. presidents were dying in office roughly every twenty years since 1840 until a bullet missed Ronald Reagan's heart by an inch and George W. Bush sailed into the new millennium unscathed—that's seven out of nine, the same accuracy as the Titius-Bode law. Suppose you're trailing a serial killer, and you notice the professions of the first four victims follow the alphabet: architect, baker, chemist, doctor. Should you wait until a hitman is murdered before deciding the theory has explanatory power?

There's another major distinction that divides the Solar System into at least two parts: planets near the Sun are small and rocky, while those far from the Sun are large and gassy, though they probably have rocky cores. Most of the planets have moons, and the giant outer planets have so many moons that they seem to form miniature versions of the Solar System as a whole. The large outer planets also have ring systems. Beyond the outermost planet, we find a nearly spherical swarm of icy rocks: comets.

These are the generalities, but all good detectives know the exceptions are interesting, too. One planet—Uranus—spins with its axis on its side rather than straight up. Another planet—Earth—has a moon that's a substantial fraction of its size. A recently demoted planet—Pluto—doesn't inhabit the same plane, and its orbit crosses that of the next planet in. Yet another planet—Venus—rotates in the opposite direction to the rest.

COLLAPSE OF THE SOLAR NEBULA

Is there one story that can explain these disparate clues? Yes! It started with a diffuse interstellar cloud. Its shape was amorphous or vaguely spherical, and it rotated gently. Some external influence, probably the supernova explosion re-

sulting from the death of a nearby star, nudged the cloud into gravitational collapse.[1] Once gravity had tightened its grip, the cloud collapsed rapidly to a
small fraction of its initial size. Most mass was dumped into the center. Deep
within the murk, the density and temperature climbed until a star was born.

As the cloud collapsed, gas in the plane of rotation began to spin up (think
of what happens as you try to move to the center of a spinning merry-go-
round in a playground), while material could collapse more readily along the
poles of the rotation axis. In this way, the collapse amplified the amount of rotation. (Think of an ice-skater pulling in his or her outstretched arms and spinning much faster.) The collapse didn't deposit all the gas onto the star because
heat within the gas eventually created pressure to resist further shrinkage. As
a result of this, a large, diffuse cloud is finally transformed into a relatively
small and rapidly rotating disk.

We tend to forget about it, but rotation is an important property of matter.
Gas in the universe is never completely smooth and uniform. Random variations of density grow when gravity acts on a gas. Swirls and eddies get amplified. Rotation in a vast cloud of gas that then shrank gave rise to the beautiful
pinwheel of the Milky Way and other spiral galaxies. If stars formed via perfectly radial collapses, they'd mop up every last piece of material, leaving nothing left to make planets, and we wouldn't be here.

FROM DUST BUNNIES TO PLANETS

The next part of the story may sound like magic, but it's equally firmly based on
rock-solid physics. It takes us into a world of snowflakes and Russian visionaries, where particles no bigger than dust motes in a beam of light get transformed into something as substantial as the Earth.

Forty years ago, a Russian mathematician named Victor Safronov collected
his life's work into a book. He'd calculated what happened in a rotating disk of
microscopic gas molecules, dust particles, and ice grains. Even though the material is orbiting quickly, nearby particles move at nearly the same speed, like
cars next to one another at a racetrack, so they can stick together. They cluster
into delicate structures similar to snowflakes. As they grow larger, they increasingly attract one another by gravity rather than by relying on collisions.
This accelerates the growth process. Motes turn into molehills, molehills into
mountains. Finally, mountains turn into planets. (See Fig. 57.)

This steady process is called accretion, and it's very efficient. In the inner
Solar System, it took only a few million years for chunks of icy rock to build up
to fifty miles across, the size of the largest meteoroids and comets. Material
swept up in a series of zones moving out from the Sun gave rise to the small

Figure 57. An imagined view inside the solar nebula just as planets are beginning to form. Rocky material is starting to grow by accretion but most of the nebula is composed of gas and tiny dust particles. Deep inside the nebula, most of the mass has collapsed into a young star that is just visible inside the obscuring material.

number of planets we see today. Planet building took fifty to one hundred million years, only 1 to 2 percent of the age of the Solar System. Radioactive material trapped in meteorites shows that the process was rapid.[2]

COLLISIONS AND CATASTROPHES

Safronov's work also pointed to the importance of collisions. The chunks of rock created by accretion can wreak havoc with a planet after its creation. Due to the Cold War, Safronov worked in isolation for a decade. In the 1970s, his book finally reached western scientists by a circuitous translation route. Soon after, several research groups proposed that the Moon had been formed by the impact of a Mars-sized body on the Earth, a theory referred to as the "Big Whack." The theory neatly explains several puzzling aspects of the Moon—its large size relative to the Earth, its unusual orbit, and its lack of an iron core.

There was initial resistance to the idea that catastrophes have shaped the Solar System, mostly because the explanation seemed like a too convenient deus ex machina. The reputation of another Russian visionary, Immanuel Velikovsky, also made scientists squeamish about explaining aspects of the Solar System with collisions. In the 1950s, Velikovsky had achieved notoriety with his book *Worlds in Collision*. Whereas Safronov was a sober mathematician, Velikovsky was a wild-eyed dreamer, and his book was a witches' brew of mythology, psychology, and pseudoscience. Velikovsky made the study of impacts and collisions seem disreputable.

With the help of recent computer simulations, the geography of the Solar System can be understood. Most features are explained in terms of gradual evolution and the interplay of gravity and radiation.

The collapse model explains why the planets orbit in a plane and rotate in the same direction that they orbit the Sun. In the inner Solar System, accretion grew four rocky planets, and the remaining gas and dust was blown out by the young Sun. In the outer Solar System, rocky cores grew to five or ten times the Earth's mass, then accreted shrouds of hydrogen and helium to make the gas-giant planets. At the distance of the asteroids, the gravity of nearby mighty Jupiter disrupted the orbits and prevented the formation of a planet. Jupiter also tossed icy rocks from the inner Solar System into lazy looping orbits far from the Sun, where we know them as comets. These processes acted in miniature to give the giant outer planets moons and rings.

To this smooth broth we add the spice of collision and catastrophe. Chaotic and unpredictable events give planets their quirks and "personalities." Thus, we explain the Earth's moon, the tilt of Uranus, and the capture of Pluto from the outer reaches of the Solar System—with regrets to traditionalists, astronomers are unlikely to reverse their decision to demote Pluto and other more distant asteroid-sized rocks to the status of interplanetary bodies.

THE PRIMEVAL EARTH

What about the Earth? Nearly one hundred million miles from the adolescent Sun, it was a toasty 650 kelvin, or 380 °C. When it's very hot, molecules exist only as vapor. As it gets cooler, they can condense into liquids and then solids, just as water molecules turn into raindrops, then ice crystals.[3] At 380 degrees, it was too hot for icy molecules like water, methane, and ammonia to condense from vapor. It was also too hot for carbon to condense onto grains, so it floated in space as pure soot (Fig. 58). The Earth acquired its organic material and water some time after it formed.

The freshly minted Earth was a bizarre place. Accretion had delivered so much energy that the entire surface was a molten sea of magma. Iron and nickel, both heavier than the silicates of the mantle, settled into a liquid metallic core. The atmosphere was made of rock vapor.

Within about fifty million years, the Moon had been gouged out

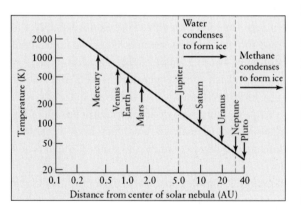

Figure 58. The temperature inside the solar nebula declined with distance from the newborn Sun, and this affected the types of material that could condense into solids and participate in planet building. At the distance of the Earth, the temperature was too high for carbon-rich compounds or water to condense. The planet formed mostly from silicon-rich minerals and iron and nickel that subsequently sank to the core.

by a giant impact. Thereafter, the impact rate declined rapidly as debris in the inner Solar System was swept up by the terrestrial planets or ejected by the influence of Jupiter. Within one hundred million years, the Earth acquired the organic material that is currently locked up in living organisms. It came from comets, which are rich in carbon compounds, as we learned in 1986 when the Giotto mission glimpsed the sooty nucleus of Halley's Comet. Water had several sources, but very little came from comets.[4] Some was released by hot rocks as the Earth cooled. Most was probably delivered by meteorites from the outer edge of the Asteroid Belt.

The conventional wisdom is that the Earth was still an unforgiving place 150 million years after it formed. This is indicated by the name geologists give to the first half-billion years: the Hadean, or "hell-hole," eon. A crust was in place, but it was likely soft and tacky, like asphalt on a hot summer day. Seams of upwelling magma and volcanoes were everywhere. Giant impacts occurred one hundred times more frequently than they do today. Water was circulating through the crust in hydrothermal systems and in the pore spaces of rocks, but it was too hot for oceans. The planet was shrouded in steam.

However, the Earth's oldest rocks tell a different story. The chemistry of their formation points to relatively cool temperatures and the presence of liquid water as many as 4.4 billion years ago. There's also new evidence from crater records on the Moon and Mars that all terrestrial planets suffered a "spike" in the impact rate 3.9 billion years ago. The cause of this "late heavy bombardment" is hotly debated but may be associated with the delayed formation of Uranus or Neptune or the migration of Jupiter to its current position. We have a new story for the early Earth in which it may have been habitable very soon after its formation (Fig. 59).

The implications for the early history of life are difficult to evaluate. The largest and earli-

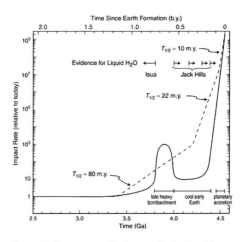

Figure 59. The rate of major impacts declined rapidly after the Solar System formed 4.5 billion years ago, as material was swept up into planets or ejected from the inner Solar System. The current low rate was reached 3.5 billion years ago. (The solid line is an approximation of the crater data; the dashed lines are models of physical processes that might explain the decline.) Water was clearly present 3.8 billion years ago in rocks from the Isua formation, and it might have been present as far back as 4.4 billion years ago, as indicated by zircons from the Jack Hills of Western Australia.

est impacts probably sterilized much of the Earth's surface by melting it. Even after the crust solidified for good, some impacts must have disrupted local experiments in biochemistry. It's not clear whether any life could survive the torrent of projectiles from the era of late heavy bombardment.

On the other hand, subsurface environments may have been immune from the mayhem. The earliest tentative evidence of life is 3.8 to 3.9 billion years ago, but the ingredients were in place five hundred million years previously, and the temperature and impact rates were not prohibitive. Also, impacts could have moved primitive life from Earth to Mars and, with higher probability, from Mars to Earth (one rock per month). If cross-contamination happened more than once, we may never know where life in the Solar System started.

Did life struggle through the turmoil like an engine sputtering to a start on a cold day? Did it get independent starts, maybe with different information-carrying molecules, with all but one of the experiments obliterated by impacts or by the reproductive success of the winners? Might we all be Martians? The evidence for answering these major questions is frustratingly scant.

It's even harder to say what early Earth history implies for terrestrial planets in other solar systems. Planet building is a random process. The initial conditions might have led to a different number of planets, with very different masses and positions. The availability of organic material and water is sensitive to the layout of the solar system. We might expect other planetary systems to show a wide variation in their architecture and number of habitable planets. We'll not know until we look!

IMPACTS

MARK BAILEY IS an unusually cheerful eschatologist. As an astronomer, he made his name calculating the intricacies of comet orbits. Some comets recur in the night sky like familiar friends. Others have orbits so elongated that they've appeared only once in recorded history. Bailey spends more time on a third category: those in Earth-crossing orbits. They're rare but numerous enough to potentially account for mass extinctions in the history of life. During the biggest one at the end of the Permian era, about 250 million years ago, 95 percent of all species were extinguished, leaving no subsequent fossil record.

None of this affects Bailey's appetite. He's the director of Armagh Observatory in Northern Ireland, and he likes to take lunch with colleagues in the local pubs. Outside, the town has the weary feel of a place that has spent too long in a war zone. Inside, there are no tense or sullen faces because everyone is tucking into steak-and-kidney pie and piles of greasy chips. Guinness flows freely.

WEIGHING THE ODDS

Bailey tosses off the frequency of impacts of a particular size—fireballs every year or so. Those are boulder-sized projectiles disintegrating in the upper atmosphere. Every century or so, a rock the size of a large house hits the ground at twenty to thirty thousand miles per hour. The Siberian impact at Tunguska was one hundred years ago. Bailey pauses with a fork halfway to his mouth and smiles—that means we're due. Every ten thousand years, a kilometer-sized rock takes out an area the size of Belgium. Bailey grins.

Then there's the big one: a comet or asteroid that could vaporize a mountain range. The most recent such impact was a mere sixty-five million years ago. Bailey drains his Guinness and then smacks his lips in satisfaction. A rim of white foam clings to his mustache.

One in ten billion—Bailey is reassuring as he quotes the odds of dying due to an impact from space. The probability is the same whether the projectile is a small meteor with only your name written on it or a huge asteroid that takes you out along with the population of an entire continent. By way of reassurance, he notes that you're a million times more likely to die in a car crash and a thousand times more likely to die in a plane crash. You're even more likely to die from botulism than from a meteoric collision. He glances down at his steak-and-kidney pie thoughtfully.

Knowing the odds are low isn't as consoling as it should be. People are unduly obsessed by grisly but unlikely outcomes, like dying in a plane crash or a terrorist attack. In literature, nobody tapped into our subconscious fear of impending death better than Edgar Allan Poe and his ominous ravens. In Poe's "A Descent into the Maelstrom," an old man and his brother spend hours anticipating their almost certain deaths in a huge whirlpool. Maybe the best illustration comes from Thomas Pynchon in his sprawling, brilliant novel *Gravity's Rainbow*. In London during World War II, German V-2s are raining down randomly on the city. The lead character is obsessed in a nightmare by the frozen instant of time at which a V-2 is poised just above him, about to hit his head. Like a meteor, a V-2 arrives supersonically from above, so you'd never see or hear it coming.

Figure 60 shows the relationship between the size of a piece of space debris and its rate of arrival somewhere on the Earth. This kind of distribution—a linear correlation between two quantities plotted as logarithms—is called a power law. What the plot really means is that debris comes in all sizes, and there are many more bits of small debris than large debris. With lots of small chunks of rock floating around in space, the number of chunks arriving per second is high, and the average time between chunks arriving is low. The large stuff is rare, so the arrival rate is low, and the time between arrivals is high.

The power-law distribution is a reflection of the accretion process in the

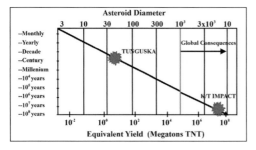

Figure 60. A plot of the average time between Earth impacts, or their frequency, and the size of the projectile. There are many more tiny pieces of space debris than large pieces, so large impacts occur very rarely. A vast majority of the pieces smaller than one meter burn up in the upper atmosphere. This is the arrival rate anywhere on the Earth's surface, not in any particular region.

early Solar System. It took many tiny chunks to make a few mighty planets. Since the planets formed, the Asteroid Belt has acted like a grinder; rocks collide and break into smaller pieces, some of which get scattered into Earth-crossing orbits.

We can step through the distribution in Figure 60 and see the effects on life, now and in the past. Every year, thirty thousand tons of space junk rains on the Earth, but most of it falls on the oceans, and most of it is made of particles too tiny to recognize. If you go out at night far from city lights, every few minutes you'll see a streak of light somewhere in the sky. Shooting stars, or meteors, are produced by tiny bits of interplanetary debris, no bigger than a grain of sand, burning up in the upper atmosphere. Whizzing in at up to 150,000 miles per hour, they carry more energy than a well-flung fastball.

Projectiles the size of a boulder arrive much less frequently, but they carry a lot more energy. About once a year, objects this size disintegrate in the upper atmosphere with the force of a small atomic bomb. In the early 1960s, meteors brought us to the brink of nuclear war. The United States and the Soviet Union each saw these occasional fireballs as evidence that the other side was cheating on the Nuclear Test Ban Treaty. Declassified documents reveal tense exchanges over the hotline between Washington and the Kremlin.

Fireballs and shooting stars are harmless light shows in the sky. We should feel relieved and grateful that the atmosphere fries up most incoming projectiles. But some fraction of the meteoroids between the size of a rock and a house reach the ground and cause damage. At the upper end of this range are chunks of rock the size of a football field, weighing more than one million tons.

TUNGUSKA

The scene was a remote Siberian forest on the morning of June 30, 1908. Out of a sunny sky there was a deafening explosion, and the forest ignited. An eyewitness forty miles away described the event, as reported in the book *Giant Me-*

teorites by E. L. Krinov: "I felt great heat as if my shirt had caught fire and there was a mighty crash. . . . I was thrown onto the ground 20 feet from the porch." Reindeer herders close to the impact were not so lucky. Even three hundred miles away, villagers report a "deafening bang" and a fiery cloud on the horizon. Seismographs registered the event in London.

Figure 61. The Tunguska event occurred in a remote region of Siberia in 1908. Since there was no impact crater, the projectile must have exploded miles above the Earth's surface, flattening thousands of square miles of forest and causing barometric pressure to change and seismographs to register the event all across Europe. It is not clear if the projectile was a comet or a meteor.

By sheer good fortune, the Tunguska rock fell in one of the most remote parts of the world (Fig. 61). If it had landed on a major city, it would have killed hundreds of thousands of people. A Tunguska event occurs every few hundred years, so impacts on this scale have played a role in human history.[5]

Just how big a role was the subject of research by Victor Clube, one of Mark Bailey's thesis advisors. The two men make a study in contrasts. Bailey is short and gregarious, with an easy laugh and sunny eyes; Clube is tall and avuncular, with the intimidating diction of an Oxford don. Clube and his colleague Bill Napier, a scientific civil servant who writes thrillers in his spare time, scoured the historical and mythological record for evidence of catastrophes from space. Their work is very controversial because it depends on a close reading of texts from many cultures and not purely on physical evidence. Few scientists have enough breadth for this kind of scholarship.

Not all space junk arrives randomly. Meteor showers happen when the Earth passes through debris strung out along the path of a comet. For example, the Orionid shower that peaks on October 21 marks the place where we pass through the orbit of Comet Halley. One of the more spectacular meteor showers is usually the Leonids on November 16. During a good meteor shower, you might see a couple of shooting stars each minute, but occasionally the showers are so intense that meteors fall too fast to count. The 1966 Leonids peaked at forty meteors per second, and the 1833 shower was even more spectacular (Fig. 62). When the debris is particularly thick,

Figure 62. In this woodcut of the famous Leonid meteor shower of 1833, as seen from the United States, the night sky is lit as bright as day by the torrent of meteors. Eyewitnesses stated that the rate of meteors peaked at a few thousand per minute. When the cometary debris is as thick as this, the larger chunks can cause substantial damage.

there are almost certainly substantial impacts to go along with the light show.

Clube and Napier scoured two thousand years of fireball records from Chinese court astronomers and found that every hundred years or so Earth's orbit passes through cometary debris thick enough to cause major damage in some parts of the world.[6] There's evidence of a peak in the debris rate around the time of Christ and again in 1000 C.E., so it may have contributed to the apocalyptic thought of the time.

The biggest reservoirs of space junk are the Asteroid Belt, the Kuiper Belt, and the Oort cloud. Asteroids orbit in the region between Mars and Jupiter, the site of a "failed" planet (Fig. 63). The Kuiper Belt is a similar arrangement of rocky debris beyond the outermost planets. Comets live in a spherical zone that extends thousands of times farther than Pluto. They spend most of their lives in the deep freeze far from the Sun, gathering warmth as they periodically swing close to the Sun on highly elliptical orbits. Jan Oort was a Dutch astronomer who hypothesized the existence of the unseen comet cloud in the 1950s. He lived an austere life in a wooden house with spartan furnishings and no central heating. The author of more than three hundred scien-

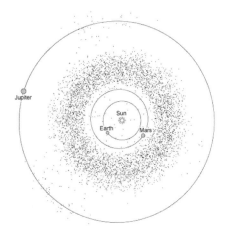

Figure 63. The space distribution of asteroids between Mars and Jupiter, the principal store of interplanetary debris that can potentially hit the Earth when collisions or gravitational influences send them on inward trajectories. Asteroids range in size up to the thousand-kilometer-diameter Ceres. The smaller group of asteroids that precede and follow Jupiter in its orbit are called the Trojan asteroids.

tific papers, he published on topics from comets to cosmology until well into his nineties.

THE BIG ONE

Imagine yourself on a comet in the Oort cloud. It's a small mountain the shape of a potato—a mixture of rock, ice, and sooty carbon compounds. The Sun is a distant dot in the sky, barely brighter than the other stars. You've spent million of years creeping along on the slow part of your orbit. Gradually, your path loops into the Solar System. Picking up speed, you whip past Uranus and Neptune. With a timing slightly earlier, you would have passed close enough to Jupiter to be flung back out into the comet cloud. Slightly later, and you would have plowed into the Asteroid Belt and been ground into harmless boulders. Instead, your vector is aligned toward a milky-blue planet, third one out from the Sun. Drawn by the Sun, you pick up speed.

Impact occurs at twenty-five thousand miles per hour. The explosive force is ten million times greater than that of the 1980 Mount St. Helens event, the equivalent of one hundred trillion tons of TNT. Almost instantly, trillions of tons of rock are vaporized and flung into the upper atmosphere. Tidal waves one hundred feet high circle the globe. The shock to the crust triggers massive earthquakes. As debris fills the sky, the Sun dims, and the photosynthetic food chain is broken. Many are killed intantly, and within a few years, all the large creatures of the land and sea will be dead.

Twenty years ago, nuclear physicist Luis Alvarez and his paleontologist son, Walter, caused a paradigm shift when they argued that the Cretacious-Tertiary (K-T) extinction event sixty-five million years ago was caused by a city-sized rock slamming into Earth. It had always been assumed that evolution was gradual.[7] But several lines of evidence pointed to a catastrophe—a very rapid decline in the fossil record and the concentrated layer of shocked quartz and iridium. Shocked quartz can be produced by volcanism but not as a global layer at a specific time. Iridium occurs in terrestrial rocks but it's more abundant in extraterrestrial material and the K-T layer shows a "spike" in iridium concentration. The clincher was the discovery of the "smoking gun," a crater of the correct age on the seafloor off the northern coast of the Yucatán peninsula (dramatic volcanic activity may also have played a role).

There have been at least seven mass extinctions in the history of life on Earth (Fig. 64). The mother of them all was the Permian event 248 million years ago, when 90 to 95 percent of all species disappeared.[8] Impacts have been suggested as the cause of several extinctions, but only the K-T event has all the extra supporting evidence. Going back in time, it becomes harder to identify craters, and the fossil record is too spotty to be sure that the extinction was instantaneous.

Returning to Figure 60, we can see that a ten-kilometer impact occurs on average every hundred million years. Astronomers anticipate that the process of planet formation will always be inefficient. If accretion does not wipe the plate clear, there will be

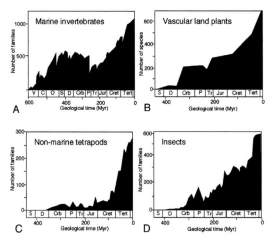

Figure 64. Changes in the number of families, a large grouping of similar species, since animals emerged in the Earth's oceans six hundred million years ago. There have been at least seven mass extinctions, with the most significant events occurring at the boundary between geological eras. The K-T event is named after the boundary between the Cretaceous (abbreviated "K" to avoid confusion with the Carboniferous Period, which is abbreviated "C") and Tertiary periods.

leftovers to create a potential hazard for any terrestrial planet. The catastrophic events are extremely rare, but the time frame of biological evolution on Earth is billions of years, so impacts probably play a role in the evolution of life elsewhere in the cosmos.

Reassuringly, we do have a plan if our luck runs out.[9] To warn of a full hit by a comet or a meteoroid knocked into Earth-crossing orbit by a collision in the Asteroid Belt, a fleet of small telescopes has been built. They scan the skies nightly for objects moving among the fixed stars and can detect everything bigger than about a kilometer across. What then? Astronomers will crunch the numbers to calculate the orbit. Even if a collision is likely, we'll have a couple of years' notice. Time enough to dust off the nuclear arsenals and send out missiles to do our bidding. Not a direct hit. A gentle kiss shot, glancing it off to the side. Nice.

HUNTING SPACE JUNK

Guy Consolmagno has to pee, badly. He's part of a group of planetary scientists and geologists conducting a "sweep" of a desolate eight-thousand-foot-high plateau in Antarctica. They're looking for meteorites. It's worth making such a long trip because anywhere else on Earth a rock is just a rock, but here small rocks from space are easily seen on the blue and pristine ice. The team members move stiffly in their all-weather suits. The air is startlingly clear. It's a balmy summer day, with the temperature nudging thirty below zero.

Meteorites are rare, so a day's haul might be only a dozen specimens. Anything the size of a fingernail would be a significant find. Consolmagno is an expert on the relationship between asteroids and meteorites and on the evolution of small rocky bodies in the Solar System. He took vows as a Jesuit brother after two years in the Peace Corps and a stint teaching physics. His home base is the Vatican Observatory near Rome, where he's the curator of the famous meteorite collection that the Catholic Church improbably owns. Consolmagno discussed the prospect of water and life on Europa in his master's thesis—the first prediction has been proved correct, and the second may prove to be prescient. He's also the informative and avuncular "Brother Guy" in occasional science pieces on BBC Radio 4.

But just now he is distracted by his bladder. It's a difficult decision. At thirty below, spit hits the ground as a solid lump, and sensitive skin shouldn't be exposed to air this dry and cold. On the other hand, he has a higher calling, and sometimes sacrifices have to be made for science.

ROCKS FROM THE SKY

Two hundred years ago, the idea that rocks fell from the sky was considered superstition. Then, Jean Baptiste Biot was sent by the Académie française to investigate strange phenomena reported in a small village. By a combination of eyewitness accounts and physical evidence, Biot demonstrated that some rocks come from far beyond Earth. Thomas Jefferson was initially a skeptic, saying it was "more likely that Yankee professors lie than stones fall from space." But he was convinced by the evidence and later used this example to argue for the scientific method.

Gathering meteorites is important because they have such interesting stories to tell about the context for life. They embed a chemical memory of the Solar System as it formed. Early in the Earth's history, they delivered most of the water on which life now depends. They also brought much of the nitrogen, phosphorus (essential to the backbone of DNA), and even some amino acids. They also, occasionally, bring death.

Guy Consolmagno is aware of all this as he crunches across the blinding white ice. The work is slightly tedious once the excitement at working on this frozen lid of the world has worn off. Several hundred miles away, another group of scientists uses a different strategy, pushing wide sticky rollers across the ice to collect the many meteorites that are as small as grains of sand. They will unroll the sticky felt from the rollers and scrape any debris off in a clean room, then return home with a tiny vial of space dust.

Consolmagno hopes he'll be as lucky as the U.S. scientist who was riding in his snowmobile on a different part of the ice field when he spotted the now famous Allan Hills meteorite from Mars, subject of feverish speculation that it contains Martian microbes. Consolmagno is a patient man, but he knows that the odds of finding a Mars rock are slim. It's far better to go to Mars and bring one home.

METEORITES FOR THE MASSES

This work is not just for professional astronomers. Ace meteorite hunters set off anywhere in the world at a moment's notice when there's a report of a new impact. Bob Haag rolled into Portales, New Mexico, a few years ago with a pocket full of hundred-dollar bills. He'd heard of a rare iron-rich fall less than a day before. He hung wanted posters in the barbershop and the Wal-Mart, and soon an army of townsfolk was scouring the desert, bringing him fragments. Some were still warm. For one specimen, he shelled out five thousand dollars to a kid on a bike. Haag, the self-styled hippie from Tucson, says he has the "best job in the galaxy."

Although not trained as a scientist, a rock hound like Bob Haag has an eye to rival that of the best geologist. Marvin Killgore was a plumber and contractor when his passion for meteorites turned him into a full-time collector and trader. He has learned how to use an electron microscope and has approval from the professional organization of planetary scientists to officially classify meteorites. Haag and Killgore share the lifestyle and adventurous spirit, if not the screen looks, of Indiana Jones. It can be lucrative work. Haag's collection is valued at twenty-five million dollars, and Martian meteorites are worth their weight in gold.

Anyone can join in the fun. Even if you can't afford a trip to Antarctica, you can collect interplanetary debris: all you need is a bucket, a magnet, and a microscope. A fine rain of micrometeorites falls on houses and fields and on us. Put the bucket under a rain spout and wait for the space dust to be washed off your roof by rain. Pull out the leaves and twigs and spread the rest of the debris on a plastic sheet. Run a strong neodymium magnet across the material to gather all the magnetic particles. Most of what you have gathered will be terrestrial debris, so you'll need to look at the particles under a high-powered microscope. Micrometeorites are small and rounded, with tiny pits in their surfaces as evidence of their fiery trips through the atmosphere. Over 20,000 tons of space dust reaches the Earth every year. This story of the Solar System is contained in a space no bigger than the head of a pin.

COSMIC INFLUENCES ON LIFE

THE EVOLUTION OF LIFE on Earth has clearly been influenced by impacts from space junk, but the Solar System is also part of a larger cosmic environment. The Sun is a very steady star, but even small variations have a substantial effect on the climate—when the Sun catches a cold, the Earth sneezes.

Looking beyond the Solar System, we live at the edge of the Orion arm of the Milky Way. Random motions of nearby stars can bring them close enough to jostle the Oort cloud and send comets cascading into the inner Solar System. We've completed eighteen circuits of the Milky Way since the Earth formed; perhaps half a dozen times during each orbit we dip in and out of the disk of the galaxy, passing through dust clouds and near star clusters. When a nearby star explodes, it can wreak havoc on the Earth by bombarding it with radiation and high-energy particles.

None of these cosmic influences is unique to our planet. Similar phenomena will affect planets around other stars as well; this information is used in astrobiology to guide the expectation of how life might evolve elsewhere.

THE FICKLENESS OF STARS

Shakespeare didn't know it, but he was using artistic license when he had Caesar declare, "I am constant as the northern star." In fact, the brightness of Polaris varies by 20 percent every four days. Most astrobiologists assume life needs a sustaining star. If so, the variability of the energy source will affect the way life develops.

Surveys with modern electronic detectors show that only a couple of percent of stars vary by more than 10 percent over a decade. Even fewer are as variable as Polaris. Main sequence stars, which fuse hydrogen into helium, are generally quiescent—the Sun has varied by only 0.1 percent over the past decade. Studies of similar stars tell us the Sun was much more active in the distant past, but all the real pyrotechnics occur near the end of stars' lives, when they change their fuel. Stars more massive than the Sun are very active, but their lives may be too short for their solar systems to host complex life.

The Sun's magnetism causes the most visible short-term effects on the Earth. At alternate peaks of each eleven-year sunspot cycle, the magnetic field of the Sun reverses, triggering complex changes to the Earth's climate as the flux of radiation varies. Climate is strongly affected by high-energy types of radiation such as ultraviolet rays and X rays and cosmic rays, which are energetic charged particles such as protons. The solar cycle correlates with regional famines over several centuries.

A skeptical detective will raise his or her eyebrows at the prospect of implicating the Sun in long-term climate change on Earth. The "suspect" has been observed with photos for only 150 years. For 250 years before that, counting sunspots was the only way to measure solar activity. And before the invention of the telescope in 1609, all measures were indirect: as the Sun varies, so does the flux of radioactive particles and cosmic rays hitting the Earth, and the concentration of those tracers is measured in sediment and ice layers going back thousands or millions of years. That's not observing the suspect; it's inferring actions based on their effect on the behavior of someone else.

Orbital variations are generally presumed to cause ice ages, but the century-old Milankovitch theory predicts a four-hundred-thousand-year cycle that *isn't* observed and can't successfully explain the one-hundred-thousand-year cycle that *is* observed. Support for the Sun as the cause comes from radioactive beryllium of solar origin in ocean sediments, which can be used to track ice ages over several million years. However, the interplay with geology and the feedback into the atmosphere is complex, so the mechanisms haven't been nailed down yet (Fig. 65).

The Earth experiences unpredictable changes in the Sun's radiation, in addition to the solar cycle. There are instances of radioactive-element deposition in the layers of the ice pack that point to times when the Sun flared dramati-

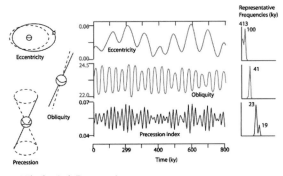

Milankovitch Frequencies (from SEPM # 40)

Figure 65. Three aspects of the Earth's orbit of the Sun that are very likely to cause climate variations. Shown here are the effects (left), the typical variations over 800,000 years (center), and the periodic signals most likely to imprint on climate (right). Variations in eccentricity in the Earth's orbit of the Sun follow a 100,000-year cycle. The tilt of the Earth's axis takes 41,000 years to complete a cycle. The top-like wobble of the Earth's axis follows a 23,000-year cycle.

cally. We know it's possible because stars like the Sun undergo brief super-flares when they brighten by factors of one hundred thousand for a short time. Such events are random and completely unpredictable. When the Sun acts up in this way it's bad news for Earth's creatures. The most extreme flares, every million years or so, scorch the atmosphere, and they may send the mutation rate skyrocketing. Even mighty *Deinococcus radiodurans* might not be able to keep up.

The ultimate threat to life is the death of the sheltering star. Relax—Sol won't exhaust its hydrogen for another four billion years. The end is not sudden, like a light switch flipping, because the Sun is constantly stirred by convection, and it uses its fuel from the inside out. The energy output will become more erratic until, like a guttering flame, it goes out. Without energy release to keep it "puffed up," the Sun's core will collapse to a new hotter state and begin helium fusion. With a new lease on life, it will shine a thousand times brighter. Its outer envelope will expand through the inner Solar System and engulf the Earth, turning it into a lifeless cinder.

Are we having fun yet? After fire comes ice. The Sun goes through a brief red-giant phase and then runs out of nuclear fuel, its gravitational muscle not strong enough to ignite fusion beyond carbon. Thereafter, it collapses to a dim white dwarf and spends eternity as a slowly cooling ember. Earth turns into a frozen rock, unable to leave the graveside yet no better off than if it were alone in deep space.

It's a bleak, if distant, prospect. But cheer up—the end will actually come ten times sooner. The Sun has been getting brighter over its entire life as its configuration changes to burn a little faster and hotter. A billion years from now, that trend will boil away the oceans. And a mere five hundred million years from now, warming will turn the planet into a global desert, and the carbon-dioxide level will drop too low for photosynthesis to operate. It spells the end of the biosphere, unless we or our descendants or some other enterprising species figure out how to reengineer the planet.

STAR DEATH

When a massive star explodes, it brightens by a factor of billions to rival an entire galaxy. In damaging forms of radiation, a supernova is a trillion times stronger than the Sun. Some of the prettiest objects in the night sky are supernova remnants, like the Crab Nebula, which was recorded by Chinese astronomers in 1054 C.E. and depicted in rock art by Native Americans, and the Vela Nebula, which is the wispy residue of a star that died eleven thousand years ago, before humans began leaving a historical record.

A massive star explodes somewhere in the Milky Way every few hundred years. Every few thousand years, the dying star will be close enough—hundreds of light-years away—to be visible in daylight. If it's much closer, it can cause real damage.

The core of a dying star reaches a temperature of billions of degrees and releases a flood of high-energy particles. If a supernova went off within twenty-five light-years of Earth, three kinds of radiation could affect life here. Lightweight neutrinos interact weakly with matter and are generally harmless—trillions from the Sun pass through our bodies every second, and we don't feel a thing—but a supernova produces such a large neutrino flux that it would cause mutations. High-energy gamma rays are mostly absorbed by the atmosphere, but the 1 percent that reach the Earth's surface cause cellular damage. Worst of all, energetic cosmic rays can destroy the ozone layer for long enough that ultraviolet radiation disrupts the base of the food chain.

It's enough to make you look warily at Betelgeuse or Antares the next time you go out at night. Luckily, these monsters are too far away to hurt us when they die, and there's no star near us that's massive enough to die as a supernova. Over the history of life on Earth, however, there's tantalizing evidence for high levels of radiation or damage to our atmosphere from dying stars, both of which can cause decimation of species.

An intriguing story about the extinction of mammoths has been pieced together by Richard Firestone of the Lawrence Berkeley Lab. He and his team find peaks of radioactive carbon in Icelandic marine sediments and high concentrations of radioactive potassium in tusks and human artifacts from 41,000,

34,000, and 13,000 years ago. The early date is hypothesized to be when a supernova went off, and it took an additional seven thousand years for the iron-
rich grains to reach the Earth. The latter two dates bracket the time span when
the Earth was subject to the shrapnel; mammoth tusks from those dates were
peppered with iron grains that impacted at millions of miles per hour. Time will
tell if this theory displaces the other explanations for the demise of mammoths
and other large mammals: disease, climate change, and excessive hunting by
humans.

Stars move around as they all orbit in the Milky Way. A team of astronomers
at Johns Hopkins University has traced the motions back in time to find out if
dying stars may have been influential in the past. Two million years ago, we
were much closer to the Scorpius-Centaurus stellar association, which was
fizzing with massive star formation and supernovae. There's an excess of radioactive iron in the Earth's crust at that age, coincident with a mini-extinction
of mollusks, plankton, and other marine organisms. A number of other mini-
extinctions have been attributed to stellar death.

THE ULTIMATE CATACLYSM

A hypernova is an even rarer form of stellar cataclysm. In the 1960s, satellites
discovered brief, nonrepeating pulses of gamma rays that were not associated
with any known star or galaxy. Astronomers recently solved the puzzle of
"gamma ray bursters" by showing that many of them are massive dying stars,
billions of light-years away. They can be seen at incredible distances because
they momentarily outshine the entire universe in gamma rays. Gamma-ray
satellites detect a hypernova about once per day somewhere in the universe.

Unlike a supernova, which explodes in a roughly spherical shell, a hypernova concentrates its lethal radiation and high-energy particles in two beams
that shoot out at close to the speed of light from the poles of the collapsing star.
It's this concentration of energy that allows a hypernova to be seen across the
universe. If this type of cataclysm happened within a few thousand light-years
of us, and we were unlucky enough to be in the beam of radiation, life on Earth
would be knocked out by a one-two punch. Gamma rays would destroy molecules in the stratosphere, simultaneously removing ozone while creating a
smog of nitrous oxide and other chemicals. Radiation levels a hundred times
higher than normal would kill exposed life, but then the smog would trigger an
ice age, killing additional species.

There's circumstantial evidence that gamma rays from a hypernova caused
the first of the major extinctions in the fossil record, 440 million years ago. Two
thirds of all species were eradicated in the Ordovician event. The hypernova hy-

pothesis explains the extinction and also a puzzling half-million-year ice age that began at the same time, after an unusually warm period. Unfortunately, the Milky Way has rotated nearly two times since the Ordovician, so any traces of the hypernova will have spread out or been carried far away.

Astronomers may be overexuberant when they connect astronomical events with mass extinctions. There are other plausible explanations for all of them, and the models of how a supernova or hypernova affects the environment are still fairly primitive. The gold standard of evidence for an external agent causing a major change to life on Earth is still the K-T impact.[10]

LIVING WITH RADIATION

Even though it's difficult to prove causation for individual astronomical events acting on the biosphere, there's a solid statistical basis for discussing them. John Scalo of the University of Texas has made models of the radiation environment of the Earth over cosmic time (Fig. 66). Everything from the Sun to distant stellar cataclysms plays a part. The biggest radiation spikes will cause mass extinctions. The implication for life beyond the Solar System is that *any* habitat will be subject to radiation stress.

If we think of the likely effects of radiation on terrestrial organisms, we probably start with a prejudice: radiation is bad, and varying-radiation environments are very bad. But Scalo has weighed the evidence and presents a different view.

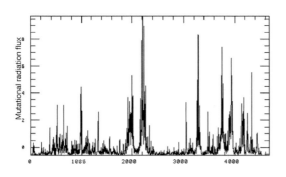

Figure 66. A model of the radiation environment of the Earth since its birth, incorporating solar flares on timescales of ten to one thousand years and distant stellar cataclysms that cause larger fluxes of radiation less frequently. The horizontal axis is time in units of millions of years. Any terrestrial planet is likely to be subject to random radiation variations such as these.

Evolution is driven by genetic diversity, which is in turn driven by mutations. DNA is damaged all the time by radiation, reactive molecules containing oxygen, and natural toxins in the environment. The average cell suffers several hundred thousand molecular lesions each day. Cells have evolved many methods for fixing DNA, and the pattern of DNA degradation and repair is part of the everyday business of a cell.

When the mutation rate is so high that repair mechanisms can't keep up, cells become dormant, commit suicide, or reproduce uncontrollably (which describes the disease of cancer). The overwhelming majority of mutations have no significant effect.

Natural selection acts on the tiny fraction of mutations that affect an organism's reproductive success by making it better or less suited to its environment. If the mutation rate is too low or the repair mechanisms are too efficient, evolution progresses slowly. In fact, some of life's earliest evolutionary advances included mechanisms to repair DNA and deal with radiation damage. Examples include lateral gene transfer and meiosis, which is the basis of sexual reproduction and diversification of the gene pool. Bacteria have had advanced methods for repairing DNA for three billion years.

An argument can be made that random or enhanced doses of radiation would promote evolution by accelerating the development of repair mechanisms and radiation resistance and also accelerating the genetic diversity on which natural selection can act. Such speculation is supported by simulations that show gains in biological complexity after bursts of mutations. The largest radiation events would overwhelm all biology's defenses, but only for organisms on or near the surface. Water is an effective shield, so creatures living deep in the ocean could ride out the environmental impact.

LIFE IN OTHER NEIGHBORHOODS

Our Sun is an undistinguished middle-aged star living in an unremarkable suburb of the Milky Way, but there are many different potential environments for life. What would we find if we were transported to a different star in a different part of the galaxy? When we set our expectations for life beyond the Earth, we're bounded not only by the history of this planet but also by the nature of our particular cosmic environment.

Our first stop isn't far from home as the crow flies, only 1,300 light-years. We're in Orion, a bustling region of star formation that traces a spiral arm of the Milky Way. The journey through a wormhole has dumped us near a hot, young star, one hundred times brighter than the Sun. Cobwebs of gas drape across the sky, and most stars look slightly red due to the gauze of dust. The four Trapezium stars blaze brightly in the Orion Nebula, and other massive stars litter the sky. There's lots of evidence of past supernovae, and heavy elements have been ejected liberally.

With so much material for planets, most stars have a dozen or more. On the other hand, many stars live fewer than one hundred million years, and those that live longer must contend with the violent deaths of their massive neighbors. Among the habitable planets, there are many truncated biological experiments. The cosmic environment favors life with a fast evolutionary clock and life that develops underwater and in rock.

Our next hop through the wormhole drops us near the urban center of the galaxy. The density of stars is one thousand times higher than near the Sun, so the entire scene is lit brighter than the sky during a full Moon on Earth. Where stars are crowded the tightest, the sky crackles with high-energy radiation from the heart of darkness of the Milky Way galaxy: a supermassive black hole. A cool dwarf hangs like a blood orange over our heads. It has six planets on tight orbits, two of which are in the habitable zone.

Overall, this is a promising environment for life. There's a lot of iron and silicon for building planets and plenty of carbon for making life. Planets are drifting among the stars, ripped from their gravity moorings by stellar encounters. A few are shrouded in thick atmospheres and massive enough not to need the external life support from a star. With a high stellar density and many planets per star, the spread of life is guaranteed—life-bearing rocks are routinely ejected from planet surfaces by impacts, resulting in an inefficient but extensive shuttle system for microbes. The high degree of stellar concentration means that the signaling time between intelligent civilizations is short. If there's an interstellar Internet anywhere, this is the place.

Our last jump takes us thirty thousand light-years into the halo. There's dark matter here, of course, as there is everywhere, but no gas or dust and very few stars. The two-armed spiral of the disk is laid out below like a Persian rug, blue-white knots of star formation set into a sparkling yellow star field. The white dwarf nearby is hot and titanium-white. Formed a long time ago, with not much grist for planets, it has three the size of Mercury, and they're all long dead. The next nearest star is one hundred light-years away. Biology is sprinkled lightly in the attic of the galaxy. It's lonely here, and it's a shame such a gorgeous view is wasted.

There are many interesting environments in the Milky Way. Some of them are likely to be hospitable for life. Life on Earth faces stress from the cosmic environment, but it might not be any better elsewhere. Voltaire was wrong to mock; this may be close to the "best of all possible worlds" that Leibniz spoke about in his 1710 work *Theodicy*.

THE EVOLVING BIOSPHERE

IMAGINE YOU'RE STANDING at the entrance to a cave, hesitant about venturing into the darkness. There's life here—trees, the grass under your feet, perhaps birds or a bat flying out of the cave. When we think of life, we think of familiar plants and animals set against a backdrop of earth and sky.

But life is not only scattered on the surface of the planet; it has also been deeply integrated into everything around you. Trees and grass are rooted in soil that was created by microbes from pure rock. The air is rich in oxygen from the respiration of photosynthetic organisms. The cave is made of limestone, which is composed of layers of ancient marine creatures. Life and its traces are everywhere.

CHANGING SUN AND ATMOSPHERE

The biosphere is the global ecological system of living things and their interactions with earth, air, and water. If we want to understand how life might evolve on a terrestrial planet in a distant solar system, we need to know how our planet has affected life and how it has been changed by life.

We start with temperature, which at most places on the surface is comfortably in the range where water is a liquid. Sunlight at a distance of one hundred million miles has enough energy to create a temperature of $1\,°F$, or $-17\,°C$, well below freezing.

Luckily for us, the atmosphere retains heat because it acts like the windows of a car or a greenhouse, transmitting visible radiation more readily than it does infrared radiation. Visible sunlight escapes back into space, but the longer infrared waves are trapped and heat the air, especially at low elevations. Trace components of the atmosphere—carbon dioxide, methane, and water vapor—are particularly effective at trapping infrared waves, so they're called "greenhouse" gases. After accounting for the mixing of vertical layers by convection, greenhouse gases raise the temperature at the Earth's surface to an average of about $60\,°F$.

That's the situation now, but as we peer into the past, there's a problem. Over the history of the Earth, the Sun has been getting brighter. As hydrogen is fused into helium, the helium nuclei take up slightly less space than the hydrogen nuclei that they're made of, so the interior contracts and gets slightly hotter. The increase in temperature raises nuclear-reaction rates and the amount of radiation received by the Earth. Four billion years ago, the Sun was 30 percent fainter. Early Earth with the same atmosphere it has now would have been a frigid $-1\,°F$, and it would have been completely covered in glaciers until one billion years ago. Yet there's good geological evidence for oceans, sedimentary rocks, and running water over most of the Earth's history.

The best answer to the paradox is that the Earth had a higher concentration of carbon dioxide early in its history, due to gas released from the core, and

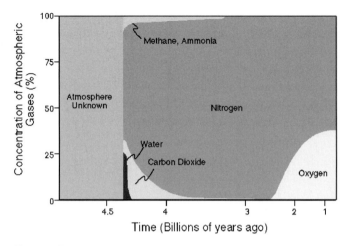

Figure 67. The Earth's atmosphere has changed substantially over 4.5 billion years. Some of the changes have been geological, like the loss of water vapor into the oceans and carbon dioxide subsequently dissolving in the oceans. The rise of oxygen is due purely to the emergence of photosynthesis and the respiration of living organisms.

most of the excess was steadily locked into sedimentary rocks and dissolved in the oceans. There's little direct evidence for this neat explanation, unfortunately, and it seems suspicious to hypothesize a declining greenhouse gas that exactly compensates for the dimming Sun and just manages to keep the Earth habitable.

One of the most dramatic changes to the biosphere was the rise in atmospheric oxygen just over two billion years ago (Fig. 67).[11] Oxygen is corrosive and reactive. Without the continual production of oxygen by living organisms, the entire atmospheric store would disappear in only four million years. Oxygen inhibits the chemical reactions that lead to amino acids, so the building blocks of life couldn't have formed with oxygen around. Descendants of the first organisms—with metabolisms that don't use oxygen—are still found in oxygen-poor environments, such as swamps and lagoons.

How did the takeover of an oxygen "economy" occur? Study of genetic trees shows that the first oxygen producers were purple prokaryotes: single-celled organisms called cyanobacteria. They also come in rainbow shades of yellow, red, green, and blue. Cyanobacteria convert light energy into chemical energy in the form of carbohydrates, with oxygen as a by-product. Photosynthesis doesn't require either sunlight or oxygen; there are bacteria that photosynthesize far from the Sun's rays near ocean vents and bacteria that photosynthesize with hydrogen sulfide instead of water, with sulfur as a waste product instead of oxygen.

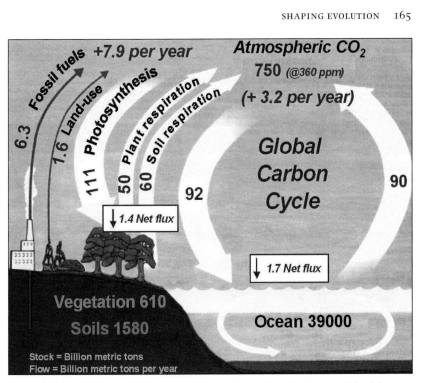

Figure 68. Carbon is life's essential element, and it cycles through the Earth, the sea, and the air in a complex cycle of chemical reactions. The biggest repository of carbon for use in the biosphere is the crust and mantle. Plate tectonics is the process that drives the deposition of carbon dioxide back into the atmosphere. A typical carbon atom might spend one hundred million years in one complete carbon cycle.

The real advantage of the oxygen economy is efficiency. Metabolic processes that use oxygen store twenty times as much energy (in the form of phosphate bonds in ATP) as anaerobic metabolic processes such as fermentation. All Earth's animals are dependent on a food chain that has photosynthetic plants and plankton at its base. That food chain is supplied by an essentially infinite source of energy: the Sun.

CARBON AND THE EARTH'S CRUST

Another deep connection between life and the planet involves carbon. Most of the Earth's carbon is stored as carbonates and kerogens, the waxy solids that form when heat and pressure "cook" the remains of plants and animals. Weathering could remove all the carbon in the biosphere in one million years. Without a fresh source of carbon, all life on Earth would have ceased long ago (Fig. 68).

The mechanism for supplying new carbon is plate tectonics. The Earth's crust is divided into plates that slide over the mantle on a "grease" of pressurized rock. Heat from the interior causes convection in the mantle—a circulating motion of magma that follows similar physical principles as thunderclouds and boiling water. At some plate boundaries, magma rises up to produce new crust, while at others crust is lost when one plate slides under another. When a plate is pushed down, the sediments that it contains are heated to over $1,300\,°F$ $(700\,°C)$, and volcanoes release the carbon dioxide (and water) into the atmosphere. Any particular carbon atom might spend up to one hundred million years trapped in carbonate sediment before being cycled back into the air—hundreds of times longer than it would spend as part of the living biosphere.

Carbon is so central to the story of life and the story of our planet that we take it for granted. Yet its story is a marvel, at once mundane and evocative. Nobody told it better than the chemist and writer Primo Levi. Levi survived Auschwitz, and his view of the worst that man does to man is a painful counterpoint to natural selection. As he wrote in his essay about the concentration camp, "Shame," from the book *The Drowned and the Saved:* "The worst survived, that is, the fittest; the best all died."

In his book *The Periodic Table,* Levi told the story of a carbon atom, and here we can build on his riff. Forged in the heart of a star, it may have drifted for billions of years among the stars before being trapped by the gravity of the solar nebula and swept up into the forming Earth. Some carbon atoms are interred so deep that they never see the light of day; this one is part of a cycle where every hundred million years or so it is dissolved in the ocean or laid down in sediment belched from a volcano. Compared to the geology, all of the carbon atom's adventures in the biosphere are lightning fast. Carbon that's not locked in rock enters and reenters the door of life every few hundred years through the process of photosynthesis. We can imagine it flowing into the lungs of birds and out of the leaves of plants, sometimes carrying information, sometimes carrying energy, winding its way through the biosphere in a story that's both magical and real.

Individual carbon atoms apply the tiny brushstrokes to life while the collective store of carbon acts as the canvas. As we've seen in the Gaia hypothesis, some chemical cycles act as negative-feedback loops: when a change pushes one way, the system pushes back the other way. The interplay between carbon-dioxide abundance in the atmosphere and rainfall-driven erosion to create carbonates may explain why the climate has been suitable for liquid water over most of the Earth's history.

Most but not all. Titanic geological forces acting over long spans of time have led to something called the supercontinent cycle. For the first two billion years of the Earth's history, there was no large continental landmass, just a set of smaller rocky platforms. It took a long time to build up a mass of granite that could form a set of large plates.[12] The plates move around and sometimes stick together, like overcooked ravioli in a pan of boiling water. Every half-billion years or so, the continents all join to form one supercontinent in a global ocean. About eight hundred million years ago, a single supercontinent called Rodinia formed and then broke up. A new supercontinent called Pangea formed about 250 million years ago, and its pieces drifted apart to form today's continents (Fig. 69).

SNOWBALL EARTH

Joe Kirschvink waves his hands to better summon up the dramatic scene he's describing. Earth is locked down in a deep freeze. Ice fields cover the land and the oceans, extending almost to the equator. Life is besieged, hunkered down in the oceans, hanging on by a thread. Outside, the modest buildings of the California Institute of Technology campus are framed by warm Spanish tile, and the palm fronds sway in a gentle breeze. Life here is good, and it's hard to imagine a time when the planet was besieged, a time that Kirschvink gave the nickname "Snowball Earth."

From about eight hundred to six hundred million years ago, there's good evidence of widespread glaciation in the geological record. Glacial deposits are seen all the way to the edge of continents and at low latitudes. In the deepest spike of glaciation seven hundred million years ago, ice probably reached all the way to the equator.

How did it happen, and how did we escape from this chilly scene? The breakup of Rodinia and the subsequent collision of the plates raised mountain ranges and increased rain and erosion, removing carbon dioxide from the atmosphere. As the plates drifted to higher and cooler latitudes, they were covered with ice and reflected more sunlight, which accelerated the cooling trend. Toss in the fact that the Sun was 8 percent dimmer then, and you have the makings of Snowball Earth. Think of it as the antigreenhouse effect.

A bizarre scene would face a time trav-

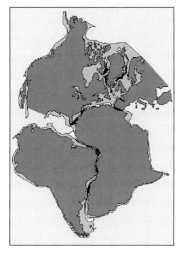

Figure 69. The breakup of the last supercontinent, Pangea, started a process of geological change that led to the current glacial cycle and climate variations that proved challenging for life. An earlier cycle of supercontinent building and breakup, seven hundred to five hundred million years ago, led to the dramatic cold epoch called Snowball Earth, when ice may have covered the oceans nearly to the equator.

eler to the pre-Cambrian. White vistas stretch all the way to the horizon, and the air temperature is a frigid –25 °F. An ice pack one thousand feet thick covers the oceans, with a narrow band of water visible at the equator. On the continents, there's little rain; weathering has slowed to a crawl. Almost all water and ice have evaporated, leaving a stark landscape of sterile brown rock.[13] There are three major epochs of glaciation spanning two hundred million years, with very little respite between them. Compared to Snowball Earth, the ice age suffered by our ancestors is a momentary cold snap.

During Snowball Earth, carbon dioxide bubbled up from undersea volcanoes and was trapped in the oceans, building up to a level unprecedented in the Earth's history, hundreds of times the present value. As it was quickly released through rifts in the ice pack, greenhouse gases began to melt the worldwide glaciers. As ice melted, more sunlight was absorbed, and the process accelerated, melting the entire ice pack in a little more than one million years. When Earth emerged from the deep freeze, the conditions for life had changed radically. The oceans were oxygen-rich after eons of being oxygen-starved. The temperature was balmy. After eons of clenched and chattering teeth (metaphorically speaking, since teeth hadn't yet come along), life could get on with living.

Kirschvink is animated as he gives a guided tour of his artifacts. He works in a basement office stuffed with rocks and core samples. It's as gloomy as a subterranean vault, and the abundant California sunshine can get in only via two small windows set high in one wall. As a full professor, Kirschvink could have a fancy office on the promenade deck, but he prefers to be closer to his lab, closer to the rocks themselves. He wears the geologist's uniform of a plaid shirt and ill-fitting pants, and his rough hands move delicately as he handles rock samples from all over the world. He can read the strata and colors and textures as easily as you or I can read a book.

Kirschvink argues for an even earlier episode of global glaciation, from 2.3 to 2.2 billion years ago. The Sun then was only 85 percent of its current brightness, so this snowball episode may have been more severe than the one just described. According to Kirschvink, life played a critical role in this ancient lockdown. Here's what he thinks happened. Despite the dim Sun, the Earth was temperate due to a blanket of the greenhouse gas methane. Then about 2.3 billion years ago, cyanobacteria evolved the ability to gain energy by breaking down water. The oxygen released by this new type of photosynthesis interacted with and then destroyed the methane in as little as one million years. Robbed of its protective blanket, the Earth's temperature plunged to –58 °F (–50 °C). Ice at the equator was a mile thick. Life hung on in a few special ecological niches and was nearly extinguished completely.

Then evolution pulled another trick. It developed an ability to metabolize oxygen. The carbon dioxide released by that familiar process is also a greenhouse gas, so gradually the Earth clawed its way back from the brink and began to warm up. "It was a close call to planetary destruction," says Kirschvink. "If Earth had been a bit further from the Sun, the temperature at the poles could have dropped enough to freeze the carbon dioxide into dry ice, robbing us of the greenhouse escape from Snowball Earth."

Not everyone buys Kirschvink's early Snowball Earth idea, but the evidence is accumulating, and he's a pugnacious and combative debater. What's fascinating is the fact that each of the Earth's near-death experiences precedes a time when evolution leaps forward with new capabilities. The more recent snowball episode immediately precedes the Cambrian "explosion" of life in the oceans. The earlier one coincides both with the first traces of eukaryotes—cells with nuclei—in the fossil record and with the invention of highly efficient oxygen photosynthesis.

VIOLENT CHANGE IS NORMAL

We're still riding a climatic roller coaster that resulted from the breakup of the last supercontinent, Pangea. As the continents separated and new mountain ranges were thrust up, carbon dioxide was scrubbed from the atmosphere, and the Earth cooled. That large-scale change initiated climate swings that persist today: eras of glaciation one hundred thousand years long with slightly warmer breaks and cold pulses every few thousand years between. Current global warming is a glitch injected into the machine by humans; overall, Earth is in a cool phase.

In addition to the long-term effects of supercontinent breakup, there are dramatic episodes of volcanism called flood basalts that may affect the course of life, not through loss of habitat but through the release of volcanic gases and their effects on climate. Large parts of Siberia, northern India, and the western United States show evidence of enormous lava flows. They can cover a million square miles with thousands of times more lava than has been released by Kilauea in Hawaii in the past few decades. Flood basalts match the timing of several mass extinctions.[14]

We've seen earlier that changes in the Earth's orbital properties can also affect climate. So which one of these traumas affects life the most: changes in the Sun, impacts from space, variations in the Earth's orbit, volcanic explosions, or large-scale geological evolution? For particular events in the biosphere, the evidence is often too ambiguous to be sure, but they all must play some role.

SHAPING LIFE ON EARTH

IN 1654, JAMES USSHER, the archbishop of Armagh and the Primate of All Ireland, published a definitive reckoning of the age of the Earth. From the genealogy in the Bible, he set Genesis at 9:00 a.m. on October 23, 4004 B.C. Aside from the implausible accuracy of the calculation, Ussher had usefully estimated the time span of human civilization.

THE EVOLVING PLANET

The Earth, however, was much older. In the late eighteenth century, James Hutton stood on Hadrian's Wall and saw carved inscriptions that had barely changed in 1,600 years since it had been built by the Romans. He'd looked out at the river valleys in his native Scotland that were etched from the same granite, and he knew they must be truly ancient. He wrote in his 1788 book *Theory of the Earth* that there is "no vestige of a beginning, no prospect of an end."

Another Scottish geologist, Charles Lyell, was an undergraduate when he returned to a beach he'd known as a child and noticed that the coastline had slightly changed. As a young man, he scrambled up Mount Etna in Italy and calculated that it must have taken millions of years to build the bulk of such a large mountain by successive lava flows. He also studied fossils, finding some that didn't correspond to any known living creature and noting that they were dispersed through equally old strata of rock. The history of life and the history of the Earth were both written in stone. His book *Principles of Geology* established a new scientific field.[15]

DARWIN'S BRILLIANT IDEA

Lyell was Charles Darwin's father-in-law, and Darwin had a copy of the first volume of his book in his sea chest as he embarked on an extended voyage on a small ship called the *Beagle.* His five years traveling the South Seas convinced him that gradual variations could generate the profuse diversity of species found on Earth. But Darwin knew his theory would rock the world and expose him to criticism, so his manuscript gathered dust in a desk drawer for fifteen years. It was Lyell who persuaded him to publish, along with the realization that he was about to be scooped by Alfred Russel Wallace, who had independently arrived at the same idea. *On the Origin of Species* appeared in 1859.

Like all great ideas in science, Darwin's theory of natural selection is compelling in its simplicity. Reproduction leads to variation, a fact well known to animal breeders for centuries. Also, living creatures tend to produce more off-spring than the environment can sustain. The probability of survival is shaped by environmental pressure. Endless changes in the world lead to small but in-cremental changes from one generation to the next, and when they have accu-mulated to the point that one group can no longer produce fertile offspring, a new species emerges.[16] After reading Darwin's book, the naturalist Thomas Huxley remarked, "How incredibly stupid of me not to have thought of that."

That's it. The fittest survive. Fitness conveys no sense of progress or "moral" superiority, it simply means a species fits the environment. When the condi-tions change, all bets are off, and species may become extinct. To some, natural selection is brutal. Not to Darwin: "There is grandeur in this view of life . . . from so simple a beginning, endless forms most beautiful and most wonderful have been, and are being evolved."

VARIATION AND EVOLUTION

Evolution is a challenging concept. The process of slow but incremental change is hard to visualize; it takes hundreds or thousands of generations for adaptation to lead to a new species. Yet there are many examples of visible or accelerated natural selection, with the beaks of Darwin's Galapagos finches as a prime one. The role of chance is also unnerving to many people. How could random variation lead to something as remarkable as an eye or flight? What use is half an eye or half a wing?

There are plausible answers to these questions. Flying has selective advan-tage for both predators and prey. From the ground up, flying emerged as an ex-tension to hopping and leaping. For tree-dwelling creatures, it developed from gliding. Transitional forms on the way to wings have been found in the fossil record. It's the same story with eyes, which first appeared 550 million years ago in ocean creatures. Predators and prey each benefit from the ability to sense light. Early eyes were no more than clusters of light-sensitive cells on the skin. The same sensors in a slight depression or pit added a directional capabil-ity. Later, transparent tissue served a role protecting the delicate cells, and when the cavity was filled with fluid it could form an image. Each of the transi-tional stages toward the modern camera eye has been identified in the fossil record.[17]

Most mutations are harmful or confer no selective advantage on the organ-ism. The few that lead to a greater probability of survival steadily take root in the population, even if the advantage is slight.[18] Life evolves new capabilities and new forms after filtering by the environment. In evolution, as in most other branches of science, it makes no sense to ask why. It only makes sense to ask how.

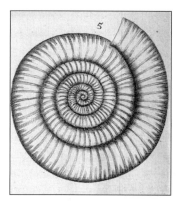

Figure 70. Ammonites are one of history's most successful life-forms, having thrived from four hundred to sixty-five million years ago with little change in their shape and function. They lived near the surface of the ancient oceans, probably ate fish and crustaceans, and were preyed upon by marine reptiles. The chambered nautilus is a close descendant.

How does natural selection play out against a backdrop of environmental stresses, some geological and some astronomical? On our restless planet, change is guaranteed, and failure is the norm. More than 99.9 percent of the species that have ever lived are extinct. The tree of life is littered with the dead twigs and branches of failed evolutionary experiments. On the other hand, some forms of life have proved to be amazingly durable and successful. Cyanobacteria, or blue-green algae, have prospered in a wide range of ecological niches for 3.8 billion years, and ammonites survived unchanged for 350 million years, ranging in size from a fraction of an inch up to five feet (Fig. 70).

THE ROLE OF CHANCE

The recurrence of certain successful forms, alongside the ephemerality of most species, has led to a tension between two ideas in the theory of evolution: contingency and convergence. Paleontologist Stephen Jay Gould was the most vocal champion of the idea that chance played a decisive role in evolution, illustrated by the opening quote of this chapter. Gould was a commanding but divisive figure in the field of evolution—masterful as a researcher and popular writer yet sometimes arrogant and contrarian in his thinking.

Gould used the Burgess Shale as his central example. The Burgess Shale is a very well preserved example of the profusion of exotic life-forms that appeared in the Earth's oceans just over 510 million years ago (Fig. 71). Most lineages didn't survive, and Gould argued there was no way to predict which of these well-adapted organisms would endure. Capricious events such as meteor impacts made evolution highly contingent. In this thinking, microbes may be common on Earth clones throughout the cosmos, but mammals and reptiles would be unlikely, and primates and humans would be improbably rare.

LIFE'S COMMON SOLUTIONS

Simon Conway Morris, a paleobiologist at Cambridge University, was a graduate student when he was first exposed to the treasure trove of the Burgess

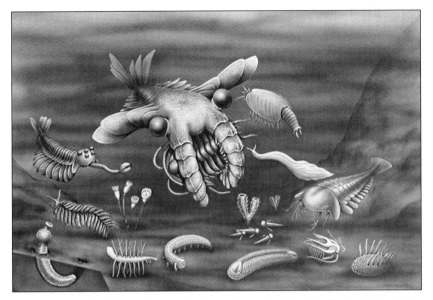

Figure 71. About 550 million years ago, an abundance of large animals developed for the first time in the oceans of the Earth. Modern arthropods are built from three basic body plans; the Cambrian explosion saw a huge number of evolutionary experiments, most of which left no survivors. The best evidence for this early proliferation of life comes from the Burgess Shale, a geological formation in Canada where the soft body parts of the animals were entombed in mud and preserved. Some of the fossils of the Burgess Shale have defied classification.

Shale. His reading of the same fossils that Gould looked at led him to a very different conclusion. Conway Morris doesn't deny the role of luck but notes that evolutionary strategies are bounded by the laws of physics and don't play out in an environment of unlimited possibilities.

Let's return to wings and eyes. Flight has evolved separately among insects, birds, mammals (the bat), and reptiles (notably the pterosaur, a flying leviathan of the Triassic). The design of wings is different in each case, but the evolutionary advantage of taking to the sky is undeniable. Vision has been discovered, and sometimes reinvented, in creatures as different as mammals, cephalopods, and insects. If you stare at an octopus, the eye you look into is eerily like your own, but it has a completely different ancestry. These are all examples of convergence.[19]

Conway Morris has identified an amazing number of examples of convergence. He acknowledges the random element of evolution but argues that life has found similar solutions to the problem of survival over and over again. As a result, he's sanguine about the inevitability of large, complex animals and even of intelligence. In a paper titled "Tuning into the Frequencies of Life," he writes, "Wherever there is life, there will, in due course, be mind. Whether it's always our mind is another question."

The British naturalist D'Arcy Thompson drew attention to convergence nearly a century ago, but some of the most intriguing examples occur at the molecular level. Two completely unrelated groups of fish use the same natural antifreeze to combat the effects of cold water. The trick is done by a protein that's coded by a sequence of the same three amino acids repeated over and over. The notothenioid fish of the Antarctic arose seven to fifteen million years ago, while the arctic cod arose at the other end of the Earth three million years ago. In another example, identical antibodies are found in two highly distinct species, the nurse shark and the camel. Similar gene circuits have been found in E. coli bacteria and yeast, showing that convergence occurs at higher levels of molecular organization.

Molecular convergence echoes the fact that heavy-element creation in stars gives a universal basis for biochemistry, as well as the fact that building blocks such as amino acids are found in a wide range of cosmic environments. There are an astronomical number of possible proteins and other biologically useful molecules.[20] Yet life selected a modest number—propagated to the level of genes, this specificity acts to constrain the functions and forms of species.

TOLERATING IMPERFECTION

Natural selection doesn't produce the best of all possible worlds. A process called genetic drift, unknown at the time of Darwin, is as important in the theory of evolution as natural selection. Genes aren't copied from one generation to the next; they're sampled, which creates some statistical error. Imagine tossing a coin, where the 50 percent probability of heads is analogous to a 50 percent probability of offspring having red hair. A few coin tosses is unlikely to give equal numbers of heads and tails. The numbers are unlikely to be equal after many tosses either, but the deviation from 50 percent is smaller. As a result, genetic drift is quicker in a small population than a large one (Fig. 72).

Genetic drift is important because it can affect any gene, and it has nothing to do with the fitness of the organism in the environment. When a population becomes small, such as after a mass extinction, there's a bottleneck effect. As the population begins to grow again, genetic drift is rapid, and it may even eliminate beneficial traits.

We should embrace the serendipity that spurred life's baroque innovation. The molecular machinery of life makes mistakes, and that's the key to its success. Imagine a microbe that evolves the ability to protect itself from radiation damage and copy itself perfectly. We might shudder, thinking these tiny clones would take over the planet. We needn't fear. A perfect replicator can't evolve, so when it meets the first bug that has learned how to eat it, it will be defenseless and totally consumed. The predator species will then die, having used up its food source. The Earth will shrug, and life will go on.

From error comes innovation. Medical researcher Lewis Thomas wrote in his book *The Medusa and the Snail*, "The capacity to blunder slightly is the real marvel of DNA. Without this special attribute, we would still be anaerobic bacteria and there would be no music."

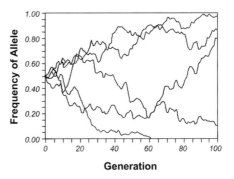

Figure 72. Genetic drift can be comparable in importance to natural selection as an evolutionary process. It is a purely statistical variation in gene expression that is independent of the role of the gene in determining fitness for survival. The graph shows the variation over successive generations in five different experiments of a gene that expresses a particular trait (an allele). The rate of drift is large for small populations; genetic drift can rapidly cause the trait to be fixed (frequency of 1) or to disappear (frequency of 0).

LANDMARKS IN EVOLUTION

LIFE ON EARTH IS a complex system. The interdependence of different components, large and small, makes it very challenging to understand the roles of chance and necessity in evolution (Fig. 73). When biologists study any single piece in the lab—organism, cell, or enzyme—they must remove it from the context that affects its function.

That we are more complicated than an amoeba goes without saying. The broad arc of evolution has moved from simple to complex, but not in a smooth or simple manner. A typical bacterium has a single cell type and

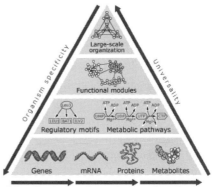

Figure 73. The biology of life on Earth is a system, in which the interplay between individual components is as important as the components themselves. A much deeper understanding of this interdependence is needed to say how biology might work under different environmental constraints, as might be found on a distant terrestrial planet.

several thousand genes. A fruit fly has 50 cell types and 13,000 genes; humans have 120 cell types and 25,000 genes. It's striking that the intricate functioning of animals such as us has required no more than a ten- to hundredfold increase in the number of genes, or information content in the genome. With twenty-five thousand genes, we could have far more cell types than we do. Evolution can generate diversity, but only a tiny fraction of the potential genetic diversity is expressed in any organism.

Let's see how chance and necessity played out in major episodes of life on Earth. Harvard paleontologist Andrew Knoll has argued that evolution follows broad directions of change that accommodate both contingency and convergence. The transitions form a logical sequence—each takes life to a new level of complexity, and each expands life's use of the ecosystem.

THE POWER OF NETWORKING

The first landmark in evolution was the emergence of a last common ancestor. We look back with hindsight at our origins and must resist any sense of inevitability. Our best mental image isn't Richard Dawkins's blind watchmaker but an untended broth where the fluctuating environment is constantly stirring the ingredients.

Simple chemical networks were sustained by free energy in the environment. With increased complexity came increased functional diversity, but it arose in fits and starts. There were many, many failed experiments. One method of replication was more successful than all the rest, and it gained traction by natural selection. Early cells didn't arise until there'd been many evanescent containers. Even the most primitive cell seems to be a marvel of function and form. What we can't see is the hundreds of millions of years of churning chemical variation that led up to it. Once success was achieved, one genetic mechanism became the template for everything that followed.

The second stage was diversification of prokaryotes, cells without nuclei. Bacteria radiated into every conceivable ecological niche. And let's not get too full of ourselves; they're still by far the Earth's most abundant and successful form of life.

Bacteria and their ancient archaean cousins reproduced by cloning: the genome is completely replicated when the parent cell splits. Diversification took place primarily by lateral gene transfer. Gene transfer is a powerful way to adapt because useful chunks of DNA are passed across different lineages. Think of it as tool swapping among microbes. Bacterial lives are short, so adaptation was refined over millions of generations, using all the combinatory possibilities of thousands of genes, each expressed in thousands of ways, depending on the action of local enzymes. Prokaryote evolution was hypereffi-cient. Microbiologist Lynn Margulis said, in the book *What Is Life?*, "Bacteria trade genes more frantically than a pit full of traders on the floor of the Chicago Mercantile Exchange."

The next stage was the development and diversification of eukaryotes. We are eukaryotes, as are all other animals and plants. Eukaryotes house their DNA within a nucleus, but they are primarily distinguished from prokaryotes by mitochondria—little powerhouses that govern oxygen respiration and cre-ation of ATP—and chloroplasts, which control photosynthesis in plants and algae. Margulis, who was cocreator of the Gaia hypothesis, noticed that mito-chondria resembled aerobic bacteria that had been ingested by the cell, and chloroplasts resembled cyanobacteria that had been similarly consumed. What if the modern complex cell had emerged as a result of cooperation among ancient bacteria?

The theory is called endosymbiosis. Bacteria have always formed colonies, as we know from the 3.5-billion-year-old stromatolites that constitute the most reliable evidence for early life. In endosymbiosis, prokaryotes entered mutually advantageous or symbiotic relationships. Out of this endless experimentation, most relationships were failures, but in a few cases chimeric organisms sur-vived and prospered.[21] Gradually, the ingested bits of genetic apparatus lost their independence and became part of a harmoniously functioning cell. Mar-gulis and her son Dorion Sagan (whose father was astronomer Carl Sagan) ob-served in *What Is Life?*: "Life did not take over the globe by combat, but by networking."

When did this happen? Prokaryotes ruled for the first 1.5 billion years of life, but starting around 2.7 billion years ago there's evidence for cholesterol-related compounds that are specific to eukaryotes. Actual fossil cells are not found until 1.9 billion years ago. Eukaryotes emerged when there was little free oxygen available. Many used hydrogen as an energy source; for five hundred million years, the oxygen and hydrogen "economies" coexisted. But photosyn-thesis provided a compelling energy advantage, spurring a rise in oxygen in the atmosphere that coupled to an early Snowball Earth. Even if this episode stressed the existing forms of life, the advanced adaptive mechanisms of eu-karyotes were already in place, and subsequently they flourished.

LIFE SURGES

Multicellularity is a natural extension of symbiosis. Just as people moved from a nomadic existence to form cities where people did specific jobs that served the whole, so organisms gained by aggregating and specializing. It's only a small conceptual step from stromatolites (bacterial mats) and lichen (symbiotic colonies of algae and fungi) to true multicelled organisms.

The timing of branching points in the tree of life comes from molecular "clocks" tethered by the evidence of paleontology. Eukaryotes diversified 1 to 1.2 billion years ago. By the time plants and animals diverged in the tree of life, about eight hundred to nine hundred million years ago, there were ten different cell types. The first signs of multicelled creatures are fossil worms a billion years old, at which time there were already fifty different cell types. All the ingredients were in place to develop more complex life-forms, yet there were no large creatures on land, and nothing in the oceans larger than a thumbnail. Then came an extraordinary surge in evolutionary diversity called the Cambrian "explosion."

It began 550 million years ago, close on the heels of Snowball Earth. Within a mere fifty million years, more than 80 percent of the internal and external skeleton designs that we see today appeared (Fig. 74). The Earth's oceans indulged in a paroxysm of evolutionary creativity not seen before or since. Some creatures were truly bizarre. One had five eyes and a nozzle for a mouth, another walked the seafloor on fourteen struts. These are among the fossils that surprised and delighted the young Simon Conway Morris. The Cambrian oceans were like an alien world (Fig. 71). Over the last half-billion years, animals have continued to spread and diversify in a lurching progress shaped by mass extinctions and geological upheaval.[22]

What caused this unprecedented surge in life? Joe Kirschvink, of Snowball Earth fame, has made a strong case that all hell broke loose on the Earth in the

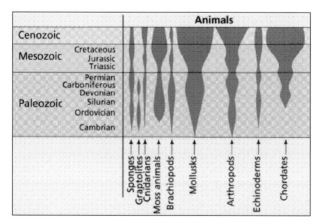

Figure 74. Starting about 550 million years ago, multicelled life on Earth began to rapidly grow in size and diversify in function. This shows the change in the number of species of major classes of animals since the Cambrian explosion. Evolution since then has not been steady but has been shaped by mass extinctions and dramatic changes in the environment. Humans are chordates.

narrow interval from 530 to 515 million years ago. The Rodina supercontinent broke up, continents raced apart and reformed another supercontinent, Gondwanaland, the ocean chemistry changed abruptly, and the rate of evolutionary diversity jumped by a factor of twenty.

Most rocky planets have magnetic fields tethered in the iron that sinks to their cores. Planets like the Earth, with a liquid core, also undergo slow wanderings of the polar axis, but the Earth went through an additional instability in which the polar axis moved rapidly. (Think of the oscillations of a spinning water balloon.) The rapid motion of the continents was one outcome, which in turn led to rearrangement of the global climate. The biosphere was disrupted, and the resulting fragmentation and isolation of populations was the spur to evolutionary diversity. Geology drove biology.

THE EMERGENCE OF BRAINS

The most recent transition in evolution is the growth of intelligence. Humans have unusually large brains for their body mass, but there's a continuum, and while we have an edge, we don't blow away the competition (Fig. 75). Among vertebrates, brains and central nervous systems developed in tandem, driven by the need to process sensory data, which in turn was driven by natural selection in a world where predators and prey were always jockeying for survival.

Before the Cambrian explosion, a creature with a patch of heat-sensing cells on either side of its head had to handle only a few bits of information each second. Compare that with the demands of the modern eye, with six million cones for color vision and 120 million rods for sensitive black-and-white vision, generating data at a rate one hundred billion times faster. It's enough to keep a big brain busy (and maybe give it a headache). Brains are demanding. We use 25 per-

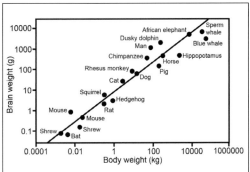

Figure 75. There is a relationship between brain mass and body mass for all animals, shown here for a variety of existing and extinct species. Elephants and certain whales have larger brains than humans, but we are exceptional in our ratio of brain to body mass, as indicated by distance away from the sloping line. Intelligence is not just defined by brain mass, however; it depends on the nature of the sensory input and the complexity of the internal connections and pathways.

cent of our metabolism to run our brains, which is an enormous investment of energy.

Since our ancestors separated from chimpanzees in the tree of evolution about four million years ago, hominid brains have increased in size twice. Evolution from 3 to 1.5 million years ago left *Homo erectus* with a brain twice as big as a chimpanzee's. A more rapid phase of evolution ended with the emergence of *Homo sapiens* about 150,000 years ago and the emergence of tool use. Our brains are now three times the size of a chimp's. Our preeminence on the planet conceals how close we are in evolutionary terms; as far as DNA goes, we're nearly twins.

As with earlier phases of evolution, we see the influence of environmental and genetic variations. Our ancestors developed as the Earth was emerging from the grip of a monster ice age that started forty million years ago. They were buffeted by sharp changes in climate for several million years. Why did it take the anatomically modern *Homo sapiens* so long to develop agriculture and civilization? Perhaps progress was delayed because the Earth plunged into a mini–ice age around 110,000 years ago, emerging just ten thousand years ago (Fig. 76). Genetic tracers show that sometime between fifty thousand and one hundred thousand years ago, human population dropped to about ten thousand. We were hanging on by our fingertips. It's not clear if this was due to a single event, like Indonesia's Mount Toba explosion, which was thousands of times more powerful than Mount St. Helens's in 1980, or a series of glacial stresses.

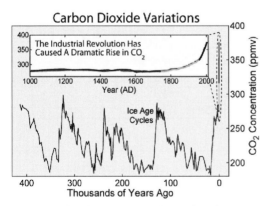

Figure 76. Air bubbles in ice cores and tracers in sediments give us the history of atmospheric carbon dioxide over the span from early humans to recent Neanderthals. Deep ice ages occurred through much of that span. We are currently in a warm phase, which is almost certainly being exacerbated by rising carbon dioxide levels due to human industrial activity.

As a result of this bottleneck, human genetic diversity is low, and most of it is expressed in the African population from which we emerged. Much as we like to judge books by their covers, human differences are only skin-deep. The mutation of a single gene 2.7 million years ago may have facilitated early growth in human brains. Another gene strongly controls brain size; new variants of this gene appeared thirty-seven thousand and then six thousand years ago. These times correspond to the emergence of art, music, and religion, and the start of the first great human civilizations. We are still evolving, even though it doesn't always seem like we're acting any smarter.

Brain size evolved to an impressive level in cetaceans as well as in primates and for similar reasons. Lori Marino of Emory University has used fossils to construct the history of brain size in marine mammals. After a wave of extinctions thirty-five million years ago, new whale species surged in brain size as they developed the echolocation skill. The dolphin line went through a second increase twenty-five million years ago, leaving them with brains that are no smaller than ours.

Are these stages in evolution unique or inevitable? What's the timescale for each one when the planet is not the Earth but instead is Earth's cousin in a distant star system? Does convergence rule, ensuring that alien species will have familiar function and form? Or is contingency king, and our imaginations are not supple enough to imagine life on distant worlds? These questions motivate the search for life in the universe.

5.

LIVING IN THE SOLAR SYSTEM

What's interesting about Mars is, if it has life, then there's probably life everywhere. That's what keeps astronomers going.

—**Seth Shostak**, senior astronomer at the SETI Institute

Her feet crunch into the red crust. She hops tentatively, testing the feel of one-third gravity. Then a second hop, higher and more playful. Nice. Looking back she sees her spacecraft resting in a shallow depression. Squat and small, it looks like a toy. A shadow of anxiety crosses her mind. With all the trade-offs, cuts to NASA, the needs of national security, it was this or nothing: a single person sent to Mars to bring back samples. A robotic clone of her spacecraft stands ready for launch, but she knows the odds of a successful rescue are long. She's one hundred times farther from Earth than anyone in history. Utterly alone.

Yet there is nowhere else she would rather be. The geologist gets to work. Her trained eye scans the alluvial plain and settles on one particular outcropping. She moves toward it in an awkward loping motion. It's difficult to judge distances through the thin atmosphere laced with dust, where ochre rocks shade into an apricot sky. An hour later, she is there, with only the sound of her breathing for company. Along one slope of the outcropping, a raised seam, split like a wound, exposes the layers below. Perfect.

She assembles the core sampler. Soon, its tiny drills are biting down into rock, and minutes later she has extracted four vertical feet of the Martian soil, its layers neatly arranged. The geologist attaches a miniature PCR assayer to the far end of the core. It dissolves samples of the rock in its reaction chamber. The green light goes on. There is DNA there, or something like it. Soon, the display will light up with the colored patterns of a nucleotide sequence. She watches intently.

She almost doesn't notice the pale, granular crystals higher in the sample, from a more recent layer. They are literally seething with life.

· · ·

THE GOLDILOCKS PLANET. It's not too hot, and not too cold. It has lots of water and a protective atmosphere. It's stabilized in its orbit by a large moon and protected from excessive debris by a sturdy sentinel, Jupiter. Welcome to our home—Earth—the only place we know with life.

Everything we've learned about the universe warns us against assuming that we are special. The continuing Copernican Revolution has revealed that our star is one of billions of stars in the Milky Way. The process of star formation naturally leads to rocky planets. Water and carbon are widespread in the universe. Yet there are attributes of the Earth that make it particularly hospitable for biology and the evolution of large, complex organisms. This has led to an idea called the Rare Earth hypothesis, which proposes that microbial life may be common in the universe, while advanced life is exceedingly rare. But is the Earth just right, or are we indulging in one of Kipling's "Just So Stories"— telling ourselves a complex tale of how we came to be and then imagining it couldn't have happened any other way?

Earth is so fecund that we assume it's the only place worth living in the Solar System. But three billion years ago, Venus and Mars were temperate and wet, and each may have hosted primitive life.

Mars still exerts a strong grip on the popular imagination. It teases—parched and dust-swept on the one hand, but with hints of recent running water on the other. Robotic probes have visited Mars several times in the past few years, and a small armada is being readied to scour the planet for signs of life. The big prize is a manned mission to bring back samples, but even the richest country in the world is blinking at the price tag, so a robotic sample-return mission will probably come first.

Venus and Mars are in or near the traditional habitable zone—a slender range of distances from a star where a planet can have liquid water on its surface. But two of the most promising sites for life, or at least sites for insights into prebiotic chemistry, are in the remote zone of the giant planets. Jupiter's moon Europa has a fractured ice pack that covers an ocean whose depths may be kept tepid by heat leaking out from rock. Saturn's moon Titan is even farther from the Sun; it has an atmosphere as thick as Earth's and a surface shaped by ice and organic rain.

If we're guided by the range of extremophiles on Earth, that makes five potential living or formerly living places in our Solar System alone. Life could be found on rocky worlds close to a star or large moons kept warm by radioactivity or tidal forces. As many as a dozen small worlds in the outer Solar System may have the basic ingredients for life: water, an energy source, and organic material. Perhaps life doesn't need a star at all?

Searching for life's traces in the Solar System, we're reminded of the stark truth about outer space. It's numbingly, jaw-droppingly big. How can it be so hard to go and get a Mars rock? Space travel is expensive, but after fifty years of superpower rivalry the private sector is finally getting a chance to participate. Entrepreneurial zeal may soon propel us to exciting new types of exploration, including interstellar travel.

It will take resolve and vision to explore the Solar System for life. In the search, we can be guided by the known range of terrestrial life, but we mustn't be blinded by it, or we'll miss important discoveries. The Earth continues to need our attention badly, but nearby worlds beckon with promises of teaching us about our own origins.

HOW SPECIAL IS THE EARTH?

IN THIS FANTASY, one hundred years pass, and we learn how to build small robotic space probes. A legion of them is launched from Earth orbit at half light speed. They're each programmed to travel to a different long-lived star like the Sun and home in on its most Earth-like planet. The project is called Artemis, after the twin sister of Apollo. As Artemis rode her silver chariot across the sky and shot arrows of moonlight to the Earth below, so these five thousand metallic voyagers travel far and send their signals home. Each probe descends on a one-way trip to an alien world, collecting images up to the moment of impact.

The architects of this project have a flair for drama; they resist the temptation to present the data on each new planet as they trickle in. Rather, they collect the signals with an orbiting satellite dish and wait until the probes reach their different destinations and the first thousand results are ready to beam to the Earth. Humans have waited millennia to find out if they're alone; another decade makes no difference.

Banks of projection TVs are set up in city centers around the world so the public can watch all the descents simultaneously. On the Artemis web site, a counter clicks off the number of terrestrial planets inspected . . . 991,992. . . . The magic number approaches, and crowds begin to gather. Nobody—not even project engineers—has seen any of the video. The screens all flicker to life. . . .

And show what? A thousand barren worlds scarred by craters and volcanoes? Hundreds of watery planets, their continental masses greened by vegetation? Diverse topographies shrouded by dense and toxic atmospheres? How many of these worlds are familiar enough to be our kin? And as each screen zooms in on a surface, what do the hushed masses on Earth see? The varied forms of uplift and erosion—rocks, rocks, and yet more rocks? How many of the thousand screens show the thrilling signs of landscape shaped by civilization and how many, just before impact, show creatures looking up at the visitor falling from the sky?

THE RARE EARTH HYPOTHESIS

If the Earth didn't have properties to allow intelligent life to evolve, we wouldn't be here to ponder our existence. In that sense, we can't draw too many conclusions from the nature of our planet. Another perspective is that we simply don't know enough about terrestrial planets in general to infer the status of our particular terrestrial planet. A third point of view says that the Earth has attributes and a particular history that facilitated the evolution of large and complex life-forms. If this set of circumstances is unusual, we may be very lonely in the universe.

Peter Ward and Don Brownlee have developed this thinking into the provocative Rare Earth hypothesis. They draw a strong distinction between microbial life and more advanced multicelled, and especially intelligent, forms of life. They consider the former to be abundant given that organic ingredients and planets are universal and that life on Earth happened quickly and spread almost everywhere. They consider the latter to be rare due to the special conditions they believe have to hold for intelligence to arise.

Most people, scientists included, are overly invested in the status of intelligent life in the universe. Space bugs don't threaten us (except in our dreams). But if the universe is littered with brains, the destiny and special role of humanity is called into question.[1] We struggled for four billion years to make it from pond scum to bipedal sophistication. Now that we have a shiny new bike, we want to be the cool kid on the block. It would dent our self-image to discover that many kids on the block had bikes long ago and have moved on to fancy cars and spaceships.

Ward is a geologist and Brownlee is a planetary scientist. They work in nearby buildings at the University of Washington, but they make a study in contrasts. Ward is lean and hyperkinetic, an avid diver who had to have hip replacement after too much pounding on the basketball court. Brownlee is older and avuncular, with a soothing Midwestern accent; he is bemused by the furor that their work and subsequent book kicked off. What's the big idea, and why do they think intelligent life is a fluke?

Ward and Brownlee make a list of the factors required for advanced animals to emerge. They argue that each is intrinsically rare and that all must hold for complex life and intelligence. First is a large, nearby moon to stabilize the climate by damping swings of the axial tilt. Next is an impact environment that promotes biological diversity without cashing in all the chips. Also, a planet of the right mass must be in the slender habitable zone, with a larger Jupiter-like planet farther out to protect from excessive impacts. The planet needs plate tectonics and liquid water. Last, the sheltering star must have enough heavy elements that planets could be constructed when it formed, and it has to live in a region of the galaxy where stellar mayhem does not eradicate life.

It sounds like an onerous laundry list. But looked at in more detail, each aspect has a corresponding counterargument. The key assumption of Rare Earth is that *all* of these conditions must hold to permit intelligent life.[2]

HABITABLE ZONES

Figure 77 shows some of the issues that go into determining whether or not a planet is habitable. Most of them involve messy physics, and they are poorly understood in our Solar System, let alone in actual and hypothetical solar systems across the galaxy. This makes it difficult to say how special or inevitable our situation is.

The hazard environment is a good example. Earth's impact rate would be higher without Jupiter in its current position, but there are plausible architectures that would give a terrestrial planet a lower impact rate. The Moon was the

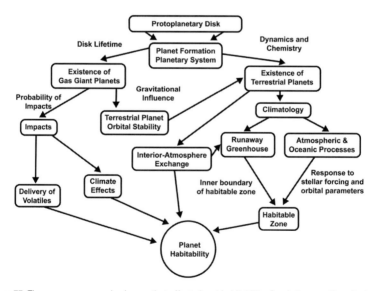

Figure 77. There are many complex issues that affect planet habitability. One is the question of where and how planets form out of the disk of gas and dust around a newborn star. Another is how planets interact with one another when a solar system forms. The habitability of a planet also depends on the impact environment and the interplay between geology and atmospheric chemistry over the lifetime of the planet.

result of a catastrophic impact, yet it acts to stabilize the spin axis of the Earth and hence its climate. Simulations show that impacts produce large moons fairly often, and even without a moon a planet can be quite habitable when it's tipped on its side or wobbles slowly. We've seen that evolution can be spurred by impacts or radiation events if they clear out ecological niches and promote genetic diversity. Is there an optimum impact rate for a life-bearing planet? If there is, there's no evidence that the Earth's situation is optimum.

The idea of a habitable zone lends itself to thinking that there's a "sweet spot" for life. The traditional habitable zone is the range of distances from a star where water can remain as a liquid on the surface of a terrestrial planet, a definition that enshrines surface water as essential for life. Most of the universe is cold, dark, and empty. Far from a star, the temperature drops to a level where air freezes. Close to a star, flesh and metal boil. We're huddled near our campfire, the Sun. We think life elsewhere must also live near a fire.

In our Solar System, the habitable zone extends from 0.7 to 1.3 times the mean Earth-Sun distance (called an astronomical unit, or AU), and it includes just one planet: the Earth. But the Sun was dimmer in the past, and the earlier habitable zone would have left the Earth in the deep freeze without some compensation from greenhouse gases.

Habitable zones don't last forever; they must be defined in both time and space. The Sun will persist for another four billion years, when it will exhaust its hydrogen in a series of sputtering spasms. The onset of helium fusion will turn it into a red giant. As the hot outer envelope expands by a factor of a hundred, it will engulf the Earth and fry life to a crisp. However, long before that, the steady increase in the Sun's brightness will boil the oceans and turn Earth into a barren desert. That gives us about five hundred million years—plenty of time for our descendants to become independent of our star. At the moment, we're not even close. With our current dependence on fossil fuels, we resemble hunter-gatherers more than future space voyagers. We control nuclear fusion—something stars do with mocking ease—at great expense and for only tiny fractions of a second.

It's often said that the Sun is a typical star. That's true in a general sense but not numerically. In fact, 95 percent of all stars are less massive than the Sun. These stars are smaller and less luminous than the Sun and have correspondingly shrunken habitable zones. Jim Kasting of Penn State is the "dean" of habitable zones and has thought about them longer and harder than anyone else. His calculations include the feedback effects of greenhouse gases in a typical terrestrial-planet atmosphere, which regulate temperature and make the habitable zone a bit more generous.

Looking first at the relatively rare stars more massive than the Sun, there's less time available for life to develop, and the habitable zone moves out to a distance where terrestrial planets are unlikely. A star half again as massive as the

Sun lives only three billion years, and the inner edge of the habitable zone is just past the equivalent of the orbit of Mars. A star four times the Sun's mass lives four hundred million years, and its habitable zone doesn't begin until near the orbit of Jupiter. A planet at an Earth distance from such a star couldn't retain oceans because they'd boil away.

At some point, considering even heavier stars, we assume the time available for biological evolution shrinks to the point where the curtain goes down before anything interesting can happen. But this assumption isn't founded on any evidence or any theory. Evolution on Earth progressed in fits and starts, and we've no idea how long it might take to develop complex life under different physical conditions.

For the more abundant low-mass stars, there's a different problem. Time's not an issue; a star half the Sun's mass lives fifty billion years. But the outer edge of the habitable zone is inside the orbit of Venus. A star one quarter of the Sun's mass is a dim glowworm one hundredth of the Sun's brightness (Fig. 78). The habitable zone of this kind of red dwarf is a wafer-thin region huddled close to the feeble star. A planet would have to be no more than a few million miles away to keep water liquid.[3]

If a star is too massive, it will be short-lived and have a habitable zone beyond where the rocky planets live in our Solar System. If a star is too puny, there will be plenty of time for biology to develop, but terrestrial planets must live in a tiny habitable zone, and they would suffer extremes of climate due to one side always facing the star. Astronomers have therefore focused their search for distant planets on stars similar to the Sun.

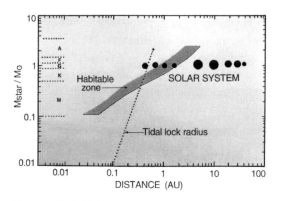

Figure 78. The habitable zone around stars of very different masses. Our Solar System is shown at a position corresponding to the Sun, a G-type star. For the rare stars more massive than the Sun, the habitable zone moves out toward where giant planets lie, and small rocky planets may not be able to form there. For the most abundant stars less massive than the Sun, the habitable zone shrinks to inside where Mercury lies in our Solar System. Planets around stars less than 40 percent of the mass of the star are tidally locked to the star, producing a potentially extreme variation of temperature from one side of the planet to the other.

WHAT WE MEAN BY "HABITABLE"

By defining the Earth's properties as the sweet spot of habitability, we virtually guarantee that Earth clones will be rare and other situations deemed less promising. Maybe we're overlooking some diamonds in the rough.

We've assumed that Earth-like planets get created only at Earth-like distances from a star because that's what happened in our Solar System. We haven't yet detected terrestrial planets around other stars, so we have no idea of the range of distances where they might form. We've assumed nearly circular orbits, but if eccentric orbits are the norm, planets will loop in and out of their habitable zones, and their climates will vary dramatically over months or years. We've assumed single stars, but most stars are in binary systems. If the stars have wide separation, planets may act as if the other star doesn't exist, but tight binaries have unstable orbits, which would make life very interesting, to say the least.

It's a good idea to loosen the grip of star chauvinism. Extremophiles on Earth teach us that life can exist above the boiling point and below the freezing point of water, deep within rock, and under the sea, where there's no light. Life's energy source might be geothermal energy from a sufficiently massive planet or tidal heating of a moon orbiting a planet. Planets can get ejected from their solar systems due to gravitational interactions; those massive enough to generate internal energy but with atmospheres not so thick that they're hot and smothering might well host life.

What we're really doing is making a distinction between an "animal" habitable zone and a "microbial" habitable zone. If Earth biology is a guide, microbial habitable zones might range from the vicinity of a white dwarf to the depths of interstellar space. The Rare Earth hypothesis says that larger and more complex life-forms take longer to evolve. It's obvious that animal habitable zones are rarer than those that can support microbes. The question is, How much rarer?

RARE OR INEVITABLE?

Rare Earth advocates and many biologists are struck by the 2.5 billion years it took for advanced animals to evolve. Three times longer, they argue, and the Sun would have snuffed out before we got to tools and toasters. Yet NASA planetary scientist Chris McKay argues that human-like intelligence could evolve in as little as one hundred million years given different but plausible conditions. Instead of being the smartest kid in class, we may be the slowest to raise our hand.

Larger animals have higher energy usage and are less robust when there are large environmental changes. It's also true that only a tiny fraction of species have intelligence—on Earth, a billion years after the first animals developed and a half-billion years after the first rudiments of a brain evolved. Are high-functioning animals really so much rarer that we're operationally alone in the vast Milky Way galaxy?

Even if intelligence isn't inevitable, there's evolutionary logic in the path that led here. The critical advance was the first cell because that switched on the mechanisms of genetic diversification and natural selection, where characteristics of the collective (the species) are determined by the viability of the individual (the organism) in a particular environment. It's a compelling theme of biology—the hypnotic bass line at the heart of the music. All subsequent evolution is variation and development of the theme as layers are added to make the stunning song of life. No cosmic tunesmith is needed; the song emerges by blind experimentation.

On the Earth, function and form developed hand in hand. Cooperation spawned specialization. Then specialization spawned complexity. Prokaryotes, eukaryotes, multicelled organisms: each stage shaded conceptually into the next. The endless evolutionary arms race led to better senses, which drove bigger brains to process all that information. Before you know it, nature had invented existential angst. Damn.

The Rare Earth argument is logically flawed because it's circular and a tautology. It's circular because life has profoundly altered the ecosystem. Many of the attributes of our planet's atmosphere and geology that make it noteworthy arose *because* there is life here. It's a tautology because we can view the road to complexity on only one planet. Evolution might proceed quite differently on Earth's distant cousins. Also, some of the items in the Rare Earth laundry list are not independent of one another, which reduces the "specialness" of our planet.

In the end, we just don't have enough information on the diversity of planets and the pathways of life to make an informed judgment. However, the proposition has led to a useful debate on the necessary and sufficient conditions for complex biology. Meanwhile, although the Earth is undeniably special, there are other places nearby that can teach us about life.

MYTHIC MARS

THOUSANDS OF YEARS BEFORE we knew their true nature as worlds in space, the wanderers of the night sky captured our imaginations. The five naked-eye planets—Mercury, Venus, Mars, Jupiter, and Saturn—are deeply embedded in mythology and astrology, and they give their names to the days of our week in Romance languages. The most potent of these symbols is Mars.

THE MARS IN EACH OF US

In ancient Babylon he was Nergal, not just the god of war but also the god of the scorching midday Sun, the bringer of plague, epidemics, and disaster. Seeking power, he forced the Queen of the Underworld to share her domain with him. In Hindu texts, he was Kartikeya, born to fulfill a prophecy. The gods were terrorized by a demon, so they forced the great ascetic Shiva to father a child by creating the illusion of a woman so beautiful that Shiva ejaculated at the sight of her. His fiery sperm was nurtured by the wives of the seven stars of the Big Dipper. Seven days later, Kartikeya was born, and as the god of war he slew the demon.

A thousand years later, he was Ares to the Greeks, a violent and spiteful god who took pleasure in combat. His parents hated him, and his sister Athene called him "a thing of rage, made of evil, a two-faced liar." The Greeks refused to honor this bloodstained coward; no sacred places were built in his name. He was remembered only at the battlefields, where he brought pain and death, accompanied by his attendants Phobos and Deimos—Fear and Panic. Everything he touched was soured by his disposition. He took for his wife the sweet nymph Harmony but fathered with her the Amazons, a warlike tribe of women.

The Romans followed the Greek lead, and so he became Mars, their god of war. The third month in our calendar was named after him because that was the time in each year when the Romans readied their legions for battle. To the Norse of that era, he was the god of battle Tyr, a one-armed man. The Goths sacrificed their captives and hung the arms of the victims in trees as a token offering. In Old English, his name was Tiw, given to the second day of our week.

How could a pale red dot in the night sky become the incarnation of bloodlust and a symbol of violent sexual energy through the ages? As Joseph Campbell said, myths are a culture's dreams. They provide metaphors for our internal struggles and our rites of passage and the means by which we transmit our collective experience. Our technology "bubble" means we've lost touch with nature and the sky. Perhaps we've also lost touch with a part of ourselves.

Campbell followed Carl Jung in recognizing the power of archetypes as the clearest manifestations of the swirl of dreams and fears and longings in our unconscious. Mars is associated with aggressive forms of maleness but, more than that, it's the planet that mirrors our weakness, our insecurity, and our rage. Mars is the archetype that must be faced if humans are ever to deliver on their magnificent promise as a species.

MARS ATTACKS

In the middle of the evening of October 31, 1938, the New York area witnessed extraordinary scenes of panic and mass hysteria. Families fled their homes and gathered in nearby parks. Churches filled with sobbing people. Roads were jammed with traffic, and telephone lines became unavailable. In one city block in Newark, more than twenty families rushed from their homes with wet towels over their heads, anticipating a gas attack. Others loaded their furniture into trucks and fled the city.

Moments earlier, a program of dance music on the radio had been interrupted by a "news flash" describing a series of gas explosions on the planet Mars. Over the next half hour, bulletins from reporters on the scene told of a "meteor" landing near Princeton, New Jersey, the fact that it killed 1,500 people, and the discovery that the meteor was in fact a metal cylinder that disgorged strange creatures from Mars armed with death rays. The reporting was so realistic that it fooled many people, all of whom missed the initial announcement that they were listening to a radio play, an adaptation of H. G. Wells's *War of the Worlds.*

New York and New Jersey were in no danger that Halloween night. Perhaps the mayhem was exacerbated by ominous news from Europe, where the storm clouds of war were building. It was certainly boosted by the skill of Orson Welles, the producer of the radio play.[4] But the seeds of fear had been sown into our subconscious long ago, and they had been cemented a few decades earlier by the idea of a civilization on Mars.

CANALS ON MARS

Percival Lowell drove his construction team hard. They gulped the thin air and worked overtime pouring cement and erecting an iron frame. Lowell was an aristocrat from Boston who'd made his money in commerce. But his passion was Mars. He was racing to complete a twenty-four-inch telescope on a high plateau in northern Arizona in time for Mars's closest approach to the Earth in fifteen years. By 1894, he was making observations with his new telescope every clear night and recording the strange linear markings, or "canals," that he saw in his notebook (Fig. 79).

Lowell was a victim of the power of suggestion and wishful thinking. He thought the canals had been built by a dying civilization to carry water from the frozen poles to the equatorial regions.[5] He wrote in his book, titled *Mars,* "Without seas and mountains, life would tend the quicker to reach a highly organized stage. Thus Martian conditions make for intelligence." But by the time of the next Mars approach, others made superior observations and didn't see what Lowell had seen. Alfred Russel Wallace, developer of the theory of evolution at the same time as Charles Darwin, reviewed the evidence and wrote sternly, "Not only is Mars not inhabited by intelligent beings as Mr. Lowell postulates, it is absolutely uninhabitable."

Since then, Mars has remained in the popular imagination—the subject of classic science fiction by Ray Bradbury and Robert Heinlein and many others, as well as fodder for TV shows and movies every few years. No other place has caused such a fever of speculation about life beyond Earth (Fig. 80). So who was right: Lowell or Wallace?

EXPLORING THE RED PLANET

NICOLE'S ALARM CLOCK goes off forty minutes later each day. On this particular morning, she encounters her two roommates in the kitchen. They try not to get in one another's way as they microwave coffee, pour cereal, and grab yogurt from the fridge. Sarah's off to French class. Ingrid has biochem lab. Nicole Spanovich heads for the basement of a building near the University of Arizona campus, to the site of a remote-operations center for the Mars Exploration Rovers. Nicole's alarm clock is unusual because it keeps Mars time.

The Opportunity and Spirit rovers each traveled many miles from where they landed, giving the public the vicarious thrill of watching a vehicle on a distant world bump over rocks and climb into craters. Few people could have been aware that the person giving the commands to the rovers was an undergraduate student. Spanovich has a central role in the eight-hundred-million-dollar mission, coordinating the display of mission status and routing data to the science teams. The schedule is relentless. Each day's data must be studied quickly, so as to make the best decision on where to

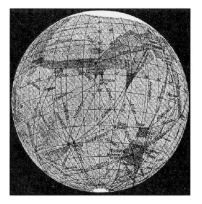

Figure 79. This hand-drawn figure based on Percival Lowell's observations of Mars was published in a book he wrote in 1906. It shows the well-known polar caps but also a crisscross pattern of markings that he described as being so straight that they had to be artificial. Lowell was somewhat mistaken in his observations—the surface features are not actually linear—and he was very mistaken in inferring that the canals are the work of an advanced civilization.

Figure 80. Mars is half the size of the Earth and about 10 percent of its mass. For its size, Mars packs a good punch, with canyons as long as the continental United States and volcanoes higher than Mount Everest. The surface is cratered and covered with windblown dust. Ice caps made of frozen water and carbon dioxide expand and shrink with the seasons. Mars has two small, potato-shaped moons, Phobos and Deimos.

send the rovers next. Soon Spanovich will be getting up at midnight for her shift, so she smiles wryly when one roommate complains about her 8:00 a.m. class.

MARS UP CLOSE

We're in the midst of an exciting phase of Martian exploration. For the past decade, robots have been exploring the surface, and spacecraft have been mapping from orbit. Thirty years ago, when the first Viking lander parachuted out of a dusty, peach-colored sky onto a rugged Martian plain, it seemed to put to rest a century of wild ideas. Wallace was right: Mars looked rocky, barren, and dead. Now we're not so sure. However, Lowell wasn't right either; Viking's biological experiments did not detect any unambiguous signs of life. The truth is tantalizingly in between.

Let's see what we've learned about Mars in recent years and what we hope to learn from the small armada of spacecraft on the drawing board. Remember that Mars still tests our fledgling space program to the limit. Billions of dollars of hardware aimed at Mars have gone missing, failed to perform, or crash-landed. Ed Weiler, the associate administrator of NASA, once said, "Mars has been a most daunting destination. Some, including myself, have called it the death planet. Why do we say that? Two-thirds of all missions that have flown to Mars have failed. Just getting to Mars is hard, but landing even more so."

Mars looks very different from the Earth, both from orbit and from the surface (Fig. 81). The smooth and relatively uncratered lowlands in the north are divided from the heavily cratered southern highlands by the huge Tharsis plateau. The western end of the plateau has the Solar System's largest volcano, Olympus Mons, and the eastern end turns

Figure 81. A close-up of the Martian surface, with the Sojourner rover, the "Little Rover That Could." The Sojourner rover was no bigger than a backpack, and it could travel at a speed of only two feet per minute. What it lacked in speed, it made up for in agility and endurance. Sojourner's power lasted twelve times longer than expected, and it took more than five hundred pictures of the rock-strewn surface. Scientists drove the rover from 120 million miles away using a computer screen, a joystick, and 3-D goggles.

into a huge fracture called Valles Marineris, which dwarfs the Grand Canyon. The planet is littered with enormous volcanoes, but they don't follow the pattern that would indicate plate tectonics. Mars is geologically dead.

Mars is also very cold and dry. The only cause of erosion is dust carried aloft in the sparse atmosphere. Dust storms can rage for years and cover a significant fraction of the planet; on small scales, they sculpt the surface into patterns reminiscent of the Earth's deserts (Fig. 82). The atmosphere is gaspingly thin, with the pressure you

Figure 82. Sand dunes at the bottom of the Endurance crater, photographed by the Opportunity rover in 2004. The Martian surface is covered with a thin red soil created by billions of years of meteoric bombardment. Small particles are carried aloft by winds in the thin atmosphere and sculpted into formations like these.

would experience at three times the altitude of a passenger jet. It's composed almost entirely of carbon dioxide, plus a whiff of nitrogen and a trace of water vapor. Mars has an axial tilt, so it has seasons. The polar caps are made of water-ice and frozen carbon dioxide (or dry ice), and the carbon dioxide evaporates and condenses with the seasons, causing the polar caps to shrink and grow. A balmy midsummer day on Mars reaches 32 °F (0 °C); at the poles, it can get down to a numbing –112 °F (–80 °C).

A KINDER, GENTLER MARS

The big surprise on Mars is water—not now, but in the distant and maybe the recent past—lots of it. Mars shimmers like a mirage, with water that seems so close we can almost taste it.

In June 2000, NASA called a press conference to release images from the Mars Global Surveyor. There was so much media speculation and leaked information that the event was moved forward a week. The results were stunning. Mars had numerous places with gullies and runoff channels. Liquid, presumably water, had been seeping out of the ground and flowing across the surface.

Figure 83. This image from the Viking Orbiter shows a network of valleys in the southern highlands of Mars; the view is about 250 kilometers across. These dendritic features are typical of running water. Some of the craters superimposed on the channels are three to four billion years old, showing that Mars was wet enough for running water in the distant past.

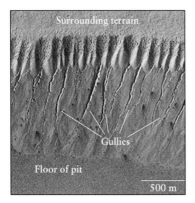

Figure 84. Evidence of running water on the red planet comes from Mars Global Surveyor images of gullies like this. Liquid has flowed from just below the edge of an escarpment to near the bottom. The markings show channels carved by the liquid; the water itself quickly froze or evaporated in the chill Martian atmosphere.

Many of the features flowed out over sand dunes, which are ephemeral due to the action of wind.[6] That meant the flowing water was geologically recent.

The gullies added to the well-established evidence for water in Mars's distant past. Viking had seen extensive valley networks in the southern half of the planet (Fig. 83). The valleys are mostly in areas with lots of craters, and some of the big craters occurred after the valleys formed, which means they are very ancient, 3.5 to 4 billion years old.[7] They look just like river valleys on Earth, and they were carved slowly, so the water that made them was around for a while. Other geological features look like they were formed by the action of glaciers. For this ice to move, it must have been relatively warm; ice too far below its melting point is brittle and will not flow. Mars was warmer and wetter early in its history, which implies that it had a thicker atmosphere.

If Mars was warmer in the past, then the existence of seas and standing bodies of water is understandable. But how do we explain the gullies, where runoff may be going on even as you read this? If you could pour a glass of water onto the Martian surface, it would quickly freeze and then slowly evaporate or sublime into the atmosphere. So how do the gullies form?

The best guess is that water erupts violently from below the surface. Despite the cold, water can be liquid under pressure hundreds of meters below the Martian surface. At the edge of an escarpment, water from these aquifers is trapped by an ice plug. When the ice plug melts or is dislodged, a small flash flood carves channels in the down slope. The water freezes or evaporates, and as the pressure is relieved a new ice plug forms. In this way, water can occasionally make an appearance on the surface (Fig. 84). We've not found the smoking gun, or

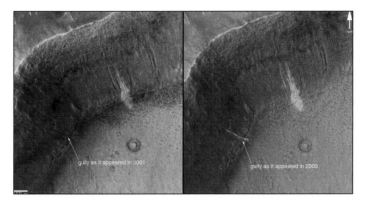

Figure 85. Dramatic indication that water bubbles up from the subsurface of Mars all the time comes from a pair of images of a crater in the Terra Sirenum region, taken fewer than four years apart. The new light-colored deposit is unlikely to have been caused by dry material. Planetary scientists' best estimates are that fifty to one hundred swimming pools' worth of water coursed down the side of the crater before freezing or evaporating.

rather the "spritzing rock," by watching this happen in real time, but in late 2006 NASA released images from the Mars Orbiter Camera that showed the resurfacing of a gully in images taken fewer than four years apart (Fig. 85).[8] This is not just recent on a geological timescale—water is bubbling up all the time!

The god of war has had some of his edges softened. As far as we know, life needs water. As long as we thought Mars was a frigid, arid desert, there was little prospect for biology there. But if Mars was kinder and gentler in the past or has water underground right now, the odds get interesting again.

THE LITTLE ROVERS THAT COULD

The evidence for Mars as a watery world was clinched by the Mars Exploration Rovers, which brings us back to Nicole Spanovich. She was a twenty-one-year-old college senior when she began running the remote-operations center for the Opportunity Rover in Tucson. Even after a year on the job, she got a shiver when she sent commands one hundred million miles to the rover and then watched on a monitor as it climbed over the rim of a crater. Another woman, Zoe Learner, was only twenty-two years old when she became a full team member at JPL's Mission Control for the rover program.

It's fitting that the Net Generation of people born in the 1990s is taking the helm. The future of space exploration is telepresence. Machines will be our eyes and ears on distant worlds. Robots make sense for space exploration because we'll shed no tears if they get damaged, and their abilities will be spurred by rapid advances in computing and electronics. Mars exploration will become just like a video game.

Early in 2004, a spacecraft hurtled out of the Martian sky and inflated just before landing. It bounced as high as a four-story building and then bounced another two dozen times before coming to rest in a cloud of red dust. Nineteen days later, on the other side of the planet, its twin landed in the same way. The landing bags deflated, and two small robots, each no bigger than a go-cart, rolled off their landing platforms at the speed of a baby crawling and began to explore a strange world. Say hello to the Opportunity and Spirit rovers.

Twins always have unique personalities, and so it is with the rovers. Opportunity is the hazard-prone overachiever. It got a lucky break on landing when it careened in a different direction than its planned trajectory and ended up at the bottom of a crater with rare exposed areas of Martian bedrock. Opportunity got to work and soon had chemical evidence that rocks in the outcrop had formed by evaporation in a watery environment.[9] It found numerous small spheres (Fig. 86), layering, and rocky fissures—all consistent with formation in water but not consistent with a volcanic origin.

Opportunity has the luck of the brave. A swirling Martian storm blew away dust that was covering its solar panels just as its power level was getting dangerously low. It once got stuck hub-deep in sand and was freed after weeks of effort, and it has a balky shoulder joint. But its early discovery of geologic evidence for flowing water has been a highlight of the entire mission. As Steve Squyres, Cornell professor and lead scientist for the rovers, said, Opportunity had discovered "the shoreline of a salty sea on Mars." But conclusive evidence may have to wait for a sample-return mission.[10] This doughty rover recently passed twenty kilometers on the odometer, forty times what it was designed for, and all without the benefit of a mechanic and regular maintenance. It's the longest-lived Mars mission ever.

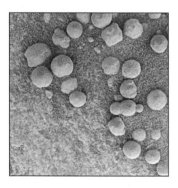

Figure 86. These small spheres, or "blueberries," as they were called, are a few millimeters across and were found by the Opportunity rover in a depression in a rock. They are rich in the mineral hematite, suggesting formation in water. The triple berry near the center is particularly interesting since aggregations like that form on Earth when water percolates through porous rock.

Spirit, meanwhile, is the hardworking twin. It landed in a monotonous plain of basaltic lava, so it didn't have Opportunity's early success, but then it roamed into foothills, where the geology is much more interesting. It has clambered up a hill as tall as the Statue of Liberty, and it has worn out the diamond bits on its drill. When its front wheel started binding, it dutifully spent some time exploring in reverse until the problem was fixed. Through Spirit's eyes the science team has gotten a vivid sense of what it would be like to be on Mars. Spirit's perseverance was finally overcome by Mars's soft soil in 2009; its roving days are over.

Squyres had his life transformed by these intrepid little vehicles. The mission plan called for ninety Martian days of exploration for each rover. Instead, they're well over two thousand. "We're so past warranty on them," Squyres admits. He's curtailed his travel because each day he has to mastermind the next day's excursions. "On the one hand, we're tired," he says. "On the other hand, there's no thrill in science that matches the thrill of discovery." He has almost no time to analyze the data returned or the quarter of a million images. The schedule is relentless, but he doesn't want it to end. "Exploration will never be complete. Whenever the rovers die, tomorrow or two years from now, there will always be something wonderful and tantalizing just beyond our reach that we'll never get to."

LIFE ON MARS

THE RED PLANET IS FULL of surprises. But recent space missions leave us with three important questions: why did Mars dry out, could it have hosted life when it was wetter, and might it host life now?

WHERE THE WATER WENT

If Mars had been warm enough to have surface water four billion years ago, when the Sun was 25 percent dimmer than it is now, it must have had a carbon-dioxide atmosphere thick enough for a strong greenhouse effect. On early Mars, there were oceans, salty seas, and glaciers, there was rain and erosion, and the air was thicker than ours is now. The feeble Martian gravity meant that some carbon dioxide leaked into space. The rest dissolved in oceans and was incorporated into rocks. Since Mars doesn't have plate tectonics, there was no means to recycle carbon dioxide into the atmosphere, so it steadily disappeared, taking its greenhouse warming with it.[11]

How much water did Mars have, and where is it now? Mars got its water in the same way the Earth did—from a mixture of small asteroids and comets. This suggests that Mars started with water equal to 10 percent of the Earth's oceans, or a layer half a mile thick over the whole surface. Such a large amount

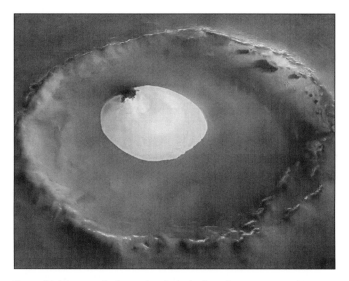

Figure 87. A large patch of water-ice sits in the floor of a crater on a northern plain of Mars, in this image from the European Space Agency's Mars Express mission. The ice is present all year, and traces of ice are also seen on the inside rim of the crater. Larger bodies of ice have been found, covered and protected by a thick layer of dust.

of water is consistent with erosion features and the evidence for ancient seas. Most of the water escaped into space.[12] But detailed calculations suggest that 10 percent remains in an unevenly distributed permafrost, which has a layer of liquid water of unknown thickness below it. The Mars Reconnaisance Orbiter discovered underground glaciers half a mile thick covered with volcanic ash far from the Martian poles, and sometimes ice is out in plain view (Fig. 87).

MARTIAN MICROBES?

Long ago, Mars had the main ingredients for microbial life: heat, water, carbon, and a stable climate. Unfortunately, the best-studied chunk of ancient Mars that we have in our hands has been mired in controversy. The Allan Hills meteorite, ALH 84001, is 4.5 billion years old, dating from the earliest days of the Martian crust. It spent eons there, only to be blasted off by a meteor impact. After drifting in space for about ten million years, it landed in Antarctica thirteen thousand years ago. The evidence that the meteorite came from Mars is compelling. Tiny air bubbles trapped in the rock exactly match the atmospheric concentrations measured by Viking, and they differ from the proportions of the same isotopes found on the Earth.

ALH 84001 was a hot news story in 1996, when a team led by David McKay at NASA's Johnson Space Center claimed it contained evidence of ancient life. The three dozen known Martian meteorites are a mixed blessing. They avoid the cost of a sample-return mission, but their points of origin are uncertain. Also, they've suffered a violent ejection from the Martian surface, a long trip through the harsh environment of space, and a time when terrestrial contamination could creep in. There's a limit to what we'll learn from these unusual rocks. It's particularly unfortunate that none are sedimentary rocks.

There were several lines of evidence for life in ALH 84001. Sometime after it formed, the rock gained carbonate globules that indicate it was exposed to water. The meteorite contained two materials—ring-shaped carbon molecules called polycyclic aromatic hydrocarbons (PAHs) and magnetite crystals—that are commonly (but not uniquely) associated with biological activity on Earth. Finally, the team discovered sausagelike forms that they argued were tiny fossil bacteria.[13] More convincing evidence for life awaits future missions (Fig. 88).

It sounds good, but the hedging and caveats started very quickly. The discovery team later agreed that the interior of ALH 84001 had not been free from contamination. Other groups argued that most of the meteorite's carbon dates from its time in Antarctica. PAHs are found in deep space, so they could have entered the rock as space dust any time in the past 4.5 billion years. The magnetite crystals remain the best bit of evidence, since they have a distinctive hexagonal shape that is identical to the form of the mineral found in bacteria on Earth. Joe Kirschvink of Caltech has showed that many of the magnetite-crystal properties are inconsistent with a nonbiological origin.

Ironically, the fossil evidence is the least compelling. The shapes are suggestive, but there are no cell walls, and they're so much smaller than terrestrial bacteria that biologists doubt they could hold a viable amount of genetic material. Similar elongated forms could have arisen by crystal growth or from molten splatter due to the impact that sent the rock Earthward.

It adds up to a resounding Scottish verdict: not proven. If we want to know more, we'll need more than a few dozen random rocks tossed in our direction. It would not be a great surprise if Mars had microbes in its wet past, but it was

Figure 88. The Mars Science Laboratory (MSL) is a NASA rover scheduled for launch in November 2011. It's mission is to determine the planet's habitability. The size of an SUV, the rover has a payload of ten instruments, built in six countries.

probably never profoundly alive in the Gaia sense, because it shows none of the global alterations to atmosphere and surface that are found in a fully fledged biosphere.

WANTED, DEAD OR ALIVE

Is Mars alive now? We just don't know.[14] The Mars Exploration Rovers were designed to measure basic chemistry and trundle around looking for cool rocks. They couldn't test soil for biological activity. The Viking landers had technology that was a generation older, and the results of their tests were stubbornly ambiguous. One scientist has even argued that Viking killed the life it tried to detect because its experiments were predicated on metabolic processes like those of life on Earth.

The exploration of Mars is undergoing a sea change. For decades, NASA built large and complex spacecraft costing billions of dollars. Scientists would be gray and near retirement by the time their visions yielded a payoff. In the 1970s, some hacker programmers from MIT approached NASA for a Mars mission based on many robotic ants connected by a neural net, and they were laughed out of the room. In the 1990s the mantra became "Faster, Better, Cheaper." A seventeen-billion-dollar-per-year federal agency will never turn on a dime, but at least it's trying to respond to the times.

The next wave will be much more technologically advanced than any previous missions to Mars. They will target places where evidence for water is strongest. The topsoil is sterilized by UV radiation, so the probes will have to drill for samples and conduct biochemical tests. Alan Waggoner is a biomedical engineer who is developing fluorescent dyes that bind to DNA, lipids, carbohydrates, and proteins. Richard Mathies, a biophysicist at Berkeley, is perfecting a biology lab on a silicon chip that will detect amino acids with one thousand times more sensitivity than Viking could. Both researchers test their techniques in Chile's Atacama Desert, where the bleak terrain is fifty times drier than California's Death Valley.

First the dry run, then it's on to Mars. A reconnaissance mission reached Mars in 2006. It's currently mapping surface features down to the size of a dinner plate and using a sounder to locate subsurface water. It has even spotted the Mars rovers hard at work! Phoenix arrived in 2008, landing near the northern pole. It found conditions suitable for life and observed a hydrological cycle in action. Next up is the Mars Science Laboratory, scheduled to land in 2012. This three-ton facility will have a suite of instruments for chemical analysis and a beefed-up rover able to vaporize rock samples with a laser. If there actually are any Martians, they'll be getting pretty worried. In that same

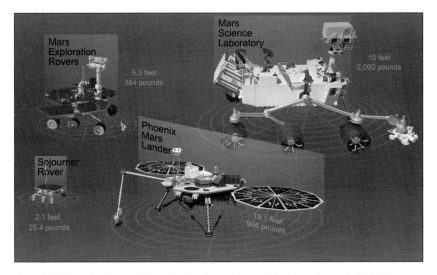

Figure 89. This family portrait of Mars missions shows the growth in size, weight and complexity. Sojourner landed in 1997, the Mars Exploration Rovers in 2004, Phoenix in 2008, and Mars Science Laboratory is scheduled for 2012.

year, Mars will also get its first telecommunications satellite, to serve the growing interplanetary Internet (Fig. 89).

Sample return is the Holy Grail, but even space visionaries flinch at the price tag. Sending astronauts would cost a jaw-dropping $150 to $200 billion; doing it with robotic probes is a steal at five to ten billion. In the late 1990s, NASA plans called for a launch in 2003, with sample return by 2008. Recent versions call for a launch no earlier than 2020, so sharp-eyed observers have noticed that NASA is moving backward. President Bush's Earth, Moon, and Mars initiative might accelerate the schedule, though it is currently in limbo, as will the European Space Agency's plan for a sample-return launch in 2018. Planetary exploration is spurred by a healthy mixture of cooperation and competition between the United States and Europe.[15]

Many people sympathize with Agent Mulder from TV's *The X-Files*, who had a poster in his basement office saying "I Want to Believe" (Fig. 90). Mars has dashed our hopes before, and it may dash our hopes again. Lowell died convinced that a Martian civilization existed. There was a false detection of chlorophyll there in the 1950s, followed by confusing results from Viking twenty years later. We must be patient and persistent—the issue will be decided only by more and better data.

Regardless of the outcome of the biological experiments, future exploration

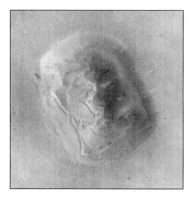

Figure 90. Believe it or not, this is the infamous "face on Mars." Seen at the superior resolution of the Mars Global Surveyor, the object of popular obsession turned into an unremarkable geological structure, a mesa like others on Earth and Mars. Irrational belief systems can be persistent—there are still people who believe that the face was built by aliens, just as there are people who believe in UFOs or that the U.S. government faked the Moon landings.

will make Mars as familiar to us as the Moon. When the first sample heads home, we may be faced with a loss of innocence. If Mars hosts familiar biology, it will mean that the Earth is not unique. If it harbors microbial life with a genetic basis different from ours, it will be one of the most dramatic discoveries in the history of science.

GREENING THE RED PLANET

It sounds outrageous, even ridiculous. At a time when we can barely handle the cost of a sample-return mission, people dream of greening the red planet. They met at a conference in 2001 on the topic "The Physics and Biology of Making Mars Habitable." One of the conference organizers, planetary scientist Chris McKay, recognizes that you have to start small. "I'd like to see NASA send a seed to Mars and try to grow it into a plant," he said. Growing a flowering plant in ambient Martian conditions would be a powerful symbol of humanity's expansion beyond Earth.

But the vision is much grander: transformation of the entire planet to allow us to live there. The act of making another planet Earth-like is called terraforming. The real problem is temperature. Mars has its thermostat stuck at $-67\,°F$ ($-55\,°C$), so some way has to be found to warm it up. The obvious ploy is to start a runaway greenhouse effect by evaporating the carbon dioxide that's frozen in the polar caps. But Mars can't release its carbon dioxide unless it's warmed up. It's a classic catch-22.

An MIT undergraduate named Margarita Marinova came up with a possible way out of this impasse. She's another of the startlingly young women working at the forefront of Mars research. With Chris McKay, she proposed using artificially created perfluorocarbons (PFCs) to initiate the warming. PFCs are supereffective greenhouse gases that last a long time. They also have no effect on living organisms or the ozone layer. How long would this take? Marinova did rough calculations. A hundred factories making PFCs, each with the energy of a typical nuclear reactor, would raise the Martian temperature by a degree every fifteen years. With an assist from evaporating carbon dioxide, it would take five hundred to six hundred years to bring the entire planet above the freezing point of water.

Warming could also be achieved with a mirror the size of Texas aiming light at the South Pole. This sounds impossibly grandiose, but the two hundred thousand tons of aluminum that would be required are only five days' worth of Earth production, and mining and manufacturing could be done in space. With the pole raised in temperature by only 9 °F (5 °C) the carbon dioxide would evaporate and take Mars to the tipping point of global warming. The phenomenon that's so dangerous on Earth works to the advantage of terraformers.

All this work is just preparation. Mars will have been turned into a cousin of the chilly pre-Cambrian Earth, suitable only for the hardiest of extremophiles. Familiar plants and animals couldn't survive there.

Two further huge steps are required. The first is the creation of a self-regulating anaerobic biosphere. There are several candidate organisms for the first Mars colonists. One type of cyanobacterium with the unmanageable name *Chroococcidiopsis* is found at such extremes of cold, dryness, and salinity on Earth that it's often the sole survivor. The cyanobacterium called *Matteia* can dissolve and bore through rock, fixing nitrogen and liberating carbon dioxide. Then there's a familiar friend: *Deinococcus radiodurans.* We imagine that Conan the Bacterium would be the first to volunteer for a tough assignment. Naturally occurring microbes could be augmented with genetically engineered varieties. The goal would be to establish the biosphere and release enough oxygen, nitrogen, and carbon dioxide to raise the atmospheric pressure from its current 0.7 percent of that of Earth to about 2 or 3 percent of Earth's sea-level pressure.

The second step is to introduce plants and boost the atmosphere to a breathable level. Generating trillions of cubic meters of air isn't trivial! Many changes will occur simultaneously. Water will carve out rivers and cause erosion. Soil will begin to form, transforming the surface, which is currently a meteorite-pulverized form of rock called regolith. When Mars has a new and complex set of biological-chemical cycles in play, different from those on Earth, it will be very difficult to predict the actual conditions. Chris McKay thinks the first step might be done in little more than a century, but the second could take thousands or even tens of thousands of years.

Terraformers often neglect the cost and difficulty of shipping all these "starter" microbes and plants from the Earth. An intriguing alternative is to build a self-replicating oxygen factory. The single hundred-ton "seed" unit would make oxygen by heating rock, which contains oxides of silicon, iron, calcium, titanium, and aluminum. Then it would use the metals mined from that same soil to construct replicas of itself. A NASA study showed that a factory with power consumption of one megawatt and replication time of one year could generate a breathable atmosphere in a couple of hundred years.[16] And as a by-product, the network of factories would generate one thousand trillion tons of refined metals and become a billion-megawatt distributed power source that's self-repairing and available for other industrial use.

THE VALUE OF PLANETS

Some people's jaws drop in horror at the prospect of terraforming. Mars coated with genetically engineered microbes? Mars covered with strip-mining factories? "It's macho. It makes gods out of geeks," was the comment of biologist Geneva Andreadis, who attended a meeting sponsored by the Mars Society on the topic. "It's better to be partners with a planet; terraforming is a hammer."

To be fair to the planetary engineers, many have been thinking about the ethical implications for a long time. McKay, who was the first person to use "terraforming" in the title of a published paper, back in 1982, proposed a system of planetary ethics based on three values. The first is preservation, where humans don't alter the Earth or other parts of the cosmos just because they have the technology to do so. The second is stewardship, where humans can use and change natural systems but do so wisely, for maximum long-term benefit. The third is intrinsic worth, which holds that human use isn't the ultimate measure of value.

We've only ever known the Earth, where life is a web and altering one aspect often has unintended consequences. Off Earth, we might decide the rules are different. Are we entitled to terraform Mars if we find it to be a dead world? If life on Mars is hanging on by a thread, is it OK to nurture it and let it evolve in new directions? Let's hope that we colonize or alter planets only as part of a plan of exploration and learning and not because we've soiled our planet and need a refuge or because of the urge for mastery and control.

Survival has always been part of the wiring of life. If we need to leave our home to continue the species, that's not against nature. It's unlikely that the universe is stuffed with rare and delicate life-forms—there's a lot of dead and rocky real estate out there. If we colonize any place beyond Earth, it will be a profound transition, our "coming of age" in the Milky Way.

EARTH'S EVIL TWIN

THE MYTHOLOGY IS AWRY. Mars is the Roman god of war. The red planet is named for fire and blood, for iron—which is hard—and maleness. Our nearest neighbor and twin, Venus, is the goddess of beauty. The milky twin is named for love, for copper—which is supple—and femininity. That's the story woven into astrology and alchemy for two thousand years. But Mars in fact is a docile place with moderate weather and almost no geological activity. Venus, on the other hand, is a toxic hellhole.

THE GODDESS OF LOVE WELCOMES YOU

"Abandon all hope, ye who enter here." It's a suitable admonition and warning for those wanting to explore our nearest planetary neighbor. Through the 1960s, the heedless superpowers attempted missions, and the roll call is a litany of failure. Venera 1: contact lost. Mariner 1: launch failure. Sputnik 20: failed flyby. Sputnik 21: failed flyby. Venera 1964: launch failure. Zond 1: contact lost. Venera 2: contact lost. Venera 3: contact lost. Venera 1965: launch failure.

After a total of sixteen attempts, in 1967 the Venera 4 probe became the first human artifact to reach another planet, but it crashed on the surface without returning data. Three years later, the persistent Soviets landed Venera 7 on the surface. It transmitted data for exactly twenty-three minutes before giving up the ghost, its circuits fried.

We should have known better. Dante told us what to expect in the *Divine Comedy.* We approach Venus with good intentions—its celestial sphere is only seven steps from paradise, for those who did good deeds for love. Our ferry through the atmosphere is driven by Charon. We leave behind the outcasts and those who merely wasted their lives. By the time we reach the ground, we've also passed the pagans consigned to Limbo.

The surface is godforsaken. Condemned sinners have been judged by Minos, and their souls are trapped here forever. There's a violent storm, and the lustful wander in it lost, never to touch one another again. The three-headed dog Cerberus is forcing gluttons to eat the toxic soil. The greedy and the indulgent are pushing huge rocks. Heretics are trapped in flaming tombs, and blasphemers and sodomites have lava raining down on them. The surface is scarred by geological clefts, which are used as pits to hold thieves, sorcerers, perjurers, hypocrites, panderers, seducers, alchemists, false prophets, and corrupt politicians. Everywhere, it's hot enough that skin and flesh burn.

Misfortune finds you on the surface of Venus. The Sun barely penetrates thick yellowish clouds. The dusky sky is scarred by lightning and sulfuric acid rain. A carbon-dioxide atmosphere crushes down one hundred times harder than on Earth. It's so hot that paper would ignite and lead would melt. Whoever named it after the goddess of love had a sorry history of relationships.

GREENHOUSE

Venus and Mars represent the divergent paths of a habitable planet. Three billion years ago, when life was well established on Earth, Venus and Mars had more temperate climates. Extremophiles on Earth thrive in conditions approaching those of present-day Venus and Mars, but the former is a poisonous furnace and the latter is an arid desert. Why did Venus go bad, and should we remove it from the list of sites for life?

Our evil twin is within 5 percent of the Earth in size and density. It has the same trove of carbon dioxide as the Earth, but ours is dissolved in the oceans and locked in rocks, while on Venus it forms a dense, choking shroud. Both planets are geologically active (Fig. 91). The biggest difference is water: parboiled Venus has only 0.01 percent of the Earth's water. Four billion years ago, when the Sun was dimmer, Venus received only 40 percent more sunlight than the Earth does today. Is this enough to make the difference between Hades and Valhalla?

Scientists think so. You may have noticed the temperature display shown in many transcontinental jets; five miles up, it's a frigid −22 °F (−30 °C) outside. Above the Earth, the temperature drop-off is so rapid that water is frozen out and trapped in the zone that causes weather. Now imagine sunlight increases by 40 percent. This would raise surface temperatures by only 18 °F (10 °C), but it would increase the amount of water vapor in the troposphere (the weather layer) by a factor of five. However, water vapor is a greenhouse gas, so it traps more radiation, which raises the temperature, which injects more water into the atmosphere, which raises the . . . and on and on. It's a positive-feedback loop.

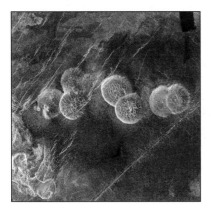

Figure 91. Venus is richly volcanic, having been almost completely resurfaced by lava flows about six hundred million years ago. These volcanoes are several kilometers across and are similar to volcanoes found on the deep seafloor of the Earth. When the pressure is too great for explosive volcanism, small domes form instead of cinder cones.

Hotter water molecules can diffuse higher into the atmosphere, where they are broken by UV radiation into hydrogen and oxygen. The hydrogen floats off into space while the oxygen gets bound into the crust. This runaway process would boil off the oceans in a few hundred million years, and that's what we think happened to Venus.

It gets worse. As the surface heated and the oceans evaporated, rocks released carbon into the atmosphere, where it combined with oxygen left over from the water vapor to form carbon dioxide, another greenhouse gas. Without water to dissolve it, the carbon dioxide built up, trapping more radiation and raising the temperature even more. This second runaway process is the classic greenhouse effect. The result: 900 °F (480 °C).

This brings us to global warming. Earth's biosphere is fragile—models by Jim Kasting indicate that a 10 percent increase in the Sun's luminosity would unleash a runaway greenhouse effect. Life on Earth won't make it until the Sun uses up its fuel; long before that, in about a billion years, the Sun will brighten by 10 percent, and the oceans will turn into steam. If we're not careful, the same thing could happen much sooner. All the evidence shows that the Earth is warmer, though whether this is part of natural climate change or caused by human activity is a matter of "heated" debate. Venus is a sobering reminder of what could happen if we let our climate get out of control.[17]

LIFE AT THE EDGE

Venus is nasty, but could it be alive? David Grinspoon, curator of astrobiology at the Denver Museum of Nature and Science, has made colleagues pay serious attention to the idea that life might exist in the Venusian clouds. Grinspoon is an iconoclast in other ways; he tours with a rock band and has hung out with UFO devotees and alleged alien abductees to better understand nonscientific belief systems that have traction in the pop culture. He's been inspired by his mentor, Carl Sagan, who had a playful and provocative side to his science.[18]

Grinspoon doesn't have a specific suggestion for the type of organism that might survive the swelter of Venus. However, he knows of terrestrial extremophiles that handle conditions just as toxic, and he knows that Venus has energy input from UV light and lots of complex chemistry connecting the atmosphere and the surface. He's really making the point that life is a *process,* and Venus has plenty of geological activity and energy flow in the atmosphere to drive all sorts of processes. Life there would be strange because Venus isn't like any place on Earth. The Venus Express mission reached our evil twin in 2006 and will continue to use penetrating vision beyond the visible spectrum to unravel more of its secrets in the next few years.

MYSTERIES OF TITAN

IN THE WASTELANDS BEYOND the Asteroid Belt, the Sun recedes to a pale yellow dot, and its feeble rays can't even keep water liquid. There are a dozen large moons here, six of which are larger than Pluto. They're major worlds with distinct personalities. Saturn's Titan is perhaps the most promising for life.

SIRENS OF TITAN

Titan is aloof and alluring. It reminds us that the search for life should never be limited to planets—the major moons of the Solar System have atmospheres and surface chemistry and even geological activity. Despite being less than half the size of the Earth, Titan has ten times its mass of atmospheric gas. Pressure at the surface is nearly twice that of the Earth. Nitrogen is the major component, as in our atmosphere. The surface isn't visible, concealed by a murk of organic smog. Titan's chemistry is likely to include many of the reactions that occurred on the primeval Earth just before life emerged.

We got our first close look at Titan in mid-2004, when Cassini swung by on a tour of Saturn and its rings and moons. Cassini is a huge and complex spacecraft, a joint venture of NASA and the European Space Agency. It weighs nearly three tons and is the size of an SUV. Cassini was built in the days before NASA adopted the mantra "Faster, Better, Cheaper" (or as wags had it, "Crappier, Stingier, Riskier"). The mission took a meandering journey, looping twice through the inner Solar System for a slingshot boost from the Sun's gravity. It flew through a skinny, ten-thousand-mile gap separating the F ring of Saturn from the faint, gossamer G ring. Saturn's rings were seen by Galileo as blurry streaks girdling the giant planet. In detail, they are startlingly thin. Imagine a sheet of paper big enough cover a football field—that's the relative size and thickness of the orbiting rubble of rock and ice.

Cassini has given us our best views of the moons of Saturn. There's Mimas, using its gravitational muscle to clear out the large gap in the rings first seen by Cassini himself. Now we see Enceladus, as bright as bright gets, its surface coated with water crystals ejected by geysers. Next up is Iapetus, one side mysteriously dark as soot. Then there are Janus and Epimetheus, sharing an orbit and playing a never-ending game of tag, one moon then the other taking up the chase. Last is little Hyperion, orbiting Saturn like a tumbling sponge.

PATIENCE AND ROMANCE

Space scientists require the patience of Job. Imagine you'd worked for twenty years on a planetary probe. After a couple of billion dollars and three billion miles of travel, you face your final exam in a JPL control room filled with expectant faces. When Cassini released the small Huygens probe toward Titan's surface, team members knew that its camera would beam back data for only an hour before the battery quit. A single error or miscalculation made years before could derail their dreams. It's showtime.

Nobody knows this feeling better than Carolyn Porco. She first glimpsed Saturn when she was thirteen, looking through a friend's small telescope from a Bronx rooftop. The hook was set. "I was a thinker, a seeker," she says, and when her interests in philosophy and religion turned outward, she began studying astronomy. She was a grad student at Caltech when Voyager flew by Saturn and the data started pouring in. The imaging team was shorthanded, so she offered to help out.

As a result, a gem fell in her lap. Saturn's rings are an indistinct blur through a small telescope, but Voyager showed them to be finely grooved, with intriguing patterns and gaps. Among other puzzling features in Saturn's rings, the curious spokes associated with one of the major rings attracted Porco's attention. It was an "aha" moment when she realized a deeper connection with Saturn's magnetic field. She still remembers the thrill of being the first person to unlock one of nature's secrets. The decade-long Voyager mission gave us our first close look at new worlds. "It was Homeric," Porco says as she recalls the journey. "There would be episodes of tremendous discovery, and then it was back into the boat and on to the next port."

Porco started work on Cassini as a young research scientist, and the bulk of her career has been invested in this distant behemoth of sensors and microelectronics. "Voyager was very romantic," she notes, "but Cassini is spectacular." A mission this large and complex is consuming, and the intensity of the work doesn't allow much time for reflection.

She will have been involved for more than eighteen years by the time the mission winds down, and the analogy with child rearing isn't far-fetched. There are peaks and valleys, emotions and challenges, hopes and disappointments. Her team web site proudly displays images as if they were family snapshots. Here, look at bright, shiny-faced Enceladus, and craggy, rough-edged Hyperion, and Dione, its surface braided like the face of an old man. These pictures from one billion miles away resonate with the media too; the shot of little Dione poised in front of mighty Saturn was a runaway winner of the Editor's Choice award in *Time* magazine's 2005 "Picture of the Year" competition, and it was MSNBC's Best Space Photo that same year.

CASSINI AND HUYGENS PAY A VISIT

As Cassini neared Saturn, Porco came into the public eye. She is sought after at NASA press conferences, where her eloquence and enthusiasm stand out in a field that's populated by earnest but slightly drab men. Porco established a quick rapport with Dan Goldin, the previous NASA administrator who shook up the agency and was known for his directness. They both hail from New York City.

Goldin undoubtedly saw in Porco the rare ability to go beyond technical mastery and convey the vital spark of wonder that drives the best science. She explains it this way: "Ever since launch I've wanted to give people a sense of adventure. That they were riding along with us on the spacecraft." The stunning images certainly help. "It's about poetry and beauty and science all mixed together," she says. Porco has a playful side, too. She once had a rock band called the Estrogens, and she's a keen Beatles fan: her science web site has a photo taken in London where she and other team members did the Abbey Road "walk." She's one of many scientists influenced by Carl Sagan. Like him, Porco is a visionary convinced that our future lies in space.

Porco has said, "Titan is undeniably the body fantastic in the Saturn system," but she's a holdout as far as microbial Titanian life is concerned. Titan is two hundred degrees colder than the Earth; water on the surface must be deeply frozen. Chemical-reaction rates will be snail-like. However, others have speculated that the surface may be coated with a rich organic sludge derived from ethane and methane. Below the surface, an ammonia-water mixture could be kept liquid by pressure. Perhaps there are hot springs.

As the Huygens probe fell into the pale soup of Titan's upper atmosphere, Porco was at the European Space Operations Centre in Darmstadt, Germany, providing live commentary on the landing for CNN. Huygens scientists were crammed into the control room, and other Cassini scientists were watching from around the world on live feeds. There was Linda Spilker, deputy project scientist, who persevered through the chilling comments and bruises many women still experience when they enter the world of "hard" science. Across the room was Guy Forget, the saturnine Frenchman and wine connoisseur who leads the team that built the aerosol-measuring device on the probe. Close to him was Ralph Lorentz, who built the unfortunately named Huygens penetrometer and spends his spare time on miniature flight instrumentation that he uses to design the perfect Frisbee.

These people share an extraordinary commonality of purpose: they've used the best of human technology to learn about a rocky world halfway across the Solar System. As the data finally flowed, the tension broke. There was applause and hugging. Relief emanated in waves. Engineers in plaid shirts had tears in their eyes.

We can imagine the ghosts of Giovanni Cassini and Christiaan Huygens looking on in satisfaction. The Italian astronomer and engineer discovered Jupiter's Great Red Spot and the gap in Saturn's rings that bears his name. After he moved to France, he set up and was the first director of the Paris Observatory, also serving as court astronomer to Louis XIV, the Sun King. Cassini was so devoted to his subject that he turned down an invitation to take Holy Orders from Pope Clement IX. The great Dutch scientist Huygens improved the theory and construction of telescopes and then used his finest instrument to unravel the mysteries of Saturn's rings. After discovering Titan in 1655, he said, "How great the joy of heart of him who sees things first!"

WHAT WE SAW ON TITAN

After years of suspense, what did the Huygens probe see? Its parachute unfurled in the upper atmosphere, and the probe was buffeted by winds as it fell through the soup. It emerged from the haze twenty miles above the surface and landed with a splat in the Titanian mud. Feeble sunlight had percolated through to give the terrain a dull orange glow, like asphalt lit by sodium light at night. Huygens carried a special calling card—four experimental rock songs commissioned for the mission, in an initiative sponsored by Mick Jagger.

Geology is strangely universal. Huygens saw dry lakes and shorelines, rocks

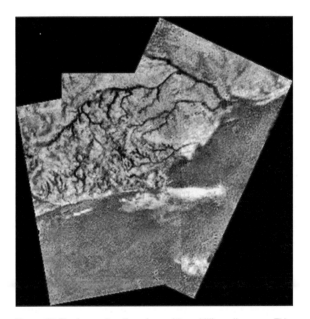

Figure 92. The frozen shoreline of a world one billion miles away. This mosaic from the Huygens probe shows a ridge on Saturn's moon Titan, where liquid methane has washed organic sediment down channels into a now-dry lakebed. The Titan topography is stunningly Earth-like, but the chemistry is utterly alien. Titan has weather and geology just as rich and complex as the Earth's.

and pebbles, channels and river deltas. The appearance is familiar, but the chemistry is utterly alien. Instead of liquid water, Titan has been formed by liquid methane. Instead of rocks, it has snowballs. Instead of silicate dirt, a hydrocarbon grit settles out of the dense atmosphere. Instead of lava, volcanoes spew out ice and ammonia (Fig. 92).

There's weather on Titan, in the form of a black methane rain that falls almost continuously. Luckily, Huygens didn't generate any sparks. As well as methane, Titan has plenty of acetylene and propane—add a little oxygen, and this world is flammable. Cliffs made of methane ice loom on the horizon. We can imagine a gloomy Norse god presiding over these exotic fjords.

To the disappointment of many, Huygens did not find ethane-methane oceans or lakes. Presumably either they dried out or the moon is between episodes of wetness. But the excitement returned in the middle of 2006, when Cassini did one of a number of close flybys of Titan. A pass of the north polar region revealed dozens of lakes (Fig. 93). At the equatorial latitudes where Huygens landed, the temperature is too high for liquid methane, but it can persist at the cooler polar latitudes. Intriguingly, the lakes seem to be located in calderas, geological features that indicate volcanism, a result that still has the planetary scientists scratching their heads. Meanwhile, the chemical and biochemical potential of miles and miles of hydrocarbon lakes is enormous.

Cassini/Huygens was a great success, but it's just a first step. The lander had only enough capability to look at the surface and taste the atmosphere for a short time. Cassini flybys will never get closer than a few hundred miles, and they can only take images. Any detailed chemical analysis will have to wait for a future mission, hopefully one carrying a rover. That may be twenty years off.

Figure 93. This radar imaging of Kraken Mare in Titan's north polar region shows part of a hydrocarbon lake that is bigger than the Caspian Sea, the largest enclosed body of water on Earth. It's possible that life exists in its depths.

COULD THERE BE LIFE ON TITAN?

Could there be life on Titan or under its surface? By tradition, the habitable zone spans the range of distances from any star where water is liquid on a planet surface. The frigid zone beyond the Asteroid Belt always seems out of the question.

Chris McKay knows otherwise. He knows that water can be heated enough to remain liquid by pressure under a planetary surface or by geological activity from the interior. He knows the largest moons of the giant planets are substantial worlds in their own rights. Even small ones can be heated due to tidal flexing by the parent planet. If life can occur outside a traditional habitable zone, there may be a number of potential sites for life in our Solar System, rather than just one. Perhaps life doesn't need a star at all? McKay smiles to himself at the thought. In science, it can pay to be radical but not too radical.

If there's biology on Titan, it will be unlike anything we have seen. The presence of hydrocarbons, water, and several sources of energy means that a chain of reactions up to amino acids and peptides might occur. McKay hypothesized a metabolism based on acetylene, where the intense cold slows down to a moderate pace reactions that would be explosive on Earth. It's plausible that energy released near hot springs could support pockets of biosphere. Fire and ice.

Titan is the most Earth-like place in the Solar System—if you swap methane chemistry for water chemistry. There may be microbes on this cold and gloomy moon, tiny golems fashioned from the mud. If life is found here, we might be forgiven for thinking it could be found anywhere.

WATER WORLD

WHEN WE THINK OF LIFE, we think of water. Three-quarters of our planet is covered by water. Life first developed in water, and land animals are still mostly water by weight. The Cassini/Huygens mission created a spike of public interest in the possibility of life beyond Earth, and NASA hopes to use that interest to gain support for a future mission to another enigmatic destination in the outer Solar System. Titan and the Earth share a thick, nitrogen-rich atmosphere and active geology, while Europa and the Earth have oceans of water. But the Sun's rays are feeble at the great distance of Europa, so these oceans are encased in a thick sheet of ice.

THE ICE FLOES OF EUROPA

The icy landscape spurs the imagination. Is it home to Hans Christian Andersen's Snow Queen, where a sliver of broken mirror lodged in your heart might make you stay forever, far from the warmth of the Earth? Or worse, is this the ninth and lowest circle of Dante's hell, Cocytus? Kept frozen by the flapping wings of Lucifer, Cocytus is the final destination of traitors. They are encased in ice according to their sins, some up to the waist, some completely.

Picture the vistas on Europa. Ice fields stretch to the horizon. Jagged crevasses crisscross the terrain, and cathedral-like blocks of ice thrust high into the sky. Angular shapes have the skewed logic of a 3-D jigsaw puzzle. You have to imagine the sound of groaning and splintering ice—there's not enough at-

Figure 94. Fractured ice covers most of the surface of Jupiter's moon Europa. The ice is cracked in many places and appears to have flowed and reassembled like a jigsaw puzzle. There's no surface water, and scientists engage in active debate about the thickness of the ice, but it's probably several hundred feet thick and may be as many as ten miles thick in places.

mosphere to carry sound. The arctic scene is lit by feeble sunlight. Jupiter hangs in the jet-black sky like a milky eye.

This little moon of Jupiter may be the most surprising world in the Solar System. We think of water as a rare commodity in space, the vital ingredient that makes the Earth special. Yet there's more water on little Europa than in all Earth's oceans (Fig. 94). Astronomers think the ice pack on Europa is hundreds of feet thick and covers an ocean miles deep. Professor of Planetary Science John Lewis and his graduate student Guy Consolmagno suggested in the early 1970s that Europa might have oceans and even life. The first prediction was confirmed seventeen years later by the Galileo mission. The idea of life on Europa was also talked up by Dick Hoaglund, an independent researcher with no NASA support. After meeting with Hoaglund, Arthur C. Clarke used the idea as inspiration for his novel *2010*.

Why is water special? It's the universal solvent, a perfect medium for complex chemical reactions. Water is the only liquid that expands when it freezes, ensuring that even at five times our distance from the Sun, Europa is not frozen solid. On Earth, land animals are mostly water, a relic of our common origins in the oceans. Water is carried to moons and planets by asteroids and comets, which are the Aquarians of every solar system. Europa is probably one of many strange and watery worlds in the universe.

VOYAGES TO EUROPA

Europa caught the eye of the Galileo spacecraft in 1997.[19] Galileo's mission was extended for several years, in part to take a closer look at the icy moon. The next step is an orbiter equipped with radar that can settle the question of the

Figure 95. A hydrobot searches for life under the ice pack of Europa, in an artist's impression. A probe to Europa would land a cryobot, which would use nuclear power to melt its way through the ice pack. Then it would release an instrument-laden hydrobot that would search for life. The mission can be tested on the Earth using the analog environment of Antarctica's underground Lake Vostok. It's currently unfunded.

thickness of the ice and water layers and work long enough to see actual motion of the ice pack. This orbiter is unlikely to arrive at Jupiter before 2026. The Europa Ice Clipper is a clever idea to drop a ten-kilogram hollow copper sphere onto the surface. On a subsequent pass, the spacecraft would fly through the debris tossed up by the impact and capture it in an aerogel, returning the material to Earth years later. The concept was deemed too risky and was not funded.

The most ambitious mission concept is a lander that would lower a probe called a cryobot onto the surface. The cryobot would use heat generated from its nuclear power source to melt down through the ice and release a hydrobot when it reached water. The hydrobot would roam around taking chemical measurements and looking for signs of life (Fig. 95). The environment is challenging for the operation of delicate machinery. Apart from the intense cold, the vicinity of Jupiter has a strong magnetic field and is subject to intense cosmic rays. The mission will be expensive and difficult, and it's not yet funded. Any future lander will likely be a joint U.S.-European venture, to follow on from the enormously successful Cassini/Huygens mission.

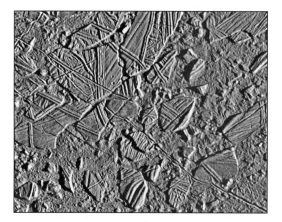

Figure 96. A close-up view of the "ice rafts" on Europa as seen from the Galileo spacecraft. The smallest features visible are less than the size of a football field. The shifting ice pack resembles arctic regions of Earth during a spring thaw. It is hoped that there are places where the ice is only tens of meters thick so that a future probe can penetrate and explore the subsurface ocean.

Planetary scientists argue about the best place to land. Some favor regions of the surface that appear a mottled red-brown. It's very unlikely that the color is due to living material, due to high radiation, but these hydrated salts may indicate organic material to be found under the surface. Others favor places where the ice is likely to be thinnest, to give the cryobot less work to do. Like pizza devotees, they have animated debates over thick crust versus thin crust. The choice of a landing site is critical because the fractured ice field presents an exceptionally hazardous terrain (Fig. 96).

Luckily, there's somewhere close to home to go practice. Just as scientists head to Chile's Atacama Desert to find a piece of Mars on Earth, they go to Antarctica's Lake Vostok to find Europa on Earth. Lake Vostok is deep in the Antarctic interior and buried under two miles of ice. As big as Lake Ontario, it's an ecosystem that has been isolated from the rest of the planet for a million years. Geologists drilled to within four hundred feet of the water but paused, not wanting to contaminate this chilly lost world. When the contamination issue is sorted out, they'll send in a hydrobot.

Europa is a prime candidate for extraterrestrial life. It has all the ingredients: water, organics, and energy from several sources—radioactive decay, tidal heating, UV radiation, and chemical gradients. But the environment is strange enough that we return to a previous refrain: what if life is so strange that we can't recognize it?

SIGNS OF LIFE

THE DETECTION OF LIFE beyond Earth centers on an awkward dilemma. The most sensitive tests, such as the amplification of DNA or RNA, are so specific to the function and architecture of our biology that they would fail if the metabolism or the information storage was very different. But a more general search based on morphology or nonequilibrium chemistry might turn up evidence with a nonbiological explanation.

Imagine you're at a lakeside with your dad. He's teaching you how to fish. The lake has lots of catfish in it, he says, and they just love worms. He baits a hook with a fat worm and sure enough, soon he's pulling in a catfish every fifteen minutes or so. That's the way it works, he tells you confidently. Every now and then, you snag a different kind of fish, but mostly it's catfish. He's tried different kinds of worms, but it's all the same. The life in this lake is catfish.

As a child, you accepted this story; after all, it was the only life you knew. But returning to the lake as an adult, you can see the limits of your dad's view. The lake has many kinds of fish, some of which will take a fly but spurn a worm. There are crabs and other bottom dwellers who never come to the surface. A handful of water has many tiny bugs in it, wriggling and darting. Occasionally, you see shadows of creatures in the water large enough to make you shiver. They're tasty grilled over a fire, but there's a lot more to life than catfish.

If all we use is worms, all we'll catch is catfish. In astrobiology, if we look for carbon-based life in aqueous environments and photosynthetic organisms powered by stars, that's all we're likely to find.

BIOMARKERS

A biomarker is something that indicates the presence of life. Jonathan Lunine of the Lunar and Planetary Lab in Tucson has thought long and hard about biomarkers. To choose the best biomarkers, you have to define life. Lunine lists the attributes in order, from those that are most likely to those that are least likely to be general: carbon; water; energy gained by exchanging electrons (also called redox reactions) or harnessing light; energy storage in phosphate bonds; a biochemical unity that involves a highly selective use of all the possible reactions; a sturdy molecule for storing and transmitting genetic information; metabolic processes that are surrounded by a membrane and therefore self-contained; organisms that are initially small; and biology defined by the range of life on Earth. The last criterion is conservative, but it leads directly to the catfish problem.

The analogous list of detection techniques ranges from those that can identify a wide range of life processes (at the expense of some ambiguity) to those that lock in on specific attributes of our biology (with the danger of missing weird forms of life).

Mineral biomarkers are the footprints and fingerprints of life, visible long after an organism has died. As an example, metabolic processes tend to create carbonate in a form with a tight crystal structure, called aragonite, rather than the looser form, calcite. Some organisms concentrate chemicals and leave the residue when they die—bacteria that metabolize sulfur often leave behind highly distinctive sulfur granules. Other times, minerals adhere to the surfaces of a cell or bacterial strand, leaving a durable cast when the organism dies. Almost all forms of terrestrial life depend on iron, and they store it in crystalline forms quite unlike iron in normal minerals. They can be identified because they usually bear the imprint of the Earth's magnetic field.

Magnetic tracers of life deserve special mention because magnetism is expected to be a common property of planets. Joe Kirschvink of Caltech uses magnetic minerals to home in on mass extinctions and provide strong evidence for a Snowball Earth. It's been known for thirty years that some bacteria navigate by creating miniature magnets from chains of iron particles. Magnetic sensing is also seen in birds, fish, and reptiles; work by Kirschvink has led to the discovery of sensory organs based on magnetite in higher animals. He's argued that detection of magnetic fields was the primal sense.[20]

Another quite general way to detect life is to look for enrichment of the rare and heavy form of carbon relative to the normal, lighter form. In photosynthesis, the heavy form is taken up more slowly than the lighter form, leading to a deficit compared to the ratio in sediments without life. A similar trick can be done with isotopes of iron.

It's also possible to measure the mass of complex molecules in a rock sample. The idea is that a set of masses will indicate components of a biological system, even though the individual molecules may not be uniquely identified. Another useful method searches for a strong excess of left- or right-handed molecules, since life does not use them in equal proportions. The most specific test of all is the amplification of DNA or RNA by a polymerase chain reaction. PCR won't give a false positive, but it will miss any form of life that doesn't use nucleic acids.

LIVING IN THE NEIGHBORHOOD

The Solar System is a richer hunting ground for life than we might have imagined. Biospheres evolve, so grim conditions don't mean that a place has always been dead. The essence of evolution is adaptation, and "grim" is a word too anthropocentric to be useful. One extremophile's toxic dump may be another's pleasure palace. When we jump out of the box and consider biology beyond the bounds of known life on Earth, it could be very challenging to recognize.

Where should we look? If life favors the most active planets, Mars may actually disappoint us by being mostly dead. Large moons turn out to be just as interesting in the search for life. Titan and Europa are compelling. But Jupiter's little moon Callisto has a varying magnetic field that almost certainly indicates an ocean under the icy crust. Early in 2006, Saturn's tiger-striped moon Enceladus joined the A-list when Cassini spotted geysers emerging from its south pole. It's unprecedented to find activity and water in a body so small and so far from the Sun. Carolyn Porco said soon after the discovery, "If we are right, we have significantly broadened the diversity of solar system environments where we might possibly have conditions suitable for living organisms."

NASA's strategy on life in the Solar System has been "follow the water." That drives the design of missions to Mars, and it informs our expectations of habitable worlds. Andrew Ingersoll of the Cassini imaging team has pointed out that "other moons in the Solar System have liquid-water oceans covered by kilometers of icy crust." In fact, careful modeling of the interiors of large moons by Adam Showman at the Lunar and Planetary Lab and others leads to an estimate of twelve to fifteen Solar System bodies with liquid water under ice or rock.[21] A combination of pressure, tidal heating, and radioactive decay within rock keeps the water liquid far from the Sun's warming rays.

In the search for life, it's also worth placing a few side bets with admittedly long odds: on Venus, of course, and Jupiter's fire-and-brimstone moon Io.[22] The last and most unlikely redoubt for life is the zone beyond the planets. More than 140 different types of molecule have been discovered in the dark space between stars, ranging up to the simpler amino acids. Reaction rates are probably too low for an interesting metabolic network to be established, but the truth is that we just don't know. Comets travel thousands of times farther from the Sun than Pluto. The old picture of a dirty ball of ice has given way to the image of a snowbank covered with soot and organic material. In 2006, the New Horizons spacecraft began its journey to recently demoted Pluto and its rocky neighbors in the Kuiper Belt. The chemistry of that environment is also unknown.

Perhaps life can leak into deep space. We know life-bearing rocks are ejected from the Solar System by impacts. In solar systems like ours, giant planets and their moons can be ejected entirely by gravity, giving rise to the idea of living interstellar "arks." With this much biological potential on our doorstep, we're inevitably drawn outward, to the limitless potential of the stars.

6.

DISTANT WORLDS

A sad spectacle. If they be inhabited, what a scope for misery and folly. If they not be inhabited, what a waste of space.

—Thomas Carlyle (1795–1881), on the plurality of worlds

The year is 2067. Earth has survived, its environment battered and bruised, and humans have muddled through, their few shining moments eclipsed by tribal squabbles over resources and religion. Many diseases have been conquered, but others have risen to take their places. The evolutionary struggle between men and microbes is at a stalemate. In an era of quantum computing, bodies of people in the western world are patrolled and retooled by medical microbots. Embedded intelligent agents ceaselessly draw on the vast web of information that permeates the air.

It was one hundred years into the space age before humans cracked the problem of interstellar travel. Fusion drives now power all major commercial aircraft, and larger craft service tourism and mining outposts throughout the Solar System. Astronomers have honed their techniques and routinely detect terrestrial planets as small as Mercury. Robot emissaries traveling at half light speed have fanned out to hundreds of the nearest stars, and a few Earth "cousins" are known.

Twenty years earlier, in 2047, one hundred people set out on a momentous journey. Fueled by the same urge that sent humans migrating across continents tens of thousands of years before, and demoralized by the loss of vision and the spiritual decay on Earth, they were the first to cut the umbilical cord and attempt to homestead a new planet. The voyage was funded by biotech entrepreneurs, who filled the spaces through a worldwide lottery, adjusted only to ensure enough genetic variation for a viable colony.

Public reaction to the venture was varied and emotional. The travelers were called fools, dreamers, and worse. Resentment was tinged with jealousy. One hundred among ten billion were to give humanity the chance for a fresh start.

For two decades, the voyagers sailed through the absolute void and silence of deep space, crammed into their little "Mayflower" with pitifully few supplies to help them at the other end. Their suspended animation was achieved with an experimental tech-nology; nobody knows if they'll be successfully revived. All they know about their new home is that the gravity is 80 percent of Earth's, the atmosphere is breathable, there's water, and the surface supports vegetation. The voyagers' space ark used all its fuel to get to its destination; a distress signal sent now would take twenty years to reach Earth, and a rescue would not arrive for forty years, by which time the settlers' children would be dead. This is a one-way trip.

The first traveler stirs as the ark enters the Procyon group. At the center of an un-familiar sky is an M dwarf with five terrestrial planets in close proximity on tidally locked orbits. One is larger than the rest, with a creamy yellow atmosphere and twin outrigger moons. The planet is both welcoming and strange as it swells to fill the ark's windows. More people stir and watch in silence as the surface comes into view. . . .

•   •   •

IF THE EARTH WERE the size of a walnut, held in front of you, the Moon would be a pea held at arm's length, and the Sun would be a glowing ball fourteen feet across, about four hundred yards away. On this scale, the Solar System is nine-teen miles across, and the nearest star is a glowing ball eighty thousand miles away. Space is numbingly, staggeringly huge.

For more than two thousand years, dreamers imagined that Earth was not unique, that distant stars also had rocky bodies orbiting them. Astronomers are armed with powerful telescopes, but finding extrasolar planets, or exoplan-ets, is exceedingly difficult. In the analogy of the Earth as a walnut lit by a star a quarter of a mile away, the nearby star is a point of light with its planet a hairsbreadth away. Imagine trying to detect the glimmer of the walnut from a distance of fifty thousand miles.

The suspense is finally over—our Solar System is not unique. Astronomers have detected more than 1000 extrasolar planets since 1995. They've mostly used indirect methods. Giant planets tug the star they orbit, and the subtle mo-tion can be detected as a repeating variation in the spectrum of the star. Plan-ets can also be found when they pass in front of their star and dim it slightly, and so the transit causes a partial eclipse. Many of the new worlds are puzzling because they're gas-giant planets orbiting their stars more closely than Mer-cury orbits the Sun. (Large planets like Jupiter are the easiest to detect.) Theo-rists are trying to explain solar systems with architectures very different from our own.

With exoplanets being found at the rate of dozens per year, the hunt is on for distant Earths. Telescopes of unprecedented stability and precision will be needed for the job. Astronomers use deformable mirrors that cheat image-smearing turbulence in the atmosphere. They look at infrared wavelengths, where a cool planet will show up more easily next to a much hotter star. They eventually plan to use free-floating telescopes in space, aligned with lasers, to form images so sharp that Earths can be seen directly.

The most ambitious goal requires new telescopes big enough to take the feeble reflected light from an exoplanet and disperse it into a spectrum. Such a new level of information can reveal whether or not a planet has surface water. Spectroscopy might also find evidence of a biosphere, in the form of reactive gases such as oxygen and ozone and in the form of biochemical tracers such as chlorophyll. This would be the first evidence that the universe is alive.

Discoveries so far have shown that planets form naturally and inevitably when stars form. Combined with the fact that stars create universal chemistry, the table has been set for life. But is anyone dining? Earth-like planets may be special, but their rarity is offset by the vast size and abundance of the universe. In the Milky Way alone, there could be enough habitable planets that we're unlikely to be dining by ourselves.

The first pictures of an Earth clone will be disappointing. They may not have the impact of Carl Sagan's "pale blue dot," as he described the Earth as seen from over the shoulder of the fleeing Voyager spacecraft at a distance of several billion miles. The first spectra of Earth clones will be noisy and scientists will agonize mightily over whether or not they contain evidence for life elsewhere. Wouldn't it be easier to visit these worlds?

Not yet. Returning to our Solar System analogy, the farthest people have traveled is the arm's length from the walnut Earth to the pea Moon, and the Apollo missions cost about $150 billion in present-day dollars. The nearest stars are eighty thousand miles away in that scale model, or one hundred million times farther. Space exploration is fantastically difficult—it's expensive to protect people from the harsh environment of space, and spacecraft are slow and inefficient because they're powered by chemical fuel. Despite this, our robotic emissaries have traveled to all of the planets, and they've explored the major moons of Jupiter and Saturn.

For decades, debate has raged among space scientists and policy makers over the usefulness of humans in space. It's a story of competing visions. NASA has an uneven track record since the glory days of Apollo. The Space Shuttle is obsolete and dangerous. The Space Station is expensive international pork, unloved by scientists and shunned by the companies that were supposed to flock to it for R&D. Some hope to send astronauts back to the Moon and then on to Mars, at a staggering cost—$150 to $200 billion. The vision of humans living and working in space still has the power to inspire.

When it comes to the stars, however, our technology is still childlike. The fastest spaceship would take tens of thousands of years to reach Alpha Centauri. But youthful technologies are growing up, the private sector is starting to flex its muscles, and it's likely that the first plausible plan for interstellar travel will occur in our lifetimes.

WOBBLING STARS

THE STILLNESS OF THE NIGHT was broken by the sound of corks popping and the first hints of birdsong as the eastern sky began to pale. It was early in the morning of July 5, 1995, at the Haute-Provence Observatory in southern France. Astronomers Michel Mayor and Didier Queloz had just woken their wives to celebrate with champagne and raspberry tarts. They had found an invisible object about half the mass of Jupiter in a rapid orbit of a bright star in the constellation of Pegasus. It was the first planet ever discovered around a star like the Sun.

Champagne moments are rare in science. Before 1995, the path to extrasolar planets was strewn with pitfalls.[1] Groups invested dozens of nights of telescope time in the search and came up empty-handed. Others made claims that were shot down or were retracted. Mayor and Queloz worked hard for eighteen months before their success.

WHY FINDING PLANETS IS HARD

News of the discovery was bittersweet for planet hunters Geoff Marcy and Paul Butler, five thousand miles away in California. Marcy had been running an experiment for eight years and had removed the star 51 Pegasi from his sample due to an error in the star catalog. Competition around the world was intense. A Nobel Prize may have been on the line. He'd been scooped.[2] Yet there was vindication in knowing that planets were waiting to be discovered. Marcy and Butler realized they could mine their data for similar objects, and by the end of 1995 they'd found three more exoplanets. It then became clear that a team led by David Latham had made an earlier observation of a planet in 1989.

After the breakthrough, the floodgates opened. The census has raced to over a thousand planets. Most orbit main-sequence stars like the Sun, but a growing number are being discovered around dwarf stars. At least fifty systems have multiple planets. Marcy's group announced 28 new planets in mid-2007 and the team working with data from NASA's Kepler satellite announced several hundred in mid-2010. The record-holder is only 2 Earth masses. A new scientific field has blossomed. But why was it so hard to detect planets for the first time, and why have astronomers been puzzled by the properties of many of the planets found so far?

The search for other solar systems operates at the limit of telescopes. A nearby star similar to the Sun would be like a hundred-watt lightbulb one mile away; such a star is comfortably detected by a modest telescope. Planets don't emit their own light, so the reflected light from an orbiting Jupiter is several billion times fainter. Even so, a large telescope can detect such a feeble signal. The real problem is resolution. At the distance of the nearest stars, Jupiter and the Sun would be separated about one arc second, which is how much the atmosphere blurs incoming starlight. In the analogy, the planet is one billion times fainter and only one-quarter inch away from the hundred-watt lightbulb a mile away. It would be hard even if you were much closer—imagine trying to spot a firefly in the glare of the floodlights of a football stadium.

Success came instead by the indirect method of looking for the gravitational tug that the orbiting planet exerts on the star. In our Solar System, Jupiter is the most massive planet, and it makes the Sun pirouette like a fat dancer. The wobble is too subtle to detect in an image but it can be detected using the signature of the Doppler effect in a series of spectra of the star (Fig. 97).

As an orbiting Jupiter tugs its star to and fro slightly, the light from the star shifts blueward then redward by thirteen meters per second every twelve years. This is sprinting speed, only 0.000005 percent of the speed of light, so it's very tough to measure.[3]

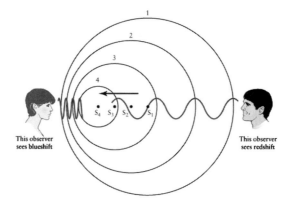

Figure 97. Christian Doppler developed the formalism for moving sources of waves in 1842. When any source of waves—light, sound, or water—is stationary, the peaks and troughs move out in concentric circles, and the wavelength is the same as seen from any direction. But when the source of waves is in motion, as shown here, the wavelength is reduced (or the frequency raised) when the source is moving toward the observer and the wavelength is raised (or the frequency reduced) when the source is moving away from the observer.

THE MAN WHO HARVESTS PLANETS

Planet hunting demands precision and patience. If Geoff Marcy had thought too much about how hard it was, he might have given up long before 1995. But he had that combination of vision and stubbornness that can open the door to greatness in science. When Marcy was a graduate student at the University of California–Santa Cruz, he was riddled with self-doubt. He said, "I felt like I was an imposter, surrounded by all these high-powered people." As a postdoctoral student, it was no better. He thought about leaving science.

Then one morning in the shower, it came to him. Rather than beat himself up over his lack of brilliance, he decided to reconnect with the joy and wonder that turned him on to astronomy as a young child: "I had to find a question to work on that I cared about at a gut level." That question was whether or not our Solar System is unique.

Marcy is soft-spoken, with a polite, almost Old World demeanor. He has a trim goatee, dark eyes, and a receding hairline—the inversion of facial hair he shares with his former grad student Paul Butler, his friendly rival Michel Mayor, and many astronomers. At Lick Observatory, he and Butler subsisted on scraps—a few nights scattered through the year where the Moon was too bright for the people who worked on faint stars and galaxies. Light is not the problem; their telescope could gather millions of photons per second from nearby stars. The critical requirement for the discovery was a spectrograph of unprecedented accuracy and stability.

Enter Steve Vogt. Based in Santa Cruz, Vogt has built a series of magnificent spectrographs, culminating in a machine that gets heavy use at the Keck Ob-

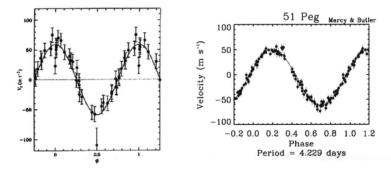

Figure 98. Doppler variations in the light of 51 Pegasi, the first Sun-like star ever to be found to have an orbiting planet. The curve is a sinusoid, which describes the periodic variation of the star's spectrum, called a reflex motion, as the orbiting planet tugs on it. A planet 40 percent of the mass of Jupiter orbits 51 Pegasi every 4.2 days, with a range of velocity variations of 110 meters per second. The curve on the left shows the discovery observations of 1995, and the curve on the right represents more recent and accurate data.

servatory, home of the world's largest telescopes. Instrument builders are the unsung heroes of astronomy. Starting with Brahe's and Galileo's, almost every major discovery in astronomy has depended on technical advances.[4] Marcy and Butler knew they would be working at the bleeding edge; they were grateful to be working with the best. As Butler put it, "When we detect planets, we're sailing in Steve's ship."

Given an accuracy of ten meters per second or so, Jupiters can be detected. Each Doppler curve is nearly a sinusoid, where the size of the velocity shift gives the planet mass, and the frequency gives the orbital period (Fig. 98).[5] Mayor and Queloz detected the first planet with a telescope five times smaller than Marcy and Butler's; they were the Little Swiss Engine That Could.

Their discovery was part of a larger sea change in the power structure of astronomy. For most of the twentieth century, American astronomy was dominant, led by private universities and philanthropists such as Yerkes, Hale, Carnegie, and Keck. Now, more than half of the world's largest telescopes are not in American hands, and the European Space Agency is competitive with NASA for space astronomy. The American share of research papers has fallen below 50 percent for the first time. This is healthy—the cosmos belongs to everyone.

After the difficulty came the surprise. Marcy and Butler had more data, and data of better accuracy, than the Swiss, but they missed the first discovery because they were looking for solar systems like ours, where giant planets far from their parent stars take a leisurely decade or more to complete an orbit. Instead, 51 Pegasi has a planet that rips around its star in a crazy four days. Imagine a planet half Jupiter's size in a sweltering orbit seven times closer to the Sun than Mercury.

Marcy and his team have increased the number of stars they survey from one hundred to one thousand, and they now use the mighty Keck telescopes in Hawaii for much of their work. Each new discovery no longer spawns a newspaper headline, but now the "real" science can be done.[6] Starting in 2010, the Kepler mission has been discovering exoplanets on an industrial scale. With more than a thousand exoplanets in the bag, what have we learned?

HOT JUPITERS

Since the invention of the telescope, we've learned the quirks and foibles of our siblings—the seven other planets of the Solar System (plus poor, misbegotten Pluto). Now we've located a small town's worth of members of our extended clan, and we can describe their general characteristics. Many of these family members are rather strange.

First, most of the planets are Jupiter's mass or larger. There's no surprise there. Planet detection pushes against the limits of instrumentation and noise. We should readily find all the planets more massive than Jupiter. The mass distribution of new planets thins out into a desert at the high end (Fig. 99). It seems that nature doesn't make planets more than ten to fifteen times the mass of Jupiter.

At the other end of the distribution, it's a different story. If you had a fishing net with holes an inch across, you won't catch fish smaller than an inch, and you can't really say anything about whether they exist or not. Astronomers are in a similar position with small planets. However, as you can see in Figure 99, the mass distribution of exoplanets piles up at the limit of detection using the Doppler method, which is a very strong hint that there are many smaller planets waiting to be found. If most of the fish in your net are only slightly bigger than the holes, you can be pretty sure there are plenty that got away.

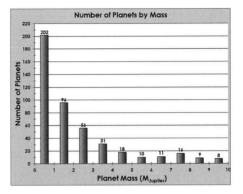

Figure 99. The mass distribution of 440 exoplanets found primarily by the Doppler method. The thinning out of the number of planets much more massive than Jupiter means they're extremely rare (unless they're on very large and slow orbits), because they're easy to detect with the Doppler method. Most surveys have difficulty detecting planets less massive than Jupiter; the piling up of the numbers around that mass is a sign that many more planets exist just below the current detection limit.

The next surprising feature of these giant planets is that they don't live where we expect giant planets to live. Figure 100 shows mass and orbital radius for nearly 400 exoplanets; Jupiter would be in the lower right of this plot at 5 AU. Almost all exoplanets are closer to their stars than Jupiter is to the Sun, and half are closer than Earth is to the Sun. These superhot Jupiters are a puzzle because we think we know how our Solar System came to be arranged the way it is, and these new systems have a very different layout.

Astronomers must be cautious in interpreting this. Returning to a nature analogy, if you camped in a forest for a week, you could watch the flowers open and close from day to night and even notice some growth over a week. But the pines towering overhead would seem immutable; you would have to return over a period of years to see any changes. Similarly, the planet hunters have finally accumulated enough high-quality data to find giant planets at the distances they occupy in our Solar System. Marcy and his team see massive planets within 20 AU around 20 percent of the Sun-like stars they survey. A

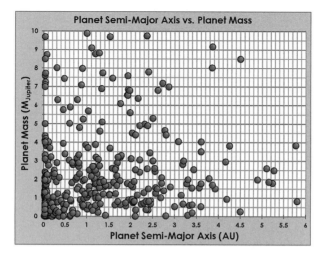

Figure 100. The architecture of distant solar systems. Nearly 400 exoplanets are plotted, mostly discovered with the Doppler method. In this diagram, Earth would be at 1 AU and Jupiter at 5 AU. Most exoplanets are more massive than Jupiter but at distances from their stars similar to the distance of terrestrial planets from the Sun.

larger percentage have less massive giant planets. That makes one in two stars with planets of Neptune or Uranus mass and larger. The fraction with Earths is unknown until after Kepler finishes its work in 2012.

Many of the new planets are highly eccentric, not in their personalities but in their orbits. Eccentricity is the amount by which an orbit deviates from a circle. In our Solar System, most planets have eccentricities of a few percent. Since planets form out of a circular disk of material, they become eccentric due to subsequent interactions by gravity. The only two planets with eccentricity of 0.2 (20 percent) or higher are Mercury and Pluto; the former has been tugged by the Sun, and the latter is a captured interloper. About 85 percent of the new exoplanets have eccentricity of 0.1 or more; their orbits are considerably more squashed than the orbits of Jupiter and Saturn (Fig. 101).

STRANGE NEW WORLDS

European and U.S. teams have improved the precision of the Doppler technique from ten meters to one meter per second, a leisurely walking pace. Multiple giant planets and super-Earths are revealed as harmonics in the Doppler curve of the most massive planet. Pythagoras was right; there is a harmony of the spheres.

Mining Doppler curves has turned up some gems. Fifteen light-years away, the star Gliese 876 has a planet seven times the mass of the Earth whipping around on a two-day orbit, with two Jupiters slightly farther out. And fifty light-years away, in the southern constellation Altar, there's a planet with the mass of Uranus, fifteen times Earth's mass, orbiting its star every ten days, plus two Jupiters on multiyear orbits. The star 55 Cancri has one twin of Uranus and two twins of Jupiter in very tight orbits, plus a fourth giant planet at a

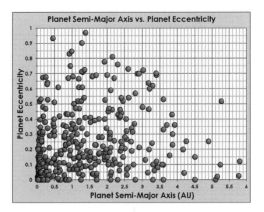

Figure 101. The distance of exoplanets from their stars is compared to the eccentricity, or the degree by which the orbit deviates from circular. Very few of the nearly 350 exoplanets, mostly found with the Doppler method, have eccentricities as low as any of the planets in the Solar System.

normal Jupiter distance. In 2007, researchers in Europe discovered a Neptune-sized planet in a tight orbit around the cool red star Gliese 436. The planet is bizarre; it's icy but it's hot because the water is frozen by gravitational pressure even as it's doused by radiation from its nearby star.

For pure exotica, it's hard to beat the pulsar planets. Pulsars are rapidly spinning neutron stars left over from titanic stellar explosions. There's a pulsar fifteen hundred light-years away in Virgo with four rocky planets in an uncanny representation of our inner Solar System, scaled down by a factor of two. The smallest planet there is only 20 percent of Pluto's mass. In fact, the first pulsar planets were found five years before the planet around 51 Pegasi. Pulsars keep such exquisite time that it's easy to notice if their rhythm is being thrown off by an orbiting planet. But pulsars have survived a supernova and their planets are so baffling that astronomers set them aside in the box they keep for "things we don't understand."

With a little effort, we can imagine the scene if we were magically transported to these strange new worlds. As we hover above one of the super-Jupiters, the parent star sizzles overhead. This giant gas ball careens around its star in a week, so close at its closest approach that the star fills half the sky. Next, we find ourselves on the surface of a twin of Uranus. Its star is also nearby, shimmering through a parboiled atmosphere. The planet is tidally locked to the star, but even on the dark side there's no respite—the hot gloom makes it seem like an antechamber to Hades. Last, we hop to one of the pulsar planets. It is a pitted, airless world like the

Moon. The neutron star is low in the sky, like a dark fist just above the horizon. The sky crackles with magnetic energy, which surges every time the crust of the rapidly spinning neutron star shifts. At this point, we yearn for home.

COPERNICUS REDUX

"These planets answer an ancient question," says Marcy, whose early insecurity has given way to the pride of a pioneer. "Over two thousand years ago, the Greek philosophers Aristotle and Epicurus argued about whether there are other Earth-like planets. Now, for the first time, we have evidence of rocky planets around normal stars."

The Copernican Revolution still has legs. Exoplanets are common, and astronomers have just scratched the surface. They haven't inspected stars that are very different from the Sun, and they can't use the Doppler method to detect planets as small as Earth, Mars, and Venus. The prevalence of hot Jupiters and elliptical orbits is disconcerting. What if our Solar System is unusual, violating the Copernican principle of mediocrity? The truth is that it's too early to tell. As the data accumulates, we're beginning to see planets in familiar settings. The process of planet formation apparently creates a wide range of architectures, of which our Solar System is just one example.

The growing census of planets leaves many questions unanswered. What is the composition of these planets? Do they have moons? How many of them are habitable? How common are Earths? Star wobble gives only the mass and orbit of an exoplanet; transits are starting to provide answers to these questions. The number of planets rises so rapidly with decreasing mass, there may be a hundred million habitats for life in our galaxy alone.

NEEDLE IN A HAYSTACK

Looking for Doppler wobbles has been a great success, but it's not the only way to find planets. Astronomers can also detect them by their fleeting shadows. This has led to a surge in the discovery rate of exoplanets.

TRANSITING PLANETS

When a planet crosses the face of the Sun, it dims it slightly by blocking some of the light. The eclipse is very subtle—Jupiter would block 1 percent of the Sun's light as seen from afar, and the Earth would block only 0.01 percent. For a gas-giant planet, the dimming is less than we'd expect from the planet's size since the diffuse atmosphere is semi-transparent. This is tough observing; imagine staring at a hundred-watt light-bulb and trying to tell if it has momentarily dipped to ninety-nine watts (Fig. 102).

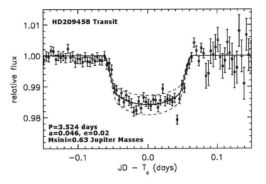

Figure 102. A giant planet transit of HD 209458, which is a Sun-like star 150 light-years away in the constellation Pegasus. The light dimmed by 1.5 percent for about two hours. The shape of the light curve allowed the researchers to deduce that the planet is about 25 percent larger than Jupiter but with a density lower than Saturn's. Its mass has been measured separately by the Doppler method.

There's worse news. The transit lasts three hours, a tiny fraction of the orbital time. And the orientation has to be just right. With distant solar systems distributed at random angles, only 0.5 percent of Jupiters or Earths at their normal distances yield a transit. Any particular target is observable for only eight hours per night, and half of the nights at most observatories are cloudy. Five of every six transits are missed.

Dogged determination and modern technology are used to overcome these long odds. CCD detectors are so efficient that it's easy to gather enough light to detect small changes in one star compared to its unwavering companions. Telescopes stare at millions of stars at a time, either by looking deep in a small area of sky or by tiling the focal plane with CCDs. Even amateur astronomers are playing their part in this hot field. They've detected several giant-planet transits, one of which was nailed with a four-inch backyard telescope.[7] By looking at hot Jupiters, the odds of success rise to 10 percent or better.

The number of exoplanets with eclipses detected has sky-rocketed since the first detection in 1999; there are over 100 transiting exoplanets, with more than half detected since 2007. These systems offer the possibility of follow-up. Since orbits repeat, it's only a matter of time—a few months or a few years—before the transit recurs. Then astronomers can turn their big guns onto the target. Several planet transits have been observed by the Hubble Space Telescope with its spectrograph. As light from the parent star filters through the planet's atmosphere, absorption lines imprint the chemical composition of gas in the atmosphere on the spectrum of the star. Careful timing gives the size of the giant planet. Such information can't be obtained with the Doppler method. Size and mass give the mean density, which is critical in distinguishing between gaseous and rocky planets.

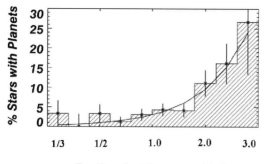

Fraction of metals compared to Sun

Figure 103. The percentage of stars with giant exoplanets depends strongly on the fraction of all elements in the star heavier than helium, generically called "metals," relative to the Sun. The overall average incidence rate of planets is about 5 percent, but when the star has three times more heavy elements than the Sun, giant planets are five to six times more likely to occur.

Such splashy discoveries make headlines, but there are hard-luck stories in the planet-hunting game. Ron Gilliland used 156 orbits of Hubble Space Telescope time to look for planets in the globular cluster 47 Tucanae, and came away empty-handed. The stars in 47 Tucanae formed long ago and far from the disk of the Milky Way, so they have a smaller fraction of heavy elements than the Sun. Without grit, it's hard to make planets. Astronomers are cavalier in their chemistry, classifying all elements heavier than helium—including the primary planet-building materials iron, nickel, carbon, silicon, and oxygen—as "metals." The average planet-detection rate of about 5 percent conceals a rising trend in the abundance of heavy elements (Fig. 103). When the star has three times the Sun's heavy elements, the planet-detection rate soars to 30 percent. Stars have been fusing and spitting out heavy elements for thirteen billion years, so if a new star contains planet-building material, it's a good bet that its cooling nebula actually made planets.

Transit detection was kicked into high gear with the launch of the Kepler satellite by NASA in March, 2009. Kepler is staring at 150,000 stars in the direction of Cygnus, taking data every few minutes for 3.5 years. The telescope is only 1-meter diameter, but the stability of the space environment means it can easily detect the 0.01% dimming of a star caused by the transit of an Earth-like planet. It will take Kepler several years to find Earths in the habital zone.

GRAVITY LENDS A HAND

Pity Shaun Hughes. He's observing on a small telescope in the Warrumbungle Mountains of New South Wales. Australian flora and fauna have their charms, but they wear thin after the thirtieth night of a fifty-night run. Hughes is looking for one star among five million that might vary. Every five minutes, the CCD dumps another image to disk. At one hundred megabytes each, it's good that memory is dirt cheap. Hughes is running software that subtracts each stellar image from the previous one and looks for anything that has dimmed or brightened. He's in a football stadium, trying to spot the one person who's going to moon him for a second.

The Moon has swept through a full cycle of phases, but Hughes has seen nothing. At this point, he'd settle for a slight positional drift, the signature of a near-Earth object. Getting his name on an asteroid might take the sting out of not finding a new planet. Luckily, a summer storm passes through, so Hughes can slink off to the pub in nearby Coonabarabran to down a couple of middies.

Hughes is looking for the signature of microlensing—a method of finding planets that's even more challenging than a transit search but has the potential to detect Earths. Microlensing stems from a prediction made by Albert Einstein in 1936. According to his general theory of relativity, mass distorts space, and then light bends as it follows space's curvature. Light from a distant object is deflected when it passes close to an intervening object. If the mass is very large, the effect is dramatic. Distant quasars are lensed into a mirage of two images or a cloverleaf. Distant galaxies are distorted into tiny arcs, as in a funhouse mirror. Less mass means less bending. When the lens is a star or planet, the light deflection is too subtle to detect. In microlensing, the image seems undisturbed, but its brightness is boosted by an easily detectable amount.[8]

Something goes in front of something and makes it dimmer. That's logical. But something goes in front of something and makes it brighter? As light travels through the universe, it is constantly being deflected by mass along the way. Analogous to a familiar optical lens, a concentration of matter can focus and magnify a bundle of light rays. In a cosmic-average sense, we don't get something for nothing; other regions of space have light that's spread out and demagnified.

How does this help with a search for planets? Lensing is an effect of mass, not light, so the temporary brightening can be seen even if the intervening object is totally dark. In other words, a background star can be lensed by a planet just as readily as by another star. The duration of brightening depends on the mass of the lens. For massive planets like Jupiter, it lasts a few days; for Earth-like planets, a couple of hours.

Lensing depends on a perfect alignment of foreground and background objects. However, two stars will generally have very different space motions. When the planet belonging to one star passes in front of a more distant star, they're like captains of ships passing in the night: one salute, and they sail into the darkness, never to see each other again. Unlike transits, microlensing is a one-shot deal.

If the brightening is brief and doesn't repeat, how do we know it's not just a star with indigestion, burping up gas or varying its output slightly? When grav-

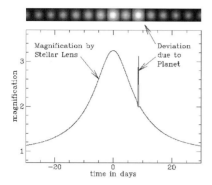

Figure 104. The appearance (top panel) and brightness (lower panel) of a distant star when a nearer star passes in front of it and the light is temporarily amplified due to microlensing. The symmetric rise and fall in brightness is a trademark of this rare effect of gravity. If the foreground star has a planet orbiting it with the correct alignment, the planet causes a briefer amplification spike. Earth-like planets can be detected this way.

ity acts on the light, it doesn't care about the wavelength. Micro-lensing's trademark is a smooth and symmetric rise and fall in brightness, equal for red and blue light. This distinguishes it from all types of stellar activity, in which the changing temperature of the gas makes for unequal variations in red and blue light.[9]

When a star begins to smoothly brighten, it's a trigger for careful observations by a larger telescope. The unseen foreground star amplifies the light of the background star; a planet around the foreground star adds a spike of amplification (Fig. 104). A more rapid spike means a smaller planet. One in a million stars is microlensed at a time, and only a fraction of those has planets. It's the ultimate needle-in-a-haystack experiment.

As with transits, microlensing surveys stare at moderately bright stars and so can be done with smaller telescopes. Shaun Hughes uses a refurbished telescope on a mediocre site that would be beneath the regard of a prima-donna observer spoiled by eight-meter behemoths. And as with transits, amateurs play a vital role in following up the "trigger" of a rising light curve. Listen to the dedication and resolve of Jennie McCormick, an amateur astronomer in Auckland, who participated in the discovery of one of the microlensing planets: "It just shows that you can be a mother, you can work full-time, and you can still go out there and find planets."

This hard road has led to the discovery of just ten planets, but already the great promise of microlensing has been revealed. Two of the planets that were discovered in 2005 are among the lowest mass of any exoplanet, and one weighs in at only 5 times the mass of the Earth. This planet is twenty-two thousand light-years away, and it orbits a dim red dwarf only one fifth of the mass of the Sun, so its surface temperature is a frigid −370°F (−223°C). The fact that such a modest planet was found in the first handful of detections is another hint that Earths may be plentiful. With transit surveys bagging hot Neptunes and microlensing surveys beginning to bag cold Neptunes, we are zeroing in on the habitable zone of planets that resemble ours.

The future is bright for microlensing (and not just fleetingly). The projects so far have been large international collaborations with cheesy acronyms such as EROS, OGLE, MACHO, and MicroFUN. The microlensers plan to refurbish and orchestrate a set of cast-off small telescopes around the world to achieve twenty-four-hour coverage of the entire sky. They'll be able to boost their currently feeble statistics and detect a dozen new planets per year. Earth clones are within their grasp. An orbiting satellite designed to stare at one hundred million stars at once will be able to detect planets as small as Mercury, but it's not yet funded. As always, the microlensers bide their time and practise patience.

PLAYING THE ODDS

Bodhan Paczynski proposed lensing to detect exoplanets long before any had been discovered. His broad, amiable face creases into a smile. He talks calmly about his plan to find distant worlds, as if he had all the time in the world. With cropped white hair and the clipped, crisp accent of his native Poland, he's the grandfather who explains something to you with great patience but expects you to eventually get it right. Paczynski is dying; he has an inoperable brain tumor. This fact doesn't cause a ripple in his placid demeanor.

Paczynski was voracious in his interests. As a professor at Princeton, he worked on everything from planets to cosmology. He was a theorist—not the modern brand of astronomical theorist, who pours equations into a computer and programs it to calculate the appearance of a piece of the universe. Paczynski was a classical theorist who might use the back of an envelope for an elegant calculation. One of his proudest achievements was purely observational. He led a group of Polish compatriots in a project to detect planets using the microlensing technique.

He has seen the ebb and flow of fascism and communism in central Europe, so Paczynski knows that nothing of value comes easy. A microlensing survey must stare at millions of stars to have a chance of success. Also, the planet passes in front of the star for a short while, and the signal never repeats. Although we have found a new planet, it evades our grasp, and our long list of questions about it goes unanswered.

Fifteen years ago, Paczynski was unperturbed by the long odds.[10] His quixotic quest finds an echo in the story of Bill Borucki, the NASA scientist who proposed finding exoplanets by transits ten years before the first one had been discovered, and who is now lead scientist of the highly successful Kepler mission. Paczynski set up a small telescope with a big-format CCD camera, and he staffed it with a handful of dedicated young observers. With a characteristic mix of brilliance and dogged determination, the group has detected dozens of interesting low-mass stars and seven exoplanets, competing successfully with larger and better-funded teams. In the end, that's all any of us can ask for—to be a momentary brightening in the darkness.

GROWING PLANETS

THE PLETHORA OF PLANETS presents astronomers with a puzzle. Did gas-giant planets form around their parent stars or did they migrate from larger distances, and how? Theorists struggle to answer these questions. It would be ironic to find out that solar systems are common while realizing that ours was more special than ever.

COMPLEXITY AND CHAOS

The theory of planet formation is still fairly primitive. Complexity is part of the problem. Newton's elegant law of gravity allows a precise calculation of trajectory only for the case of two objects. With hundreds of objects, there are tens of thousands of forces as each object acts on all the others; in this case, the gravity calculation is only a good approximation. But the planets in the Solar System accreted from thousands of chunks of rock called planetesimals, which means millions of forces must be tracked. And those planetesimals each formed from billions of dust grains, so . . . you get the idea. This complexity makes it impossible to reverse engineer planet formation by running the calculation backward to see how things looked at the beginning.

The converse is also true; a slight change in the starting conditions can lead to the formation of a totally different set of planets. Once planets form, their chance encounters can alter the architecture of a solar system. Planets collide and fragment. Gas giants eject comets and sometimes terrestrial planets, or they send debris hurtling inward. These processes are understood statistically but can't be predicted with certainty for any particular case. Chaos leaves its mark on every solar system.

Like all scientists, astronomers are riding the wave of Moore's Law, the doubling of computer power every eighteen months. Memory and disks are so cheap that they can do simulations today that were inconceivable a decade ago. To make planets, place a gas cloud into your computer, switch on gravity, and wait for it to collapse. Then, ignoring the central star, follow the complex gravitational dance of the accretion process. Other subtle forces are taken into account. Approximations must be made.[11] Now you play sorcerer's apprentice, changing the initial conditions to see what emerges. Some solar systems are re-assuringly familiar; others are startlingly different. Each of them may exist somewhere in the universe.

It takes a brilliant theorist to compete in the age of computation. Peter Goldreich still has what it takes, thirty-five years after he first stamped his mark on planetary dynamics. His early work was prescient in describing many characteristics of the new exoplanets. Goldreich has an uncanny intu-ition for gravity. He's also a black belt and has a Brooklyn accent and a boxer's nose. He can deliver cruel blows to researchers with flawed ideas, but he's un-stinting in his support of the best of the younger generation.

One of those is Renu Malhotra. Goldreich was her mentor, and she shares his preference for analytic elegance over computational number crunching. Malhotra has a calm equanimity, which came in handy when she moved from an all-girls high school outside Delhi to a university where she was nearly the only woman among five hundred men. In conversation, she has the disarming habit, common to Indians, of using the roll axis of head motion (in addition to the familiar pitch for yes and yaw for no) for an equivocal reaction. She's serene, which is an unusual attribute for a scientist.

The theorists use math to tease regularity out of the seemingly random mo-tions of small planetary bodies. Widely separated rocks in space can have their orbital periods synchronized by the ratios of whole numbers, in a phenomenon called resonance. It's an echo of the Pythagorean music of the spheres. There is synchrony in the interactions of planets and their moons—Mimas sweeps out Cassini's division, Cordelia and Ophelia shepherd a slender ring of Uranus, and our Moon turns one cheek toward us, like a doting lover. Solar systems op-erate with a curious mixture of regularity and chaos.

FORGING EXOPLANETS

The current best bet on how planets form is called the core-accretion model. It starts with the rapid growth from dust particles into rocks and then into moun-tains and then into terrestrial planets. This amazing progression in size from microns to meters to kilometers to a rock as big as the Earth is actually the most reliable part of the story—gravity depends on mass, so it acts to accelerate the growth. Harvard researcher Scott Kenyon has said, "The dust bunnies under your bed grow in a similar way. After a million years, a dust bunny can get pretty big." Forming Earths is easy.

In the outer part of a solar system, where it's cooler, when a planet "core" gets to about ten Earth masses it can grab gas from the surrounding disk and build an envelope that turns it into a gas giant. This process stops when the gas is used up or is driven out by radiation from the young stars. That's the basic idea of core accretion. But there are two problems with the theory. Data from the Galileo probe has shown that Jupiter's core may be only three Earth masses, which is too small for accretion to work properly. Also, growing a Jupiter takes several million years, but recent results from the Spitzer Space Telescope suggest that disks often don't last long enough and that giant planets may form in as little as one million years.[12]

A second idea is on the table; its strongest advocate is Alan Boss from the Carnegie Institution of Washington. Lumps can form within a disk due to gravitational instability and potentially grow terrestrial planets in as little as one hundred to one thousand years. In computer simulations, disks do become unstable, and spirals and eddies and rings are seen, but the simulations aren't yet good enough to know if a combination of gravitational instability and core accretion can explain the formation of giant planets.

If we don't fully understand the formation of normal giant planets, how on Earth do we explain the newly discovered extrasolar Jupiters with tight elliptical orbits? These hot Jupiters could not have formed where they are now because the temperatures are too high to grow a giant gaseous envelope. Giant planets migrate from much larger distances and then park in orbits close to the parent star. The mechanisms for the migration and the parking are hotly debated with no consensus in sight.

The elliptical orbits support the idea that the giants weren't formed close to the star—gas in the protostellar disk follows circular orbits. It's a good bet their elongated orbits resulted from interactions with other planets. A very nice example is the Upsilon Andromedae system, which has a very hot inner Jupiter-mass planet and two moderately hot planets two and four times the mass of Jupiter, all on elliptical orbits. (Codiscoverer Debra Fischer asked classmates of one of her young children to name them. Their choices were Twopiter, Fourpiter, and Dinky.) A team at Northwestern University made a precise model of this solar system and showed that the orbits could be understood if an unseen fourth planet came in too close and scuffled with the inner planets in a gravitational feud. The troublemaking planet was ejected, but the remaining Jupiters still show imprints of the disturbance in their elliptical orbits.

Observations show that conditions in star-forming regions are very chaotic, with unpredictable implications for planet formation. Simulations are important because they show that planets may not always have been arranged the way they are now. Sometimes they jostle one another. Sometimes they scatter like tenpins. The range of outcomes may be limited by our imagination as much as by our computation.

While theorists argued about whether planets form or persist in gravitationally complex environments, observers answered the question: they can. One in five of the known exoplanets are in binary-star systems (also called "Tatooine" planets, after the Skywalker home planet in *Star Wars*) and there's one in a triple-star system. Less than half of all stars are single; our simple sunrises and sunsets may be unusual.

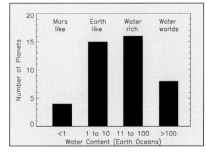

Figure 105. Terrestrial planets that developed in a series of simulations carried out in a computer. In the simulations, planets assembled by gravitational accretion from smaller pieces, and water reached the inner Solar System on debris sent inward by the gravity of giant planets. Each simulation generated several planets in the range of one-quarter to four Earth masses, and most of them had more water than in all the Earth's oceans and were at the distance from their stars where water remains a liquid.

FOLLOWING THE WATER

What's special about our planet? One answer is obvious: water. The Earth is unique in the Solar System for the liquid sheen that covers three-quarters of the surface and moistens most of the rest. Terrestrial life is made mostly of water. Water and life appear to be strongly linked.

The best simulations whet our appetite for what may be waiting out there to be discovered. Water delivery depends on the location of giant planets. Sean Raymond is a postdoc researcher at the University of Colorado, but while still a grad student in Seattle he did eye-catching work on water delivery in young solar systems. In most simulations, he and his collaborators found several terrestrial planets, ranging from one-quarter to four times Earth's mass. The majority of these were in the traditional habitable zone, where water is liquid on a planet surface. The water content ranged from bone-dry to hundreds of times the water content of all the Earth's oceans (Fig. 105). Water worlds may be very common, which primes the pump for life beyond Earth.

Raymond's more recent simulations have concentrated on mimicking the known systems of exoplanets. Hot Jupiters on close orbits to their stars were once thought to be bad news for terrestrial planets, because the gas giants act like unruly bullies, tossing the smaller planets out as they barge through the inner parts of their solar systems. But with a dozen desktop computers crunching for more than eight months, Raymond found that while the giant planets are indeed disruptive, it's not all bad news for water worlds. As the hot Jupiters park in their tight orbits, they fling rocky debris outward, where it can coalesce into Earth-like objects. At the same time, small icy bodies in the dense outer

Figure 106. An artist's impression of a distant planetary system, where terrestrial planets straddle a gas giant, all in the habitable zone of a Sun-like star. The exoplanets discovered so far are part of solar systems unlike ours; many are gas giants on very close, highly elliptical orbits of their Sun-like stars. Astronomers are only just acquiring enough data to find solar systems with architectures like ours, and we don't yet know if our Solar System is typical.

parts of the gas disk slow down and spiral inward, delivering water to the fledgling Earths. One-third of the giant planet systems discovered so far may harbor Earth-like planets covered in deep, global oceans.

Meanwhile, Willy Benz and his group in Switzerland are doing very ambitious simulations that take account of the limitations of the Doppler detection method. This allows them to predict the properties of the whole population rather than just the massive "tip of the iceberg." Benz works in Bern, proving once again that the Swiss should be known for knuckles and know-how as well as cuckoo clocks and chocolate.

Their conclusion: we're nowhere near recovering all of the exoplanets. With the precision of most existing data, only 3 percent of planets are found, and the fraction of "normal" giant planets discovered is no higher. Even with the best current precision of one meter per second, the simulations suggest that only 5 percent of all the planets are recovered. Earths remain tantalizingly out of reach, but early results from the Kepler mission suggest that the universe contains huge numbers of them (Fig. 106).

DETECTING EARTHS

THE DETECTION OF EXOPLANETS is a stunning achievement. But the planets discovered so far are often broiling Jupiters, along with icy and hot Neptunes. They seem alien and uninhabitable. How long will it be before astronomers quicken our pulses by discovering a twin of the Earth?

HOW TO FIND CLONES OF HOME

The technical requirements for detecting distant terrestrial planets are forbidding. Compared to Jupiter, the Earth is three hundred times less massive and five times closer to the Sun, so it's a much smaller lever on the Sun. This reduces a Doppler wobble of thirteen meters per second to an undetectable nine centimeters per second, the pace of a scuttling beetle. So far, the Doppler method has had the best success, but it will run out of traction in the search for planets much less massive than Uranus and Neptune. The current record holder is a planet just five times more massive than the Earth, orbiting the red dwarf Gliese 581. It's only twenty light-years away, and seems to have conditions suitable for liquid water.

In principle, it's possible to directly detect the wobble of a star being tugged by a planet. With the analogy of a nearby star as a hundred-watt lightbulb one mile away, the star appears no bigger than a small speck of dust. However, the Earth's atmosphere blurs incoming light by about one arc second, which smears the dust speck up to the size of a dime. The wobble of the star caused by a Jupiter would be a thousand times smaller angle, 10^{-3} arc seconds. Think of a dime a mile away wobbling by no more than the thickness of one of the hairs on FDR's head. That's bad enough, but the wobble caused by an orbiting Earth would be one thousand times smaller, or one millionth of the amount that the atmosphere blurs starlight.

The situation is a little better for direct detection of planets by imaging. The Earth is one-tenth Jupiter's size but five times closer to the Sun, so it reflects four times less light than Jupiter. In our scale model, Jupiter is one billion times fainter than the star, sitting at the edge of the blurred-out "dime" of stellar light. This is challenging but doable, and 2005 saw the first-ever detection of exoplanets by imaging and over a dozen have been detected. The first detections were super-Jupiters far from their parent stars, and one was orbiting a brown dwarf.[13] An Earth in this scenario is four times fainter and five times closer to the star; its feeble light would be engulfed in starlight.

Transits of Earth-like planets are also difficult to detect from the ground. Seen from a distance, Jupiter would dim the Sun by 1 percent but only one in a thousand randomly oriented systems are aligned so that the transit is visible.

Also, the transit lasts for only one four-thousandth of the orbit: thirty hours every twelve years. The closer orbit of an Earth improves the orientation and timing odds. One in two hundred systems has the right orientation, and the transit lasts one seven-hundredth of the orbit: thirteen hours in a year. But this good news is offset by the fact that the smaller Earth dims a star by much less than Jupiter, a minuscule 0.01 percent. This is what Kepler was built for; it should find 500 or so planets smaller than twice the Earth's size.

BIG GLASS

The world's best telescope designers and instrument builders have embraced the challenge of detecting and characterizing planets. Telling them it can't be done just makes them work harder. Breakthroughs will come from a mix of ground- and space-based telescopes. For light-gathering power, the ground wins—each Keck telescope collects twenty times more light than the Hubble Space Telescope. For image sharpness, there's no substitute for the vacuum of space. Image sharpness also improves when shorter waves are used.[14]

Roger Angel is the heir to William Herschel and George Ellery Hale, the architect of the Mount Wilson and Mount Palomar observatories. He knew twenty years ago that all the easy gains had been achieved. Detectors were nearly perfect. Telescopes had grown to the point where they sagged under their own weight. He came up with a way to make large, rigid mirrors that had a short focal ratio (fast, in optical parlance) so they could fit in a smaller dome, thus reducing the cost. Angel makes mirrors in a rotating oven under the football stadium at the University of Arizona.

Once a year, when the stands are empty and the cheerleaders are elsewhere, honing their routines for a new season, the huge oven starts turning. As the temperature climbs past 1000°C, chunks of borosilicate reach their melting point and flow into a parabolic form. The oven spins like a crazed carousel, lights flashing and giving off singeing waves of heat. When it has fully cooled a month later, the lid will be cracked on a new mirror, its front face only an inch thick and its back plate a hollowed-out honeycomb. When Angel cast a 6.5-meter mirror for the Multiple Mirror Telescope in 1992, it was the largest mirror made in the United States since the mirror for the mighty Palomar reflector fifty years earlier.

There are several approaches to the design of large telescopes. One is to align hexagonal segments with lasers and a neural net. This is the approach taken at the Keck Observatory. Another is to make lightweight mirrors that flop like pancakes and employ hundreds of actuators to constantly maintain the proper shape. That's the approach taken by an international consortium with the twin 8.1-meter Gemini telescopes, by the European Southern Observatory with its four 8.2-meter Very Large Telescopes (VLT), and by the Japanese with their 8.3-meter Subaru telescope.

Is it a coincidence that Roger Angel reached up to 8.4 meters for his most recent series of giant mirrors? He looks a little sheepish as he says, "There's been something of a pissing contest going on." Angel is soft-spoken and has a shy smile, but the competitive juices run deep in this expatriate Brit. He has an entrepreneur's spirit and relishes going toe-to-toe with the well-heeled European and Japanese governments. Although close to retirement age, he's embarked on his most ambitious venture ever: spin-casting seven 8.4-meter mirrors to combine into a 21.4-meter leviathan in Chile. It will be the world's largest scope by far, and the six off-axis "petals" in the design are incredibly challenging to figure. Even this is a sideshow for Angel, who's had encouragement from the National Academy of Sciences for a plan to mitigate global warming by launching an armada of Sun-deflecting mirrors to a place between the Earth and the Moon.[15]

The mirrors that emerge from the football stadium are exquisite. After spinning and cooling in the oven, they move over to the grinding and polishing machine. When Angel delivers a mirror, it's big enough to park five SUVs on, yet the surface is smooth within one-fifth of a wavelength of light. Scaled up to the size of the continental United States, its biggest bumps would be an inch high.

CHEATING THE ATMOSPHERE

Big glass is the first ingredient needed to detect and gather spectra of distant Earths. The second is a way to cheat the atmosphere. Roger Angel's colleague Nick Woolf has done a lot of pioneering work on ways to rapidly adjust mirrors to compensate for the jumble of light waves that reach a telescope, thereby dramatically sharpening the image. The technique is called adaptive optics. Woolf is tall, thin, and angular, and white hair rises from the top of his head like a brush. He exudes both a startling intelligence and an often startled air.

At major observatories around the world, adaptive optics is getting millions of dollars of investment as engineers try to wring extra performance from their telescopes. Current state-of-the-art procedure is to aim a laser at the atmospheric layer that causes image blurring, and then use the reflected light to adjust telescope optics twenty to thirty times per second.

The last ingredient is an instrument that can form star images sharp enough to detect extremely faint planets in their wings. Phil Hinz regards the scattered piles of electronics in his lab with equanimity. Hinz is the protégé of Angel and Woolf at the University of Arizona, still in his thirties. He's slight, with sandy brown hair and a voice like a whisper. The gleaming machinery will soon turn into something called a nulling interferometer. To most people, the mere thought of assembling this chaos into a smoothly functioning machine would be daunting. Hinz has been taking things apart and putting them back together since he was a kid. He's that rare astronomer who can master the complex hardware of a modern astronomical instrument—servos, high-speed electronics, and precision optics.

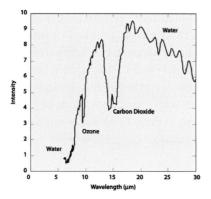

Figure 107. Mars Global Surveyor looked back at the Earth and took this infrared spectrum of the whole planet. If we could do this for distant Earths, spectral features due to oxygen, ozone, and water would act as biomarkers to indicate life. At optical wavelengths, there are additional biomarkers, such as the chlorophyll edge. In practice, the first spectra of distant Earths will be of poor quality and much harder to interpret.

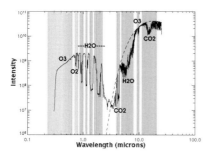

Figure 108. A simulated spectrum of an Earth-like planet from ultraviolet to infrared wavelengths. With sufficient spectral resolution, many absorption bands due to water, oxygen, carbon dioxide, and ozone are visible. Taken together, these tracers may be used to indicate the presence of respirating organisms.

Hinz plans to use it to study planets like ours, so he has a few tricks up his sleeve. The first is an adaptive-optics system on the new Large Binocular Telescope (LBT) in southern Arizona. Twin 8.4-meter mirrors on a common mount make it the most powerful telescope in the world. Atmospheric compensation is done with a delicate secondary mirror, one meter across and just a millimeter thick. Even Hinz loses his perfect equanimity when he handles the secondary—one slip and the cost is one million dollars and a year's delay to the project.

The next trick is to combine the light from the two "eyes" of the LBT to give it the sharpness of vision of a single mirror twenty-five meters across. It's called interferometry.[16] The light beams from the two telescopes must be brought to a common focus with a precision of twenty billionths of a meter, otherwise the waves would combine into an incoherent mess. In a nulling interferometer, the light adds so as to almost perfectly cancel out the central star. With the star blotted out, the much fainter planet is revealed.

A final big gain comes from shifting the experiment to infrared wavelengths. At wavelengths longer than visible light, a star has declining intensity, but a warm planet has rising intensity. The contrast between planet and star improves by a factor of one thousand. After all this effort, the dime-size star in our earlier analogy shrinks to the size of a small bead, so the planet is no longer buried in its glare. The hundred-watt star is suppressed, and astronomers look for the 0.0001-watt planet. It's challenging, but not impossible.

Planet hunting is difficult work, but Hinz is persistent. He knows the Doppler teams worked for more than a decade before they had their first success. The prize is huge. With the planet light isolated, the light-gathering power of a large telescope lets the feeble light be spread out into a spectrum. Primitive life could reveal itself in biomarkers such as oxygen and ozone, or the chlorophyll spectral edge that's characteristic of vegetation on the Earth (Fig. 107). In preparation for upcoming data, astronomers are simulating the spectra of Earth-like planets in great detail (Fig. 108). This compelling experiment can take astrobiology to a new level, and it may well be the first demonstration that we live in a biological universe.

A PLETHORA OF PLANETS

THE NEXT LEAP in our understanding of distant worlds is likely to come in space. Planet detection is easier above the blurring of our atmosphere, and the fact that we can launch only relatively small telescopes is offset by the fact that stars are seen against a much darker sky background. NASA and the European Space Agency are launching a small armada of missions to detect and characterize Earth-like planets. These missions are expensive, and the launch dates of some of them are uncertain.

CURRENT AND UPCOMING MISSIONS

Kepler is early in its mission, but looks like it will be a stunning success. With less than two months of data it doubled the number of exoplanets. Its "fishing net" has holes small enough to snare Earth-like planets. As a result it will give the true demographics of exoplanets. Early results indicate that, relative to Jupiter-mass planets, Neptune-mass planets are twice as abundant and Earth-mass planets are five times as abundant. The projected number of Earths in the galaxy is a hundred million.

COROT is a smaller French mission that launched in 2006 and has had a lower yield and some hardware problems. Meanwhile, the doughty micro-lensers are working on funding for their own mission, which would sift

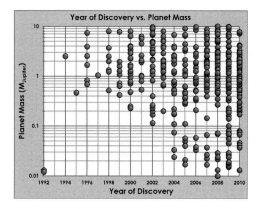

Figure 109. The steady march of increasing sensitivity for exoplanet detection is shown by the lowering in the mass that can be detected since 1995. There are now over 30 "Super-Earths" planets up to 10 times Earth's mass. One system, Gliese 581, has three super-Earths, two of which are in the habitable zone of the parent star.

through two hundred million stars in order to find about one hundred Earths. Several promising missions exist purely as concepts; they have modest funding for design studies but no growing wedge in the budgets of NASA or ESA. The supporters of these missions roam hopefully among scientific meetings, giving flashy presentations and hoping for the green light that will turn dreams into piles of shiny hardware.

SIM, the Space Interferometry Mission, is an amazing feat of engineering: by suppressing vibrations to less than one billionth of a meter, it measures positions to within the size of an atom. The instrument on board will measure stellar positions with exquisite accuracy.[17] This will allow it to see the minuscule (one millionth of an arc second) wobble of a star caused by an orbiting planet. SIM will search for Earth-like planets around the nearest 60 stars and Neptune-like planets around the nearest two thousand. SIM could see details on Lincoln's face on a penny as far away as the Moon or detect grass growing in your backyard from five miles away. Although its yield on Earths will be less, it's an important complement to Kepler, since it will find much closer and brighter planets.

THE PROMISE OF TPF

Exoplanets are booming (Fig. 109). Eventually, NASA plans to launch the Terrestrial Planet Finder. The design of TPF is still sketchy, because the relevant technologies are still being developed. A first phase will employ a six- to eight-meter optical telescope for planet imaging. Image contrast of one billion to one is attained using a coronagraph, where a central obscuration blots out light from the star. The TPF coronagraph will home in on the best 50 terrestrial planets found by SIM, providing images and optical spectra of the atmospheric gases. It has no firm launch date (Fig. 110). Five years later, it will be joined by

Figure 110. NASA's Terrestrial Planet Finder (TPF) coronagraph which has no current launch date. This six- to eight-meter telescope will block the light from a central star with a contrast of one billion to one and reveal terrestrial planets around 50 nearby stars. Spectra of the reflected planet light will reveal composition of the atmospheres.

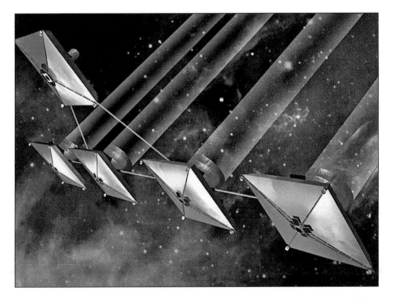

Figure 111. NASA's TPF interferometer is a set of four three- to four-meter telescopes, either flying in precise formation or physically linked. Working at infrared wavelengths, it will be able to image terrestrial planets with a contrast of only one million to one. Spectra of the reflected planet light will reveal a different set of potential biomarkers. This mission complements the coronagraph, and is similar to ESA's proposed Darwin mission.

the TPF interferometer, an array of three- to four-meter telescopes, linked by lasers and floating in precise formation (Fig. 111). Infrared wavelengths are preferred for the interferometer, because many common molecules have spectral transitions in this region, and planets are more prominent than stars at long wavelengths. It will be able to characterize terrestrial planets around five hundred nearby stars.

The twin TPF missions will cost several billion dollars each, so the smart money says that NASA will join forces with ESA and its Darwin mission, a proposed set of three space telescopes flying in formation. Beyond TPF, there is the grandiose vision of a mission called Life Finder that would have the sensitivity and resolution to see oceans and continents on distant worlds. NASA's James Webb Space Telescope, the successor to HST, is due to launch in 2016 and will be able to characterize exoplanets. But the crystal ball is cloudy.

In fact, any of these future missions could be cancelled due to funding pressure within NASA. The space agency fights for a share of the shrinking discretionary portion of the federal budget, going up against Veterans Affairs and Social Security in the ugly sausage making of the congressional budget process. Moon, Mars, and Beyond is the eight-hundred-pound gorilla that threatens to eat the lunch of smaller missions. Astronomers hope that the goal of finding Earths will resonate with the public and the legislators.

The scientists who dedicate their professional lives to these missions sometimes let the blood rush to their heads. The truth is that even their dream hardware may not be able to prove the existence of life on a distant planet. The missions will be looking for the signposts of life—biomarkers—using tracers like oxygen, ozone, methane, carbon dioxide, ammonia, and nitrous oxide. However, Earth life interacts in a complex way with the planet surface and atmosphere. Since other living worlds may not be like our own, it's important to run careful simulations and use a suite of diagnostics because of the real chance of false positives.

The first observations won't look like the elegant Earth spectrum of Figure 107; they'll be noisy, ratty traces that may tantalize as much as they inform. Science at the frontier is always this way. The rate of progress is very rapid, and astrobiologists are still learning their craft. The adventure is just beginning.

CUTTING TO THE CHASE

Planet hunters Debra Fischer and Greg Laughlin are impatient with incremental thinking. They don't want to wait decades until we can harvest distant Earths. They want to find the nearest Earth clone and take a close look at it. They are inspired by NASA administrator Dan Goldin's speculation back in 1992, before any exoplanets had been discovered: "Imagine if spectroscopic analysis revealed a blue planet with an oxygen atmosphere just four light-years away orbiting Alpha Centauri. Demand to build a warp drive would start right away!"

Alpha Centauri is not only the closest star to the Sun; at a distance of 4.4 light-years, it's also one of the best places to look for life. Alpha Centauri is actually a triple system. Component C is an unpromising red dwarf, but component B is only slightly cooler and dimmer than the Sun, and component A is almost a twin of the Sun. A and B orbit each other with a minimum separation of eleven AU, which simulations show won't disrupt the orbits of any planets within two AU of either star.

Buoyed by the expectation that terrestrial planets are commonplace, Fischer and Laughlin propose to harness a dedicated telescope to stare at A and B every night for two years. The European exoplanet hunters are already observing the system. The Doppler detection technique bottoms out around the mass of Uranus or Neptune, but combining three hundred thousand observations of two such bright stars would allow them to beat down the noise. If either star hosted a Mercury, a Venus, an Earth, or a Mars, they could detect it. That's part one.

Part two involves sending a fleet of nanobots to Alpha Centauri, powered by miniature antimatter drives. Think of cellphones accelerated to a tenth of the speed of light. Having large numbers would keep the unit costs down and ensure redundancy. With minimal guidance systems, the nanobots would home in on any terrestrial planets and beam back pictures and other information. But without high-powered transmitters, how would they beam the information back over such a large distance? Fischer and Laughlin have that covered. Wave after wave of nanobots would be launched. The leading wave would send its data the modest distance back to the next one launched, like a firemen's bucket line. Within two generations, we'd have detailed information on nearby Earths. After that, how could we resist a visit?

TRAVEL TO THE STARS

THE DISCOVERY OF SO MANY distant worlds is thrilling. But our information is still crude and fragmentary. In science, the best evidence is physical evidence. We know the age of the Solar System from meteorites that landed on Earth and from Moon rocks we brought back to Earth, both of which we subjected to radioactive-decay measurements. We've learned a lot from the Martian meteorites that have landed in our lap, but to know for sure whether Mars has been alive we'll have to go there and bring back samples. The exoplanets beckon, and using telescopes is so indirect—why can't we just go there?

THE FASTEST THING THERE IS

As a young man, Galileo speculated about the speed of light. One evening, he and a friend stood on hilltops a mile apart in the brown, corrugated Tuscan countryside. They each had a lantern with a sliding shutter. By prior agreement, Galileo opened his shutter, and his friend opened his as soon as he saw Galileo's light. Galileo then tried to measure the time it took the light to make a round-trip between the two hilltops. He couldn't. It happened so fast that he was measuring the sum of the two reaction times. Galileo crudely estimated that light must travel at least tens of thousands of miles per hour. For all he knew, the speed could be infinite.

The first accurate measurement was made by English astronomer James Bradley in 1728.[18] He realized the light from a star overhead must arrive at a slight angle because the Earth is moving in its orbit. The effect is subtle—only one two-hundredths of a degree—but his observation led to a measurement of three hundred thousand kilometers per second, or 186,000 miles per second. Now we can trivially play tag with photons using ultrafast electronics; the best modern measurement is 299,792.458 kilometers per second.

Let's go back to the Earth as a walnut. The analogy is useful because it reduces both distance and speed. In this scale model, light moves at walking speed. To walk the yard or so to the pea-sized Moon would take about a second, which is indeed the light travel time to the Moon. The ten-foot Sun is a quarter mile from Earth, a leisurely eight-minute stroll. It would take five hours to walk to the edge of the Solar System. Then we encounter an aching void of nothingness: several years of metaphorical walking to reach the nearest star. Ten thousand years to traverse the Milky Way galaxy.

Let's now put the achievements of spaceflight in perspective. The most distant spacecraft is Voyager 1, launched in 1977 and now fourteen billion kilometers away, or three times Pluto's distance. That's impressive, but it's only 0.01 percent of the distance to nearby stars, so it will take three hundred thousand years to reach them. The fastest spacecraft is the solar probe Helios 1, which reached a speed of 160,000 miles per hour in 1974. That's pretty fast, but still only 0.02 percent of the speed of light.

BIGGER AND BIGGER FIREWORKS

Why can't we do better? There's a sociopolitical answer and a physics answer. Space exploration has been spurred by the rivalry between nations. The first chemical rockets were used by the Chinese military a thousand years ago (think of large, guided fireworks). Robert Goddard flew the first liquid chemical rocket in 1926 (Fig. 112), and the next major advance was Werner von Braun's development of the V-2 rocket in World War II.

Goddard was hooked early. As a boy, he was inspired by H. G. Wells's *War of the Worlds*, and he was nearly expelled from college for firing a rocket through the basement of his physics building. Goddard suffered much professional ridicule, but by 1920 he had developed a proposal for sending a rocket to the Moon.

Von Braun also learned early about the attraction of rocketry. As a thirteen-year-old in Germany, he attached six firework rockets to a red toy wagon and lit the fuses. Trailing flames and smoke, the wagon roared five

Figure 112. Robert Goddard posing with his liquid-propellant chemical rocket just before its first flight in March 1926. His first flight traveled only 184 feet across his Aunt Effie's farm to land in a cabbage patch, but it was as seminal as the Wright brothers' flight at Kitty Hawk.

blocks into the town center before exploding and leaving a charred wreck. Von Braun was delivered to his father by a policeman and severely reprimanded. But he was hooked.

Von Braun's V-2 was too late to affect the outcome of the war, but his entire team was brought from Germany to kick off the American space program. NASA was founded in direct response to the Russian launch of Sputnik in 1957, and until the recent surge in telecommunication satellites most space launches were for military and surveillance purposes. NASA has never had any real competition to keep it lean and innovative, and the well-funded military has no motivation to develop better rockets.

The physical reason for our disappointing performance is a reliance on chemical energy. Imagine your one-ton car needed nine tons of fuel to drive across the country. Worse, your engine is so spendthrift that it burns through all the fuel within a few miles of your house. That's today's rocket. "Chemical rockets are just too slow," worries Les Johnson, manager of NASA's transportation technology program. "They burn their propellant at the beginning of a flight and a spacecraft just coasts the rest of the way."

If we send people to Mars, it will be in this slow and inefficient way, with 90 percent of the mass of the mission consumed as fuel. Rockets throw something overboard to propel themselves. The momentum of the fuel ejected backward at high speed through a nozzle is balanced by the forward momentum of the payload. That explains the "mystery" of a rocket—why it doesn't need to push against anything when it's in deep space.

Chemical rockets are barely contained explosions. We can sense the ghost of the teenage von Braun egging us on with glee. Every element of the U.S. (and Soviet) space program, from Mercury and the giant Saturn V to the Space Shuttle, has used either solid or liquid chemical fuel. As we learned shockingly with the loss of two space shuttles, chemical rockets are extremely dangerous. There must be better and safer ways.

NEW TECHNOLOGIES

NASA is working on ion engines that propel an inert gas such as xenon at speeds of one hundred thousand miles per hour from the exhaust, using solar panels or fission to charge them. Better still is to not carry fuel at all but live "off the land." Solar sails gather sunlight and use the momentum—a push equivalent to the weight of a coin per square meter of surface at the Earth's distance—to gradually accelerate through space. Plasma sails are related to solar sails. They create a magnetic bubble around the spacecraft and let the charged particles in the solar wind shove the vehicle away from the Sun. While NASA has struggled to move beyond chemical rockets, pioneers have kept their eyes locked on the future.

Robert Forward was foremost among the space travel visionaries. The perfectly named Forward, who died in 2002, lived at the juncture of what could be done and what might be done in space. A physicist with two hundred research papers to his credit, he also wrote eleven science-fiction novels that are classics for their imaginative use of speculative science. In his writing, you could find planets orbiting so closely that they shared an atmosphere, or life on a neutron star, or metabolism that worked at absolute zero. Forward was a fixture at physics colloquia around the country, with his shock of white hair, owlish glasses, and colorful waistcoats. He used skeptical audiences to hone his ideas.

Forward was known for work on space tethers and solar sails. A space tether is a cable of high tensile strength that connects two objects in different orbits. It can generate power due to its interaction with the Earth's magnetic field or be used like a whip to fling a payload toward the Moon or another planet. Forward designed an interconnecting set of tethers that could get a spacecraft to Mars with no propellant. Solar sails also need no propellant, but they lose their oomph as they move far from the Sun's light. Interstellar travel requires a sail the size of ten football fields, with a huge laser aimed at it.

Kepler dreamed of solar sails four hundred years ago after watching comet tails blown by the solar breeze. The technology is finally catching up with the vision. Sails can now be made of carbon-fiber mesh one hundred times thinner than a piece of paper, coated with a dusting of aluminum. An onboard laser would provide the push. Forward also proposed a version in which a microwave transmitter pushes on a grid of superconducting wires. To reach one-tenth the speed of light with a long, steady push—and the nearest stars in forty to fifty years—a trillion-watt laser or transmitter would be needed. Monsters of this power are used to create fusion in the lab, but only for tiny fractions of a second. They would have to be vastly reduced in size to be operable in space.

The other way to get to the stars is to harness particle power. All chemical energy—including the familiar processes of life—comes from rearranging electrons among atoms and molecules. The atomic nucleus is governed by a force trillions of time stronger than the force governing electrons. Making energy by rearranging atomic nuclei in fusion or fission is ten million times more efficient than chemical energy. Instead of a ten-story Saturn V rocket filled with chemical fuel, Apollo could have reached the Moon with a lump of uranium the size of a baseball.

The idea of nuclear-powered rockets dates back fifty years to Project Orion and Project Daedalus (Fig. 113). Project Orion was the first engineering design study of a spacecraft powered by nuclear pulses. A number of brilliant scientists and engineers worked on it in the 1960s but it was terminated due to concern over radioactive fallout and the passage of the 1963 Test Ban Treaty. Project Daedalus was a study conducted by the British Interplanetary Society from 1973 to 1978 to design a fusion-powered unmanned spacecraft to travel to nearby Barnard's Star. The project never moved beyond the design stage.

Unfortunately, the hardware needed to control fusion weighs much more than the fuel itself. Any practical design for interstellar travel would require a supertanker of fuel to get a Shuttle-sized payload to the nearest star in fifty years. There's not enough hydrogen in the space between stars to scoop up this much fuel on the way.

BETWEEN SCIENCE AND SCIENCE FICTION

Fusion and fission convert only about 0.1 percent of particle mass into pure energy. However, when matter annihilates its quantum twin, antimatter, the process is perfectly efficient.[19] If you could liberate the mass-energy in the ink in the period at the end of this sentence, it could power your home for several days.

Figure 113. The British Interplanetary Society produced a detailed design for an interstellar probe in the mid-1970s. Project Daedalus was intended for a one-way trip at 15 percent of the speed of light to Barnard's Star, six light-years away. The spacecraft was not built but would have weighed fifty thousand tons. Its nuclear fuel was supposed to be helium-3, which is not available on Earth but which could have been gathered from the atmosphere of Jupiter on the way out of the Solar System.

Antimatter isn't science fiction; the shadow twin of the stuff we're made of is created routinely but fleetingly in particle accelerators around the world. However, it takes special conditions and a lot of energy to create and store it, and there's the rub. While a modest fifty grams of antimatter could get a small robotic probe to neighboring Alpha Centauri in a decade, current accelerators create just one-tenth of a microgram per year, so we're a factor of one hundred million short. Debra Fischer and Greg Laughlin will need this tough problem to be solved to realize their dream of a mission to Alpha Centauri's hoped-for terrestrial planets.

What about warp drive? To the disappointment of *Star Trek* fans, the speed of light is an absolute limit based on currently known laws of physics.[20] Wormholes sound attractive for instantly tunneling through space-time, but there's no evidence they exist. Some scientists hope that we might one day harness the vacuum energy that is causing the universe to accelerate. Others speculate about an antigravity device. Interestingly, there's no theoretical obstacle to teleportation, which is the remote construction of an object, atom by atom, after transferring the information at the speed of light. Primitive versions of teleportation are used in quantum computing. But it's a huge technological leap to transmit and reconstruct a macroscopic object or a human.

Meanwhile, the visionaries are driven by irrepressible and infectious optimism. Every year sees new proposals for hybrids of fusion and solar sails or hybrids of fusion and antimatter. In 2005, a converted Russian ICBM launched a solar sail funded by the Planetary Society and Cosmos Studios. The rocket failed, but it was an inspiring way of turning swords into ploughshares, and the groups will try again. Robert Forward's ideas are finally being tested.

As the almost equally perfectly named Robert Frisbee puts it, interstellar

travel is a "stretch" goal. The JPL flight engineer notes, "When it was first proposed, Apollo was technically impossible. But there were all these bits and
pieces of technology being developed, from big rockets and fuel cells to life support systems and heat shields. A stretch goal helps people to focus their efforts,
take blinders off, and open up to new possibilities." He adds, "And don't underestimate the power of dreams."

OUR FUTURE IN SPACE

IF INTERSTELLAR TRAVEL IS a misty dream, and NASA continues to rely on
decades-old technology, perhaps we'll shrink back from the challenge of space
travel. After all, we've got plenty of problems right here at home. Humans have
not left Earth's orbit since 1972. Do we have a future in space?

Space travel will always be difficult and expensive. The environment is totally unforgiving, with constant hazards from UV radiation, cosmic rays, and
micrometeorites. Electrical power and data bandwidth are limited. No matter
how much redundancy is built in, there will be failures—a solar panel that
doesn't unfurl, a battery that shorts, a shutter that sticks half open. About 30
percent of missions fail completely. This worry leaves most NASA engineers
with just enough hair for a bad comb-over.

NASA GOES ON A DIET

Yet space exploration is changing. In the 1970s and 1980s, NASA designed
spacecraft like top-of-the-line Swiss Army knives. Planetary missions such as
Voyager and Cassini and space observatories such as Hubble and Chandra
were crammed with complex instruments. Given the time needed for development, ground testing, and integration, it was guaranteed that much of the
hardware would be obsolete when launched. You may own a smart phone or a
Pocket PC with a faster processor and more memory than the onboard computer that operates the Hubble Space Telescope.

In the 1990s, tight budgets forced NASA to consider the pros and cons of
high-profile, billion-dollar missions. To spread the bets and get more bang for
the taxpayer buck, smaller classes of mission were created, with a single goal
and one instrument. They have peppy names such as Scout and Explorer.
This strategy has advantages, but it sacrifices the lure of high stakes and
collective adventure, where each voyage is like a *Mayflower* heading to a
new world. Lean missions are a bargain. About twenty years ago, the average

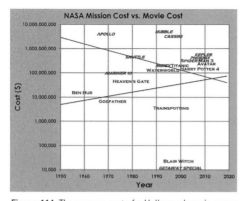

Figure 114. The average cost of a Hollywood movie versus the average cost of a NASA space mission over the past fifty years. Some prominent examples and some outliers are identified. It may be a while before space travel becomes mass entertainment, but the revenue potential matches that of movies. Rich people currently pay twenty million dollars to visit the Space Station, but several companies have plans to bring the price of an orbital flight down to fifty thousand dollars or less. Recently, the Kepler mission cost a similar amount to the movie *Avatar*.

planetary mission started costing less than the average Hollywood movie (Fig. 114). And we all know that more than one in five movies is a dud.

BRING ON THE ROBOTS

Suppose you were a young engineer. How would you design a planetary probe from scratch, without being constrained by current practice? Would you really come up with a space truck like Cassini? Nanotechnology suggests the answer. It's now possible to cram sensors, a miniature CCD camera, a processor, a motor, and a radio transmitter into a package the size of a bumblebee. These airborne nanobots could be programmed remotely and send their data back to a mother ship. The military will use "smart motes" to test for chemical agents on the battlefield; geologists want to use them for exploring remote and inhospitable terrain; and no doubt government agencies will use them (and perhaps already are using them) to spy on us.

Miniaturization and the doubling of computer power every two years mean that we can cram a lot of intelligence into a small package. What might a mission to Mars or Titan look like if we used cutting-edge technology ten years from now?

The spacecraft is the size of a suitcase, and the cost is a few hundred million dollars. It contains a set of pods, each the size of a grapefruit. Pods are dropped from orbit at different locations, and on the way down each pod releases dozens of motes. Dispersed by air currents, the flying motes sample pressure, temperature, and chemical composition. When the pods land, they launch hundreds more motes. Some crawl across the surface like robotic ants, others burrow under the surface and test for microbes. They send their data wirelessly to the nearest pod, which beams them up to the spacecraft, which beams them on to Earth. The motes are linked by a neural net, and strategy is adapted on the fly. Loss or malfunction of some motes doesn't affect the success of the mission.

In addition to micromachines, spacecraft will be transformed by new materials. Carbon nanotubes are one hundred times stronger than steel yet one-sixth the weight. The skin of a spacecraft could be made self-healing against meteorite impacts if it was made of long-chain molecules called ionomers. With materials in research labs today, the body of a spacecraft could have sensors and power generation built into it, doing away with the need for wires and batteries. Other new materials can self-assemble and "remember" their shape, giving them flexible function and replacing actuators.

Robots are already capable of doing our bidding in the Solar System and beyond. Commercial jets can already be flown by autopilot and often are—the pilot is there as a placebo. Battle is dangerous, so the military is increasingly looking toward robots to do the dirty work. We use robots to explore mines, volcanoes, toxic-waste dumps, and other places where humans fear to tread. Both the airlines and the military make extensive use of simulators for training and to save money. As we've seen with the Mars rovers, space exploration is taking on the flavor of a video game. We'll use robots as extensions of ourselves—to see, hear, and touch worlds we will never visit.

THE EXPERIENCE OF SPACE

The vision of robots exploring the Solar System warms the hearts of engineers and technologists, but it leaves most people cold. Where's the thrill? Where's the romance?

People are big and clunky, but some of our finest moments in space have come when human ingenuity gave technology an assist, like when Shuttle astronauts fixed the aberration of the Hubble Space Telescope, or Neil Armstrong took over control of Apollo 11 after the computer failed and landed in treacherous rocky terrain with only seventeen seconds of fuel in the tank.

Pinky Nelson is one of the select band of people who've stared down at the Earth from above. Now an astronomer and professor at the University of Washington, he flew on the Shuttle three times in the 1980s, logging more than four hundred hours of time in orbit. His real name is George; when you're an astronaut, it's OK to be called Pinky. Pinky is nearly sixty, but his face is smooth and lineless and framed by sandy hair. He's virtually unchanged from the eager young man who stands in a shiny spacesuit in a NASA publicity shot, clutching his helmet, an American flag draped in the background. Born in the vastness of the Iowa corn belt, he's had unique opportunities to contemplate the vastness of space.

When asked what it's like up there, Pinky says, "I'll give you my stock answer," and then gives it. But if hundreds of retellings have shaped the story into familiar lines, his eyes still sparkle as he relates the experience. It's not all checklists and experiments, he says; there's plenty of time to chill and stare out the window. Although, scientists being what they are, Pinky talked to experienced astronauts and prepared a checklist of ways to relax and enjoy himself.

His high point in space was his first flight, on the ill-fated *Challenger* in 1984. Pinky and the crew repaired the Solar Maximum satellite, and Pinky shot footage for an IMAX movie. He was the first American to move untethered in space when he did two extra-vehicular activities to test the Manned Maneuvering Unit. On a flight with five test pilots, Pinky had assumed one of them would get picked to fly the MMU. But he got the nod and then couldn't resist a little boyish taunting as he floated past the Shuttle while the flyboys were stuck inside.

Pinky's face lights up as he recalls his MMU flight; after twenty-two years, he can still reconnect with the primal experience of space. He's only two hundred miles straight up, a distance you could drive in a few hours, but it's an utterly different world.

He stands on a little perch, rocket pack at his back; Earth yawns below his feet. The Shuttle is poised above his head; stars dance at the ends of his outstretched arms. This small package of flesh and bone and gristle is treading vacuum while he whirls at seventeen thousand miles per hour. A mere thousand generations after his ancestors scratched out short lives on the African veldt, Pinky is a colossus who straddles the world and walks around it in ninety minutes. He's alone with his dreams and thumping heart. If gravity eased its grip an iota, he'd be flung halfway to the stars.

YOUR NEXT VACATION?

Astronauts have had a singular experience. For the rest of us, space is like music is to a deaf person. It can be appreciated intellectually but not grasped. We have many needs on Earth, so technology and exploration alone are not enough to motivate funding for space travel. If space is our future, our destiny, then the activity must embrace many more people than it has so far. We need a reason to be there.

The reason may be recreation. As commercial space travel takes its first tentative steps, tourism is a big driver. By the end of 2009, six men and one woman, all rich entrepreneurs, had each paid twenty million dollars for a trip on a Russian rocket to visit the International Space Station (one went twice). That's a hefty price tag for a weeklong vacation in a cramped can with fewer creature comforts than a youth hostel, but it's four times cheaper than the cost per astronaut with the Space Shuttle. If you can handle a price of thirty-five million, they'll even throw in a space walk.

In 2004, the spindly SpaceShipOne soared above the Mohave Desert to win the ten-million-dollar X Prize, given for the first reusable civilian spacecraft to reach the suborbital landmark of one hundred kilometers. This achievement is the catalyst to open up space for everyone. NASA can't launch a payload into Earth orbit for less than ten thousand dollars per kilogram. Soon after the X Prize was awarded, Virgin Galactic announced that it will use a descendant of SpaceShipOne to send thousands of people into space for about two hundred thousand dollars per head. Over 80,000 people have made down payments on bookings. When the price comes down by another factor of three to one thousand dollars per kilo, the commercial opportunities will soar.[21] Bigelow Aerospace is putting more than one billion dollars into developing a cruise ship to the Moon. The Shimizu Corporation plans to have a hotel in orbit by 2020. The United States and Russia won't dominate the next fifty years in space the way they did the first fifty—the Europeans and Japanese have ambitious plans, and the Chinese are likely to use space and lunar exploration to cement their ascent as the world's next great superpower.

The new pioneers of space are hands-on entrepreneurs like Sir Richard Branson, who likes to fly high-altitude balloons and who started a record label, which signed bands such as the Sex Pistols, before he moved on to found a music-store chain and an airline. The financial muscle for SpaceShipOne came from Microsoft cofounder Paul Allen, who's also invested in the search for extraterrestrial intelligence.

But the nonpareil of aviation and the new space age is Burt Rutan. Like his hero Werner von Braun, he started building rockets as a kid. "The ones we made were very dangerous, and the kids that played with them didn't have all their fingers, and sometimes were blind in one eye," he says. Rutan and his brother set up shop in the Mohave Desert, and their composites and canard wings have transformed the way light aircraft are made. In 1986, his *Voyager* was the first plane to travel around the world without refueling, doing it in nine days. He smashed this record down to three days in 2003.

But all along, space was the big prize. Rutan was inspired by a 1950s film in which von Braun visualized going to the far side of the Moon. "That was so important because the whole world had that sense of adventure five hundred years ago when Magellan made it round the world," he says. Rutan watched NASA go into the doldrums in the 1980s and was convinced it could not deliver on cheap spaceflight. He badly wanted to go up himself, so he did something about it. Rutan developed SpaceShipOne for just twenty million dollars. Think of what he could do with NASA's billions to spend on his visionary designs. People like Rutan believe that one day we'll live and work in space (Fig. 115). Let's hope it's not because we've spoiled the biosphere but because the Earth is too small to contain all our dreams and ambitions.

Figure 115. Our future in space. After fifty years of the space age, only five hundred people have been into Earth orbit. As commercial and private companies start to participate, the odds are good that recreation and tourism will become major reasons for space travel. In this vision, a space colony houses fifty thousand people in a realistic terrestrial setting. Artificial gravity is provided by the rotation of a gigantic space wheel.

DREAMS OF OTHER WORLDS

Remember the rapid arc of technology. When cars were first invented, they were expensive novelties, less reliable than horses. Only a bold visionary would have looked at the capabilities of airplanes a century ago and predicted sleek aluminum tubes ferrying millions of people daily. Space travel today is disappointing only because we've been so well served by the fertile imaginations of science-fiction writers.

Since the 1960s, only five hundred people have been able to experience the thrill of spaceflight, and most of them have been military test pilots. It's a shame that NASA never sent poets and writers and filmmakers into orbit to inspire us with a visceral sense of what it means to slip the bonds of Earth. The next fifty years will belong to the private sector. Purists may flinch at the prospect of billboards in orbit and zero-gravity brothels, but as we start to recreate in space, other possibilities will open up. Just as a week of skiing or lying on the beach is vacation enough for most people, there will always be some who challenge themselves against an unclimbed mountain or explore a cave that has seen no footprints. When space travel is as routine as flying on a commercial jet, we'll be ready to reach for the stars.

The problem is the length of the journey. With plausible technologies, interstellar travel is a multigenerational endeavor. Travelers will have to be in suspended animation. As Mark Ayre of the Advanced Concepts Team at the European Space Agency puts it: "We've been looking at suspended animation to cut consumables—food and water—for a trip that could take five years or longer. That's important because missions are driven by the mass of the spacecraft. The other thing is trying to avoid psychological problems. If you have people awake, you need to keep them entertained." Ayre is talking about travel within the Solar System, but the issue is the same for travel to more distant worlds.

The medical obstacle may soon be removed. In the past few years, doctors have had 90 percent success in putting mice and pigs into short-term suspended animation. The mice were given a gas mixture in which hydrogen sulfide, or "rotten egg" gas, replaced oxygen. Their metabolic rates dropped by 90 percent, and their core temperatures fell from $37°$ to $11°C$; then they were revived with no ill effects. These tests on our close mammalian cousins raise the hope that the techniques will work on us, too.

Will we ever realize the scenario that opened this chapter? Perhaps in as few as forty or fifty years a set of adventurers will be headed for distant worlds. If Fischer and Laughlin's quest to find Earths around Alpha Centauri succeeds, why not send voyagers alongside the nanobots? It's difficult to imagine the bravery of someone willing to leave friends and family and be interred—for these spacecraft would be little more than airtight coffins thick enough to absorb cosmic rays—only to face an alien environment and an uncertain future light-years from home.[22] Death awaits all of us. These travelers would be using their lives to make a bold gesture, acting out a curiosity that is quintessentially human. In a sense, they would be returning home, since we're all children of the stars.

7.

ARE WE ALONE?

> Sometimes I think we're alone. Sometimes I think we're not. In either case, the thought is staggering.
>
> —**Buckminster Fuller,** architect, designer, and visionary

If Earth's future space travelers came to this place, they would declare it a godforsaken wilderness. A trackless vista of cliffs and escarpments stretches to the horizon. Brown rock scorches under twin giant suns in a blood-red sky. The air is a thin gruel of sulfur dioxide, methane, and nitrogen. But there's more here than meets the eye. In the pore space of the rocks, something is stirring.

At a boundary layer with the deep mantle, mildly acidic water bubbles through the rock, driven by heat and pressure from the interior. Miles underground, microbes thrive in a rich brew of organics and dissolved minerals. They move by sensing magnetic fields and temperature gradients. Huge colonies begin to differentiate their functions and metabolisms to better use the available resources. Symbiotic behavior emerges.

In the course of ceaseless and random genetic variation, some organisms develop the ability to vibrate their outer membranes and sense when it is perturbed. The timing of a return ultrasound wave acts as a primitive proximity sensor. This brand of microbes maintains the spacing to garner more resources, so it rapidly dominates the colony.

As the strategy becomes more successful, the organisms with the most powerful emission or most sensitive reception must deal with a cacophony of ultrasound signals. Some do this by tuning their vibrating membrane to a fixed-frequency channel. Others learn to combine different inputs, moving beyond stimulus-response to a simple form of signal processing. The sonic champions gradually migrate to the center of the colony, where they can emit in synchrony and so increase their power and range.

These profound changes don't happen overnight. There are tens of millions of years of experimentation and dead ends before individual organisms begin acting in concert. But once it happens, a positive-feedback loop is set up that spurs even more experimentation.

Something unexpected happens. The activity of the colony as a whole creates a low-frequency sonic signal that travels easily through rock. Imperceptible to individual microbes, the hum is registered by the cooperative nexus at the heart of the colony. But there are other low-frequency signals with more remote origins. Gradually, microbial colonies sense one another throughout the vast subterranean biosphere. The signals form a primitive network.

All this would be invisible to a casual interstellar tourist. "Awareness" is too strong a word, too anthropocentric. At what point does reaction shade into intention, or signal processing shade into intelligence? We may not have enough distance from the buzz of our own thoughts to judge. But something interesting is happening here, and the planet, like the universe itself, is still young.

·   ·   ·

THE SENSE OF EXPECTATION IS PALPABLE. In the past few decades, we've learned that chemistry is universal and that distant worlds are commonplace. Earth burgeons with life in every nook and cranny. On this world, evolution has led to creatures that know of their place in the universe and are taking the first steps to venture into space. It seems unlikely that this is the only time and place in the cosmos where biology led to intelligence and technology, implausible that we are the only life-forms to experiment with space travel and interstellar communication. The information of astrobiology frames one compelling question: are we alone?

Imagine a young girl sitting in a small sandbox. Her finger is raised in front of her nose, and she inspects the grains of sand one at a time. Each is a tiny, well-formed world. One is a shiny and angular quartz crystal. Another is a black chip of basalt. A third is blue and opalescent. Now imagine that every village and town in the world is scattered with one hundred sandboxes. These sandboxes are like galaxies in the universe. There are about as many grains of sand in the millions of sandboxes as there are stars in the universe. We are the little girl, inspecting nearby planets in our own galaxy a few at a time. It's difficult to imagine what biology may have created on so many worlds.

Motivated by the generous contents of time and space, the physicist Enrico Fermi boldly reversed the earlier question. He supposed that there had been plenty of time and opportunity for any advanced civilization to explore the galaxy, so he wondered instead, "Where are they?" In a situation of little information, there are many possible answers to this question, and the way we view the answers sometimes tells us more about ourselves than about the universe we live in.

Regardless of how we parse the probabilities of life, intelligence, and technology elsewhere, the universe itself has properties that seem noteworthy because they allow for the creation of heavy elements and the persistence of stars. The fabric of physics embeds numbers that are both finely tuned and propitious for the development of carbon-based life-forms. Are these facts trivial, because we can only observe a universe with properties that would allow us to exist? Or is there a deeper meaning?

As a way to give shape to our ignorance, radio astronomer Frank Drake invented a simple equation to encapsulate our expectations for intelligent life in the universe and the prospects of communication. The Drake Equation has factors that range from well measured and astronomical—the birthrate of stars in the galaxy—to speculative, such as the longevity of an advanced civilization in a communicative phase. Undeterred by the uncertainty, some scientists are listening for artificial signals from space. The search for extraterrestrial intelligence (SETI) is one indication of how keenly we are curious about kinship in the cosmos.

We still know of only one place in the universe hosting life. But scientists are becoming persuaded that the pervasiveness of life on Earth means there will be many biological experiments out there. The universe may be littered with microbial life. Some fraction of these experiments will lead to advanced organisms with the ability to explore space and understand their place in the universe. The question of whether or not we are alone hinges entirely on whether that fraction is significant or vanishingly small.

If it's not small, we're faced with the dichotomy in the epigraph to this chapter, framed in slightly different terms by the science-fiction pioneer Sir Arthur C. Clarke: "Two possibilities exist: either we are alone in the universe, or we are not. Both are equally terrifying." The reaction of terror is understandable. We're a young race, unsure of ourselves and gingerly feeling our way into the cosmos. Not long removed from a time of hunting and gathering, we're still fearful of the night. We express our confidence as a species in our science, and it will one day reveal whether or not we inhabit a biological universe.

WHERE ARE THEY?

It was the summer of 1950, and Enrico Fermi was walking to lunch with several colleagues. The men were talking about two stories that had been peppering the nation's newspapers for months—the disappearance of trash-can lids and a spate of UFO sightings. Fermi joked that the two phenomena were connected. They talked about other things for a while, and then, during a pause in the conversation, Fermi abruptly asked: "Where *is* everybody?"

His colleagues immediately knew Fermi was talking about extraterrestrial visitors. They also suspected that the question was more profound than it appeared. Fermi was one of the preeminent physicists in history. Winner of the Nobel Prize in 1938, he was skilled as both a theorist and an experimenter, and his judgment was so unerring that his colleagues called him "the Pope." Fermi addressed complex scientific problems by breaking them into pieces and doing swift calculations in his head to get a rough answer. So-called Fermi questions form a vital part of the training of a scientist.[1]

FERMI'S PROVOCATIVE QUESTION

Fermi didn't write down his thought process, but we can guess at it. The logic is as valid as it was then. Unless the Solar System is special, there are Earth-like planets near some of the billions of stars in the galaxy. A fraction of them host life. Given enough time, life on some of those worlds will evolve intelligence and technology. With a modest extrapolation of current technology, we will be able to travel at 1 percent of the speed of light, and at that speed it takes only ten million years to explore or colonize the galaxy if we assume that travelers don't linger and move swiftly on to new worlds. That's a small fraction of the age of the galaxy and its oldest stars. There has been plenty of time and opportunity for alien civilizations to communicate with us or to visit us, yet we see no evidence of their existence. Where are they?

This conundrum has been called the Fermi paradox, because we fail to observe something that we might expect to. Most people, including many scientists, tend to have one of two gut reactions to Fermi's paradox. They think either, "Wow, there's got to be tons of life out there, and there's been time for many civilizations to evolve far past our capabilities, so the galaxy should be crawling with aliens," or, "That's crazy; intelligence is rare on Earth and will be rare elsewhere, and space is so vast that the chances of alien contact must be very low." What's the more sensible reaction?

Before we consider possible answers to Fermi's question, let's look at the nature of the argument. You can sense that it's not a normal proposition. Science deals with observed phenomena and then tries to explain them. In this case, we are trying to account for the *failure* to observe something.[2]

The paradox stems from three statements, each of which seems reasonable. First, extraterrestrials capable of communication and interstellar space travel exist or have existed in the past. Second, if they'd visited our planet, we would've seen them. Third, we haven't seen them. (We'll use the shorthand "ETs" for space-faring aliens.)

Everything hinges on the first statement. If ETs have never existed, then the contradiction formed by the second and third statements goes away—we are the first space-faring civilization. This conclusion accepts the Fermi paradox at face value, but to prove the first statement false would be a very tall order given the vast realms of space in which ETs may be distributed. If ETs exist, then attention falls on the second statement. There are many plausible reasons why ETs could exist yet not communicate or avoid our detection, so the second statement may be wrong, and the contradiction is avoided. If the third statement is wrong and we *have* actually seen ETs, then the paradox evaporates entirely.

In this tricky terrain, where evidence is absent, we have to be very careful with logic. To say "I haven't seen ETs, so they probably don't exist" is invalid, because the bar is set very high to prove the absence of something in a large and complex universe. But saying "The argument above is invalid, so ETs probably do exist" is also invalid! We're not entitled to be surprised by the fact that ETs haven't made contact unless we have a rational reason to believe they exist *and* a rational expectation that they would have made contact.

MAYBE THE QUESTION IS MOOT

The simplest option, believed by a majority of the U.S. public, is the rebuttal of the third statement: aliens exist, and they've already made contact. Fueled by reruns of *The X-Files* and *Star Trek* and by pervasive science fiction in the popular culture, most Americans are convinced that UFOs are examples of alien visitation. Many also believe that a nebulous conspiracy at the top levels of the military and government has kept the secret for more than fifty years.[3]

Figure 116. Classic grainy photograph of a UFO from the 1960s, similar to hundreds that can be found in books and on the Web. While evocative, images alone will never provide convincing scientific evidence of something as profound as a visit from an extraterrestrial species. The case for UFOs as aliens is not proven and there are many alternative explanations for much of the evidence.

There's such a vast literature surrounding UFOs, ancient astronauts, and alien abductions that there's room here for only a pinprick in the balloon of speculation. The history of UFO sightings is instructive, however. One hundred years ago, UFO sightings were inspired by the fiction of Jules Verne, and they took the form of flying airships and galleons. In 1947, a sighting by pilot Kenneth Arnold in Washington State, followed by the infamous Roswell "incident," kicked off the modern UFO era. These UFOs looked like sleek, metallic flying saucers, a frontier technology of the time (Fig. 116). People still report flying saucers, but after fifty years the sightings seem to have fallen into a rut, describing technology that seems retro and low-tech, almost corny.

Most UFO sightings still occur in the United States, despite plenty of people in other countries who could be making sightings (Fig. 117). Also, UFO sightings don't occur randomly. It's illuminating that major peaks have occurred at times of great tension in the Cold War or during pivotal events in the space program, such as the So-viet launch of Sputnik, the first Mars landing, and the Apollo Moon program (Fig. 118).

Figure 117. The geographical distribution of UFO sightings in the western hemisphere, from a comprehensive database of nearly twenty thousand maintained by amateur researcher Larry Hatch. There is a strong concentration in the United States, despite a larger pool of observers in South America. The same is true on the other half of the globe, where Europe accounts for about 90 percent of the sightings, and there are very few in Africa and Asia.

When a UFO can be identified, in most cases it corresponds to Venus, an aircraft, or a high-altitude weather balloon. Visual observations are notoriously unreliable, and in the age of digital photography an image alone will never be decisive. (Fermi discounted this possibility with his joke about the garbage-can lids.) But even when all conventional explanations, delusions, and charlatans

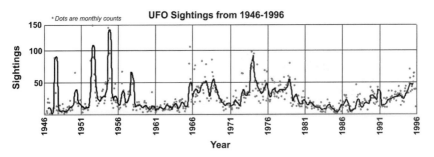

Figure 118. Fifty years of UFO sightings worldwide (mostly in the United States and Europe), from the database of Larry Hatch. Since the first peak of the modern era in 1947, many of the waves of sightings correspond to major events in space—such as the launch of Sputnik, the first Mars landing, and the Apollo program—or times of increased tension and military activity during the Cold War.

are removed, there's a persistent residue of cases with eyewitness testimony that's hard to discount. Are they real visitations?

Scientists are in a tricky situation when it comes to UFOs. Many think that the evidence of astrobiology points to large numbers of habitable planets across the galaxy, and it follows that there's a strong possibility of intelligent life out there, too. But they're utterly unconvinced by the evidence for UFOs. The sensible position is to insist on a high standard of evidence—images, eyewitness reports, and vague conspiracy theories will not suffice. Carl Sagan said, "It pays to keep an open mind, but not so open your brains fall out."

BELIEF SYSTEMS AND CONSPIRACIES

Why are some people so fervent in their belief in UFOs? Planetary scientist David Grinspoon delved into this question by "consorting with the enemy"—visiting UFO sites, investigating the bizarre phenomenon of cattle mutilation, and talking to the people who claim to be alien abductees.[4] He has written eloquently about the bubble of belief that envelops and sustains this culture, a place where skepticism and critical thought are in short supply. The gatherings have an evangelical fervor. Grinspoon also notes with disappointment that the stories are slightly shabby, like B-grade science fiction.

Physicist Richard Feynman put it best: "The first principle is that you must not fool yourself and you are the easiest person to fool." There must be deep psychological reasons why we want to believe in aliens (Fig. 119).

In western culture, aliens have been shape-shifting metaphors for the ills that lie within us. UFOs can be found in medieval paintings and tapestries and throughout the Christian tradition. Many of these archetypes were incorporated in the classic 1956 science-fiction movie *Invasion of the Body Snatchers*. At the time, this story of pods from outer space that replicate and replace humans in a small Californian town was a thinly veiled commentary on the paranoia induced by Senator Joseph McCarthy and his anti-Communist witch hunts. However, the film also embodied the tension

Figure 119. A version of the poster in the basement office of fictional FBI agent Fox Mulder, from the TV series *The X-Files*. The series ran from 1993 to 2002 and created an intoxicating blend of UFO, alien-abduction, and government-conspiracy stories. The image and the slogan epitomize the widespread public belief in UFOs, despite the lack of any hard evidence that they require alien visitations as an explanation.

between individualism and mindless conformity—people acting like "peas in a pod" and turning into vegetables.

Conspiracy theories play into the same mind-set. Superficially, it's reassuring to suppose that we're being buffered from the enormity of alien contact by the government or that a shadowy military-industrial cabal runs the world. But in fact conspiracy theories represent the ultimate abdication of personal responsibility. If you check your critical faculties at the door, you've already voted with the pod people. Ideas like this endure because they're self-reinforcing. Facts that fit the theory are admitted; facts that don't fit are excluded. This is worse than ascientific; it's antiscientific.

THE ABSENCE OF SPACE TOURISTS

Let's therefore discount for now the option that aliens exist and the evidence is under our noses. Let's consider the opposite answer to Fermi's question: they exist, but they've never visited. For this, we revisit the idea of space travel and colonization.

We've seen that the energy requirements for interstellar travel close to the speed of light are forbidding. But what if we don't try to send frail and short-lived humans to the stars but send robots instead? The convergence of several technologies suggests an interesting possibility: replicating space probes. Miniaturization and new propulsion technologies will give us the capability of accelerating small spacecraft to 1 percent of light speed. Another rapidly maturing technology allows 3-D components to be fabricated by programmable machines. The result is rapid galactic exploration.[5]

Here's how it might work. A civilization like ours sends robotic probes to the nearest stars. They explore planets and search for signs of life, then mine asteroids or rocky moons for material to make replicas of themselves. The replicas fan out to a new set of stellar systems, and the process repeats. At 1 percent of the speed of light, it would take only one million years for the probes to diffuse across the entire galaxy. They could report their findings to the home planet at light speed. There have been billions of years for a civilization to develop this capability before us. So why should we be the first? In this scenario, contact can be made even if ETs are extremely rare. It takes only one.

You'll recognize this as another form of the "mediocrity" principle. Why should we be the only kids on the block or the smartest kids? But notice the assumptions. To be surprised that we've *not* been visited, you have to assume that ETs do exist, that they have the capability for interstellar travel, that they've chosen not to communicate or travel in space, and that we wouldn't detect them if they had. If any part of that chain of reasoning is invalid, we can't conclude that we're alone.

272 THE LIVING COSMOS

OPENING PANDORA'S BOX

Perhaps some version of the Rare Earth hypothesis is correct—terrestrial planets with stable enough conditions for the development of complex organisms are a minuscule subset of all planets. Microbes are common, but the bottleneck is brains. After all, in four billion years of evolution on Earth, only a handful of species have developed a high level of intelligence. And intelligence might not often lead to technology and the capability of exploring space; think of the limitations of dolphins and orcas.

If we venture into the realm of alien sociology, the options sprout like weeds. Perhaps ETs travel widely but choose not to make themselves known to us (the "zoo" hypothesis). Perhaps most civilizations self-destruct before they become technologically advanced enough for space travel. Perhaps exploration is a human cultural phenomenon and intelligent species elsewhere see no need to venture into space. Perhaps ETs don't travel but instead communicate in ways that we can't understand or recognize. Perhaps ETs are so strange we wouldn't recognize them even if they were under our noses.

Most galling to our self-esteem, perhaps they visited and didn't find us worthy of attention. The large age of the universe means that if we're not the first intelligent and technological civilization, others may have a big head start on us. Suppose one day that astronauts visit an Earth-like planet. As they walk on its barren surface, they record with disappointment in their space log: *lichen and bacterial colonies, uninteresting.* Similarly, if we time-traveled to the Earth one billion years ago, we'd find no life-form larger than a pinhead. What if aliens can visit and have visited but are vastly more advanced than we are? What if aliens are to us what we are to a bacterium?

Fermi's question opens a Pandora's box of unconstrained possibility.[6] We accept without question that biology is just one of the phenomena that occur in the universe, like comets and black holes and quasars. But what if life has a more central role? What if life is profoundly connected to the cosmos of which it's a tiny part, so that we (and perhaps ETs too) are a built-in feature? We're nudged into this audacious proposition when we notice certain coincidences in nature.

COSMIC COINCIDENCES

YOU'RE READING THIS BOOK. Depending on how you look at it, that simple act may seem miraculous or mundane. On the one hand, you might have been born long after or long before I wrote it. You might have lived in a country where it wasn't published. You might have passed the aisle where it was shelved or driven by the bookstore that carried it. If it was a gift, the giver might have chosen a different title. With all the twists and turns, what are the odds that you'd be reading it right now? It seems highly improbable.

On the other hand, unless you picked up this book by chance and are reading this sentence randomly, it's not remarkable. You can afford books, and you like to read. You are probably interested in science, and I wouldn't have bothered to write this book if it was unlikely that there are willing readers like you. And we couldn't have this strange, one-sided conversation if you weren't reading this book right now.

Agatha Christie's fictional detective Miss Marple once said, "A coincidence is always worth noticing; you can always discard it later if it *is* just a coincidence." As we near the end of our journey in this book you happen to be reading, we'll venture beyond science into a land of deeper meanings. Cosmic coincidences make us reconsider the role of life in the universe.

FINE-TUNING IN NATURE

Physics contains some gritty and important numbers—the mass of the proton, the mass of the electron, the electric charge carried by subatomic particles, the strength of the fundamental forces of nature, and so on. If many of these numbers were slightly different, we wouldn't be here. In other words, you could tweak the fundamental basis and physics would still be functional, but the consequences of these laws playing out in the universe would not include carbon-based life-forms like us.

Atoms are held together by a strong nuclear force, which acts over a very short range and serves as glue, and a weak nuclear force, which is responsible for radioactive decay. If the strong force were a bit stronger, nuclear reaction would be so efficient that stars would quickly turn almost all hydrogen in the universe into helium and on up to iron. With no hydrogen, there's no water. If the force were a bit weaker, electrical repulsion between protons would stop any complex nuclei from forming, so there'd be no carbon created.[7] If the weak force were a bit stronger, neutrons would decay so fast that nuclei would unravel before any heavy elements could be built. If it were a bit weaker, there'd be plenty of neutrons hanging around, with the result that once again all the hydrogen would be converted into helium and on up to heavier elements, with none left to make water. We're not talking about any large changes; "a bit" here means 5 or 10 percent.

There's more. The electromagnetic force controls the ways atoms interact, and it explains light. If this force were slightly stronger, atoms would become selfish and not share electrons, and no chemical reactions would be possible. If it were slightly weaker, atoms would not hold on to their electrons, and the universe would become a sea of loose particles, with no chemistry possible. There would be no chemistry, no life.

We're not done yet. Gravity is the weakest force in nature, but in many ways it is the most important since it sculpts everything from planets to the cosmic expansion. Stronger gravity would cause bigger stars to form, which would burn quickly and be unstable; this is unlikely to be good news for life on planets around such stars. Weaker gravity is worse because stars wouldn't get massive enough to die explosively. Supernovae are needed to create some of the trace elements needed for life and to disperse carbon and other heavy elements into the regions where new stars and planets can form.

Cosmology presents us with more puzzles. For much of its history, the universal expansion slowed down due to the gravity of dark matter. But a few billion years ago we entered a phase of acceleration as dark energy took over from the weaker gravity of all that highly dispersed matter. History since the big bang is driven by the amount of dark matter and the amount of dark energy. The ordinary particles of which you and I and our familiar world are made are inconsequential in their effect on the expansion.

A universe with far less matter would have expanded more quickly in the early phase—so rapidly that gravity would not have had time to get a grip before everything turned into a cold, diffuse gas. If no stars and galaxies form, there will be no life. A universe with far more matter would have reached a maximum size and collapsed quickly under the weight of its own gravity. Since we think biology needs a long time—perhaps one billion years—to develop, a baby universe like this would be stillborn. It's the same with dark energy; if it were much stronger, the universe would have ripped itself apart before life could have any chance to form.

As it is, dark energy will cause galaxies to separate with increasing speed. This puts a damper on the idea of intergalactic communication or a universal consciousness because galaxies will eventually flee faster than light can travel the distance between them. The physicist Freeman Dyson called this a Carroll universe, after Lewis Carroll, because "it takes all the running you can do, just to stay in the same place."

IS LIFE SURPRISING?

The "special" properties of our universe lead to the anthropic principle.[8] The anthropic principle isn't a single idea; it's a web of concepts and logical arguments, and it has spawned as much controversy and confusion as anything in science. In its weakest form, anthropic reasoning is a truism: we can only observe a universe that allows us to exist. The strongest form says that the universe had to be the way it is so as to allow intelligent observers to exist.

So which is it? We can yawn and say, Of course the universe is old and large, and stars have made carbon, and chemistry is possible. If all of that weren't true, we wouldn't be here. Or we can be gobsmacked by the incredible fortune that led to our existence.

When Nick Bostrom is talking, the most outrageous ideas seem sensible, even obvious. Bostrom is a young philosopher who has written extensively on the anthropic principle, the Doomsday argument, and other esoteric issues. He runs a new institute at Oxford University, but he has the slight singsong and sculpted vowels of a native Swede. Bostrom has pointed out that we must rein in our surprise when we consider the properties of the universe. Observational self-selection is a powerful constraint on what we can see and what we do see. The fisherman using a net with one-inch holes shouldn't be surprised to catch only fish larger than an inch. The queen of England shouldn't be surprised to wake up each morning to find that she's the queen.

We can't simply point to our existence to account for the particular properties of the universe. If we relax our definition of life and intelligence, the fine-tuning presented above isn't as severe. Life might not need carbon or stars. On the other hand, fine-tuning doesn't go away entirely. The laws of physics don't rule out an exotic universe that expands and collapses in a year, or a universe that has so much disorder that time has no arrow, or a universe filled with magnetic monopoles. But they're almost certainly dead.

ENTER THE MULTIVERSE

Ideas from the frontier of physics give us a context for interpreting the universe we inhabit. The big bang started as a quantum fluctuation, with a backdrop space-time foam that may have spawned, and may still be spawning, a plethora of universes, each with different properties. The "multiverse" concept conjures up an infinite number of universes, disjoint from our own. Their properties vary so radically that life might be possible in a small fraction of them, robbing our universe of its specialness. At first sight, the multiverse seems like a bizarre contradiction of the scientific aesthetic that simple solutions are best, but it's consistent with inflationary cosmology, and there are ways it might eventually get observational support.[9]

Meanwhile, theorists are trying to complete the job that beat Einstein: unifying relativity and quantum physics in a "theory of everything." Instead of thinking of subatomic particles as tiny spheres, we should visualize them as one-dimensional strings or as two-dimensional membranes ("branes"). They roam around familiar three-dimensional space, which is embedded in a ten-dimensional landscape. We're oblivious to the higher dimensional space, which could contain vast numbers of parallel universes with different laws of physics and different numbers of dimensions (Fig. 120). The fiendishly difficult math describing this is called M-theory.[10]

M-theory provides a basis for a plentitude of shadow universes. There might be 10^{100} different physical states of the vacuum, each one pregnant with possibility. Self-selection dictates that we'll find ourselves in one with properties like those we see. The theory might even explain why space has three dimensions. Life must be unlikely in two dimensions because it can't contain enough complexity. (Think of the difference between dominoes and Lego bricks.) It's hard to imagine gravity in four or more dimensions, but if you do the math gravity falls off so rapidly that stars can't hold on to planets and the modified forces inside atoms make them unstable. Within M-theory, ten-dimensional space settles into configurations with either three or seven (don't ask) dimensions. As E. E. Cummings once wrote, teasingly, "Listen: there's a hell of a good universe next door; let's go."

We've shimmied along a branch of speculation so slender that it might not hold our weight. To return to the actual science of astrobiology, let's ignore all those possible parallel universes and ignore all the billions of galaxies beyond

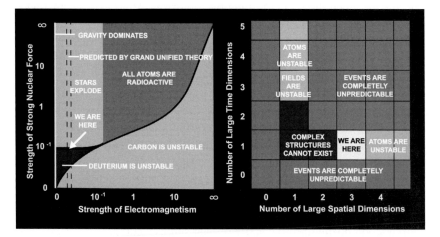

Figure 120. Fine-tuning in physics leads to a situation where if the forces of nature were slightly different, biology would be difficult or impossible. The graph on the left shows how nature would change if the electromagnetic and strong nuclear forces were different. Over most of the variation, stars or normal atoms are unstable. The graph on the right shows our universe relative to highly speculative theories with an arbitrary number of space and time dimensions. Once again, a universe with the number of dimensions we observe seems required for the existence of life.

our neighborhood. Let's regroup and organize the information—also the igno-rance—into a framework that deals only with our galaxy. This leads to the Drake Equation.

THE DRAKE EQUATION

THE YOUNG RESEARCHER WENT to the blackboard and paused. The meeting had no agenda, and he wanted to give some structure to the discussion. After thinking for a bit, Frank Drake wrote an equation on the board, not realizing that it would later bear his name and attain iconic status.

MOTIVATING THE SEARCH

It was 1961, and the conference room at the radio observatory at Green Bank held the pioneers of SETI. The young Carl Sagan was there. Biochemist Melvin Calvin was there; within days, he was to learn he had won a Nobel Prize for his work on photosynthesis. Dolphin researcher John Lilly celebrated the cama-raderie of the group by forming the "Order of the Dolphin" and distributing membership pins. MIT physicist Philip Morrison had sparked excitement two years earlier when he coauthored a paper arguing that we might be able to de-tect a radio signal beamed at us from elsewhere in our galaxy. Frank Drake fol-lowed up six months later with an attempt to detect artificial signals from two nearby stars. It was a time of great hope and expectations.

The group was well aware of the Fermi paradox. They also knew that most of the arguments ended as epistemological standoffs. Speculation was no sub-stitute for actually looking for evidence of ETs. As Morrison put it in his paper in *Nature* titled "Searching for Interstellar Communication," "The probability of success is difficult to estimate; but if we never search, the chance of success is zero." As they say with the lottery, you can't win if you don't play.

The Drake Equation is a series of numerical factors that combine to give N, the number of communicating civilizations in the galaxy at any particular time. We're forced to observe a snapshot of a long-lived and dynamic universe, so two factors involve time. The others relate to the census of habitable worlds.

The first factor is the raw material for communication, or the rate at which stars are born that will live long enough to host biology. The next six factors ac-count for the odds that there is any transmission to listen to, either deliberate or inadvertent. Two are astronomical: the fraction of stars with planets and the mean number of habitable planets per star. The next two are biological: the

Figure 121. The Drake Equation, used to guess at the number of intelligent, communicable civilizations in the Milky Way. It incorporates a set of numerical factors relating to the number of habitable planets and the nature of life on them, as well as one factor that accounts for the longevity of a civilization in the communicating state.

fraction of habitable planets where life actually develops and the fraction of these where intelligent life evolves. The last two depend on culture or sociology: the fraction of civilizations that do communicate over interstellar distances and the typical lifetime of the communicating technology (Fig. 121).

Conceptually, the Drake Equation starts with a huge reservoir of potential life: habitable zones around billions of stars in the Milky Way. Then it winnows down the numbers by making more and more restrictive requirements, until we're left with the small fraction of those stars that have a planet where a lifeform does what we've just learned how to do: communicate across space (Fig. 122).

THE BOY OR GIRL OF YOUR DREAMS

Let's use an analogy. Some years earlier, you met a red-haired girl. Something about her—maybe her jaunty walk or the slightly tacky way she chewed gum—got to you. As she turned to leave, she popped her gum loudly for emphasis, and that simple sound put a lump in your throat. What are the odds that a red-haired girl somewhere in your hometown is popping gum right now and maybe breaking someone else's heart? Perhaps it was the soulful eyes, maybe the winsome smile, but when she popped that bubble, something inside you went wobbly. The memory still aches. You moved on, life happened. Could someone else experience what you felt? Perhaps it was a boy; let's imagine boy and girl are interchangeable in what follows.

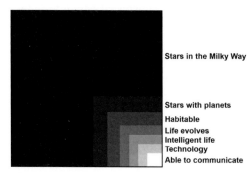

Figure 122. Conceptual view of the Drake Equation. Among the vast number of stars in the Milky Way, some fraction of stars has planets, a fraction of those planets are habitable, some of them actually host life. Among the life-bearing planets, a fraction has intelligent life, and some fraction of those has technology and is able to communicate in space. These successive reductions of an initially large number lead to a number that is potentially much smaller and in the end N could be just one: us!

You can start by estimating the rate at which red-haired girls are born in the town. Suppose the town has a population of one hundred thousand. The birthrate might be about one thousand per year. We multiply that by the fraction of new births that have red hair (10 percent, or 0.1), the fraction that are life bearing (the women, or 0.5), the fraction of those that are actually lively (maybe 0.2), and the fraction who will grow up to chew gum (perhaps 0.1). The product of all this is about one. The last factor is the "lifetime" of the phenomenon in years. It's tiny, perhaps one ten-thousandth, since even avid gum chewers probably spend only ten minutes per day blowing bubbles.

In the analogy, the product of all the factors gives $N = 0.0001$. The combination of circumstances leading to a red-haired woman popping gum is rare; you're unlikely to see it. You are alone. But the population of the world is six billion, or sixty thousand times larger than the town. In the entire world, $N = 6$, so there probably is a red-haired woman *somewhere* popping a bubble right now. As with the Drake Equation, the factors range from concrete and measurable—the population of a town—to sociological: the circumstances that lead someone to be chewing gum.[11]

Time and space are both folded into the Drake Equation. Sticking with the human analogy, we might imagine it's implausible that anyone else shares our interests, but in a country of three hundred million people and a world of six billion people, it's possible. What are the odds that any two people have whistled a particular tune? Pretty good. But what are the odds that they have whistled the same tune at the same time? A lot lower, and perhaps it's never happened.

THE OPTIMISTS WEIGH IN

The participants at the Green Bank SETI meeting dutifully plugged numbers into Drake's equation. The stellar birthrate was fairly well measured: about three new Sun-like stars per year. They estimated the product of the next four factors to be one-quarter to one-half, where there are a couple of habitable planets per star and all of the fractions are close to one. They assumed that planets are ubiquitous and that life, intelligence, and technology are nearly inevitable once you have a habitable planet. This marked them as optimists, since there were no data then, nor are there now, to support that assumption.

As a result of these choices, all the factors except the last multiply to about one, and the Drake Equation reduces to an elegant form, $N = L$. (Drake, now a silver-haired senior researcher at the SETI Institute, has this on his license plate.) This concentrates attention on an essential fact: the prospects of contact depend crucially on how long the technology of transmission continues. That's not necessarily the same as the lifetime of the civilization; it could be much longer if a civilization builds beacons that continue to transmit after its demise. In 1961, the Green Bank participants recognized the enormous uncertainty in this number and assigned it a range of one thousand to one hundred million years. They correspondingly estimated a huge range of one thousand up to one hundred million intelligent, currently communicable civilizations in the galaxy. Wow.

It's important to recall that the lifetime used in the Drake Equation is an average, since different civilizations may communicate for different time spans. As an analogy, average income can conceal huge disparities. If 90 percent of the population earns ten thousand dollars per year and 10 percent are millionaires, the average income is $109,000, which certainly sounds misleading to the majority that are struggling along on ten grand. Similarly, if 90 percent of civilizations have radio communications for a hundred years while 10 percent communicate for a hundred thousand years, the average L is 10,090 years. For SETI optimists, this is a hopeful sign since a tail of long-lived civilizations can substantially boost the average.

The researchers that gathered at Green Bank had a vested interest in doing the actual experiment, and they picked the Drake Equation fractions at the high end of their possible range. If habitable planets are very rare, and if intelligence and technology don't inevitably follow from the emergence of life, then instead of N being 10^3 to 10^8, it may be much lower. Pessimists would point to the Fermi paradox as evidence that $N = 1$. Among billions of stars in the Milky Way, we're unique. There are, of course, billions of galaxies beyond ours, and there is no reason to believe the Milky Way is special, so it's far less likely that the entire universe has no biology. But a universe with neighbors separated by millions of light-years is a pretty lonely place.

LOST IN SPACE

Regardless of how many active civilizations there are in the galaxy, N has a direct consequence for the nature of interstellar communication. If there are 10^8 civilizations out there, one in one thousand stars hosts one, and the typical distance to the nearest is thirty light-years. We could exchange signals in a human lifetime and travel there in a few centuries with the next generation of interstellar spacecraft. If the lower end of the range is right, then only one in one hundred million stars hosts a civilization. They'd be sparsely scattered through the galaxy, with a mean spacing of ten thousand light-years.[12]

Worse, if the civilization lasts only one thousand years, it will have disappeared by the time we received its signals and attempted to reply. Space might be scattered with the expanding spheres containing the runes and glyphs from long-dead worlds. If we could decode it, the information might be valuable to us as we stumble toward becoming a mature civilization. Science fiction often uses the conceit that we can join the galactic "club" as long as we survive the hazing rituals or rush week.

Wait a minute. Optimists? Pessimists? Isn't science supposed to be about logic and evidence? Sure, but when data are in short supply, scientists feel free to speculate. And when scientists speculate, their biases and emotions seep into the thinking process. The tendency for anthropocentric thinking is great but must be resisted if possible. We can't escape the human condition, but it pays to get out of our heads sometimes.

The Drake Equation is limited as a tool because it's based on guesswork and has no predictive powers. In math, when a set of numbers each has an error or uncertainty, the uncertainty in the product of those numbers is dictated by the largest uncertainty. In other words, 3.14159 times 5.5 times roughly 0.1 times anywhere from 10 to 1000 times gee-I-dunno equals . . . gee, I dunno! We can deal with this issue by simplifying the Drake Equation to separate observation from speculation. We also need to question its implicit assumptions.

POTENTIALLY LIVING WORLDS

FIRST, WE RECAST the Drake Equation into the form $N = N_* \times f_h \times f_i \times P$. The first quantity is the number of stars in the galaxy (N_*). This gets multiplied by a factor that is the fraction of those stars with habitable planets times the average number of habitable planets per system. To get the number of potential pen pals, the next factor is the product of three quantities: the fraction of habitable planets with life, the fraction of those planets with intelligent species, and the fraction of those planets where the species develops the technology for interstellar communication or travel. The last factor we'll call persistence (P)—it's the fraction of the age of the universe over which a civilization endures. Our first goal is to estimate the number of habitable planets in the Milky Way.

There are roughly two hundred billion stars in our galaxy. The number is uncertain due to the difficulty of detecting dim stars like white dwarfs and low-mass cousins of the Sun, but they don't figure into the estimate since they have razor-thin habitable zones.

COUNTING STARS WITH PLANET STUFF

What we want to know next is how many of those stars form planets. It depends on the abundance of heavy elements—to make planets, you need some grit along with all the gas. We've seen that the incidence of extrasolar planets is at least 5 percent for stars like the Sun, rises to 25 to 30 percent for stars with

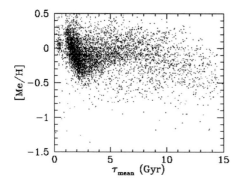

Figure 123. The abundance of "metals"—elements heavier than hydrogen and helium, relative to hydrogen—in five thousand stars near the Sun. The horizontal axis is age in billions of years. The vertical axis is a logarithmic scale, so the Sun sits at zero vertically and 4.6 horizontally. Most stars are somewhat younger than the Sun, but there are plenty of nearby stars with more planet-building material than the Sun and that are older than the Sun; they're very plausible sites for planets and therefore life.

three times the Sun's amount of heavy elements, and may fall to a very small fraction for stars with less than one-third of the Sun's heavy elements. Figure 123 shows the age and metallicity of six thousand stars within a few hundred light-years of the Sun: our galactic neighborhood. The Sun has a typical heavy-element abundance, and most stars are one to three billion years old, slightly younger than the Sun.

Let's toss out the stars in the galactic halo—that spherical cloud of mostly dark matter that contains 90 percent of the Milky Way's mass—since they've had little recycling and so are almost without heavy elements. That brings us down to one hundred billion.

Now let's acknowledge a sweet spot in the disk of our galaxy. Far out in the galactic disk, the rate of star formation is too low to make many heavy elements, so stars in the distant suburbs are probably planet-free. Close to the galactic center, there are lots of stars forming to make planet-building material, but the supernova rate is so high that it may be hazardous to evolution. This ring of suitable conditions evolves with cosmic time, spreading out from the center of the galaxy. Right now, the ring includes us and contains one-fourth of the stars in the disk. That leaves us with twenty-five billion.

A sweet spot in star mass can also be invoked by throwing out stars so massive that they live fewer than four billion years—the time it took for advanced life to evolve on the Earth—and stars so much less massive than the Sun that their habitable zones are very skinny. This is a conservative choice; it might take less time elsewhere. We're down to five billion, three-quarters of which are older than the Sun. Look at the beauty of large numbers—we've discarded 195 billion stars, and we still have lots left!

POTENTIALLY LIFE-BEARING ROCKS

The next step is to figure out what fraction of these stars has terrestrial planets and how many those stars typically have. We discard all gas giants, assuming that life can't function under high pressure at the base of such massive atmospheres. (But perhaps that's wrong?) We're forced to move beyond the realm of direct observations, since planet hunters are just reaching the sensitivity needed to find super-Earths. It's scientific guesswork.

What we actually observe is 20 percent of Sun-like stars with gas-giant planets. To conservatively project this to Earth-like planets, we could decrease it by a factor of two to 10 percent, assuming only one terrestrial planet per giant planet. Simulations agree in finding at least two terrestrial planets in a typical system. Tossing out the 50 percent of the stars that are binary or multiple, the fraction drops to 5 percent. This might be too cautious, since Frank Lin at the University of California–Santa Cruz has shown that planets may be just as likely to form in a typical binary system; stable orbits are nurtured better by two stellar parents than by one. Only the rarest close binaries would disrupt their orbiting planets. We've reached 250 million terrestrial planets orbiting half that number of stars.

Let's call this the teetotaling, short-back-and-sides, ultraconservative estimate. By pure coincidence, working with almost no data, English philosopher Thomas Wright made a similar estimate of 170 million worlds back in 1750! It also agrees with the early estimate from Kepler of 100 million. Now let's let our hair down and ask what a still plausible but more optimistic estimate might be.

The number of exoplanets rises rapidly at the limit of current detection, making it almost certain that the observed fraction of 20 percent Sun-like stars with one or more gas-giant planets is a serious underestimate. Careful models have shown that we're seeing only the tip of the iceberg of the planet population. The observations are consistent with *all* such stars having giant planets, if only we had better data. Also, it's easier to make a terrestrial planet, since large amounts of gas don't need to be accumulated (though some giants will eject their smaller brethren from the system). Gravity makes small planets more easily than it does large ones, so let's use a factor of two for that. This leads to a prediction of ten billion terrestrial planets orbiting five billion stars.

Now lets fold in the idea of habitability to the conservative estimate. By consensus among researchers in astrobiology, the minimum requirements for life are organic or carbon-rich material, liquid water, and an energy source. In a solar system like ours, if planets form with equal probability out to 3 AU, the liquid-water zone from 0.9 to 1.2 AU is 10 percent of the range. Applying that, we get twenty-five million habitable worlds in the galaxy.

This number is a lower limit, since we've played it safe at every turn by throwing out binaries, ignoring the vast number of dwarf stars, not accounting for the likely ease of making terrestrial planets, neglecting giant-planet moons, and using a strict definition of a habitable zone. Let's pause for a moment to let that sink in: twenty-five million potentially living worlds in our galaxy alone.

That's the crew-cut estimate. What about the dreadlocks or ponytail-down-to-the-butt version? Then we would accept the evidence that giant moons Europa, Callisto, and Ganymede have oceans, as well as the models suggesting that an additional dozen large moons also have subsurface oceans. The traditional habitable zone is irrelevant—moons far from their suns can keep water liquid by pressure under a rocky or icy crust, and heat can come from tidal forces or radioactive decay. We should also readmit to the census the huge number of long-lived, low-mass dwarf stars.

We gain two orders of magnitude. Several billion potentially life-bearing rocks in our galaxy alone. Do we imagine that all these biological petri dishes could remain sterile? Or do we think that microscopic life has burgeoned on many of them?

INTELLIGENCE AND TECHNOLOGY

WE NOW STEP where the ice is even thinner. The next factor in the Drake Equation is the fraction of habitable worlds where life advances to intelligence and technology. It's the product of three quantities, the first of which is the fraction of habitable planets where life actually develops. Evidence for past or present life on Mars or Europa will mean that this fraction is close to one, as life would have had multiple independent starts in a single solar system. A supporting, but weaker, argument comes from the rapid start to Earth life and the versatility of extremophiles. It looks as if life was almost inevitable.[13] Suppose we buy this argument?

If we weren't so fixated on companionship, we could stop right here. Whether we suppose that only the Earth-like water worlds spring to life or that 1 percent of the much larger number of water-bearing moons become biological, the answer is the same. Imagine tens of millions of living worlds in our galaxy. Think of all those biological experiments, each with different physical environments and each with biospheres as unique as fingerprints. Adding in all galaxies beyond the Milky Way implies a staggering million trillion (or 10^{18}) versions of life in the universe.

DEFINING INTELLIGENCE

Next up is the fraction of times that a planet with biology develops intelligent creatures. How do we define intelligence without being self-referential? If we focus on our modalities of thought, we'll risk a customized definition that may be the result of contingent evolution and the particular environmental conditions of Earth. The key is to distinguish between thought and behavior. Many animals on Earth display complex behavior as they survive and procreate. Their brains are tools that process information from the environment and act on it.[14] But it's possible to have complex behavior without cogitation.

Let's define intelligence by two additional qualities: the ability to reflect on past experience and the power of abstraction. Reflection allows for learning that goes beyond operant conditioning. Abstraction enables the manipulation of concepts not represented by concrete objects. For example, counting embeds both reflection (I see two dogs) and abstraction (I see two things, and they may be cats). These qualities enable the creativity that we like to think is exclusively the preserve of humans.

HOW WE GOT SO SMART

How did human intelligence evolve? We learn a lot from our classification in the web of life. We're from the domain of eukaryotes, which means our cells are large and have nuclei, and their function was shaped by the buildup of oxygen in the atmosphere billions of years ago. We're in the animal kingdom, which means we move, reproduce sexually, and have diverse and specialized cell types. In the evolutionary tree, the four other kingdoms—plants, fungi, bacteria, and protists—successfully populated the planet without the benefit of a centralized intelligence system. We are chordates, with internal skeletons protecting our central nervous system. About seven hundred million years ago, precursors to modern brain cells first appeared.

Water-dwelling amphibians moved onto the land and evolved into reptiles about 350 million years ago. By the late Jurassic, about 150 million years ago, all of the major vertebrates—fish, amphibians, birds, reptiles, and mammals—had well-developed brains. We are mammals, with hair and warm blood. The shrewlike *Hadrocodium wui* scurried around in a world of dinosaurs two hundred million years ago and weighed less than a dime, but its brain was unusually large for its body. Big oak trees grow from little acorns.

Mammals lost out to dinosaurs in the diversification of life that followed the Permian-Triassic "Great Dying" 250 million years ago. It was not until after the next mass extinction sixty-five million years ago that mammals grew bigger than a squirrel. Early primates were distinguished by an increased use of sight relative to smell, by grasping digits, and by complex social lives. The development of acute, stereoscopic vision was probably a spur to larger brains; in modern humans, processing vision takes 10 percent of the brain's capacity. The fossil record is very patchy, but it indicates that the hominid line split off from

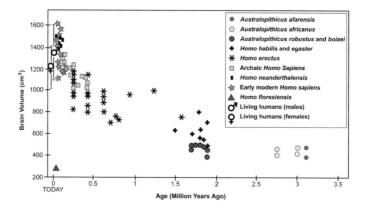

Figure 124. The steady march of increasing brain size among human ancestors, with a more rapid rise up to the modern level in the last few hundred thousand years. The Neanderthals with their slightly larger brains died out only twenty-five thousand years ago, and 2003 saw the startling discovery of *Homo floresiensis,* an isolated pygmy human ancestor with a 300 cm³ brain that may have survived until fewer than eighteen thousand years ago.

apes six to eight million years ago as our ancient ancestors began to walk upright and turned from forest dwellers into foraging nomads on the grasslands of Africa. Our brains steadily grew to eclipse those of all other species of ape (Fig. 124).

DNA analysis confirms the fossil evidence that all six billion people on this planet are descended from a small group of geographically isolated individuals who lived in Africa about two hundred thousand years ago. Staring into the eyes of a great ape, we can see the self-awareness of creatures that share 95 percent of our DNA. It's a mere accident of history that we don't share the Earth with the Neanderthals, whose brains were slightly bigger than ours and who disappeared fewer than twenty-five thousand years ago, a blink of the eye in evolutionary history.

IS INTELLIGENCE INEVITABLE?

Is intelligence an inevitable outcome of natural selection? It took nearly four billion years to develop and emerged in only a handful of species among hundreds of millions that have lived, almost all of which are extinct. Big brains are clearly not required for evolutionary success, but they convey a selective advantage by processing enormous amounts of sensory data and by allowing the cognition that leads to adaptive strategies for survival.[15]

On the other hand, large brains take maintenance. They need a high-protein diet and exquisite temperature control. About 25 percent of our metabolisms are devoted to brain function, which is a huge investment. The fragility of large organisms translates into an argument that a habitable zone must be more stringently defined to allow intelligence to evolve, since climate extremes could be fatal. (Yet even this argument is a two-edged sword, since deep episodes of glaciation may have fostered the flowering of complex animals in the first place.) The requirement relaxes for ocean life since water acts as a heat reservoir to even out temperature variations.

We accept without question that humans are nonpareil. No other species invented art or mathematics. No other species has created machines to extend the senses and do its bidding. Not since stromatolites and other early photosynthetic organisms has any species been able to alter the planet. Psychic immaturity makes us cause pain and mayhem, but there's no questioning the horsepower under the hood.

In astrobiology, it makes sense to take a broader view—if our intelligence sets the definition, then we'll only recognize creatures like us as intelligent. Evolution continues, so intelligence on Earth may not begin and end with us. If we set the bar at self-reflection and abstraction, there's evidence that we share our caliber of intelligence with dolphins and orcas, whose last common ancestor with primates was eighty-five million years ago. They have brain-to-body-mass ratios near that of humans and higher than those of other mammals (Fig. 125). Some birds show the ability for reflective behavior. And let's not be too mammalcentric. Cephalopods have large brains and amazing capabilities for learned behavior, although mollusks are much farther from us in the evolutionary tree.

THE ALIENS AMONG US

You're standing on a bleak Ordovician shore. Behind you, a vista of barren rocks stretches into the distance. There are no trees to break the horizon and no plants or animals of any kind. Your lungs work hard sucking air with a quarter less oxygen than you're used to. Earth spins faster on its axis; the day is only twenty-one hours long. The Moon looms closer in the sky. Water laps at your feet. Suddenly, the surface is broken by a huge horn, thirty feet long. Its smooth shell is mottled dull red and white and ends in a gaggle of tentacles. Another splash, and it's gone.

This strange scene is the Earth 470 million years ago. For seventy million years, since the Cambrian explosion of life, a great evolutionary battle has been taking place in the oceans. Our distant ancestors, the chordates, don't feature prominently in this tussle. The early winners were predatory arthropods. But

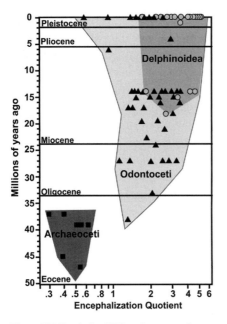

Figure 125. The skulls of 210 marine-mammal specimens were studied in order to reconstruct a history of the ratio of brain mass to body mass, called the encephalization quotient, or EQ. The first major jump in brain size in the ancestral group Archaeoceti corresponded to the development of echolocation. More recent increases in brain size among toothed whales (Odontoceti) and the dolphin superfamily (Delphinoidea) have brought them close to the EQ of humans, which is 7–8.

after dramatic climate change and a mass extinction, the small-shelled mollusks seized their chance. Starting as crudely chambered shells a few millimeters across, within fifty million years they had diversified and grown into creatures like *Cameroceras*, which just breached the surface. During the Ordovician, huge cephalopods ruled the seas.

Cephalopods are among Earth's most successful forms of life. The nautilus has survived with very few changes for four hundred million years, through several mass extinctions, by hunting in the dark netherworld along the deep slopes of coral reefs. Their cousins—cuttlefish, squid, and particularly the octopus—display some of the richest behaviors in the animal kingdom. They're shape-shifters just as strange as *Star Trek*'s Odo.

Roger Hanlon got hooked on cephalopods as a junior in college, when an octopus on the Panamanian reef startled him. Now a senior researcher at the Woods Hole Marine Biological Laboratory, he's been studying the octopus in particular in its natural habitat for more than thirty years. His best dives are in the Caribbean, and he laughs when asked if he still sees things that surprise him after thousands of hours of observation, insisting he's not even close to cataloging their rich lives. Hanlon is careful not to talk about octopus "intelligence." He says it's too easy to project our thoughts and capabilities onto these exotic creatures. He prefers to observe and interpret what he sees. Although he literally wrote the book on cephalopod behavior, Hanlon follows the dictum of noted zoologist Louis Agassiz: "Study nature, not books."

Vertebrates like us use a command-and-control architecture—the brain acts as a centralized processing unit to accept sensory input and control our limbs. An octopus, by contrast, has one centralized brain and another highly distributed brain. You can think of it as a mind melded to a body (Fig. 126).

Start with locomotion. An octopus carries intelligence in its limbs. Each arm has an elaborate nervous system composed of fifty million neurons, and each has an enormous range of continuous motion. By comparison, our simple arrangement of knees and ankles, elbows and wrists, seems quite primitive. Israeli researchers discovered that each arm has an underlying motor program with no centralized control. The brain gives an initiation command, then the smart limbs take over.

Figure 126. The octopus is far from humans on the tree of life, but it shares with us binocular vision, a large brain, and a complex set of adaptive behaviors. The distributed architecture of their brains and the impossibility of communication mean that we simply do not know how to evaluate octopus intelligence.

The octopus eye is as advanced as the human eye, although its design is different. Using a special balance organ that can sense gravity, the octopus keeps its eyes perfectly aligned regardless of the orientation of its body. It can sense polarization of light, which helps it see transparent prey such as shrimp and fish. Eye and brain combine for highly sophisticated pattern recognition, which is essential in complex terrain like a coral reef. An octopus has chemoreceptors in its suckers so it can taste what it touches and instantly reject the wrong foods. As a hunter, it's unparalleled—swift, strong, propelled by a water jet, and armed with suckers, a beak, ink that can be ejected and directed, and toxic saliva. The giant octopus grows to one hundred pounds, but the sailor's fearsome legends about them are fanciful; when faced with humans, they're curious and gentle.

The most amazing feature of an octopus may be its skin, a direct extension of the brain, used for camouflage and communication. Translucent sacs called chromatophores are filled with pigments of various colors: yellow, orange, red, brown, and black. When muscles attached to each sac contract, it expands, and the color becomes visible. Special cells lying below the chromatophores act as iridescent reflectors of many colors, allowing for optical interactions of bewildering complexity and beauty. Octopus skin can morph in texture, instantly mimicking smooth coral or a sponge or the rusted hulk of a sunken ship. A lot of processing power is needed for twenty million skin "pixels" to select from a palette of hundreds of thousands of colors in under a second. The comparison with a chameleon is insulting to an octopus.

Many of the behaviors are ingenious. An octopus can release a cloud of ink in its own shape, confusing a predator for long enough to get away. They can break off an arm as a decoy—the severed arm will continue to change colors and crawl around, making an excellent distraction. Octopuses memorize complex visual cues to navigate the tortured topography of the coral reef, a 3-D landscape that can disorient even experienced divers such as Hanlon. In captivity, octopuses are the Houdinis of the animal kingdom, able to pass through an opening no bigger than their eye. There's even some controversial evidence that octopuses can learn by observing the behavior of others.[16]

Evolution doesn't stand still. Cephalopods may develop new abilities. Humans might succumb to microbes or their own hubris. We can visualize a time in the not-too-distant future when the descendants of the octopus, as curious as ever, take their first tentative steps on land.

THE ROLE OF TECHNOLOGY

If intelligence at the level of self-awareness and abstraction is difficult to predict, the advent of technology elsewhere in the universe is even more uncertain. In one sense, technology is a natural consequence of evolution. Millions of years ago, Earth creatures invented flight, sonar, fiber optics, magnetic and electric sensing, seismic and pressure detection, and holography. Humans have been here for hundreds of thousands of years, but we matched that list only in the past century.

Tools are key transitions to technology, but they have never been unique to us; apes, birds, otters, elephants, and dolphins all use tools. (The latter use sponges to protect their noses as they forage on the seafloor, passing the technique on through families.) It would be rewriting history to suggest any inevitability about the human march to modern technology. Neanderthals used stone tools for more than one hundred thousand years without progressing to anything more sophisticated. Some early landmarks in human technology were probably accidental, such as the ability to control fire, or the onset of agriculture based on natural variants of wild rye with larger seeds that fell rather than were scattered on the wind, or the invention of bronze that spurred the first great civilizations six thousand years ago.[17]

If the issue is communication with intelligent life in the universe, it's clear we're talking about technology far beyond that used to adapt and survive. As with intelligence, let's add two attributes. One is the ability to manipulate and control matter and energy at will; the other is the inclination to use technology for space exploration. Technology isn't inevitable; octopuses, dolphins, and even some birds may approach us in cognitive ability, but they can't point telescopes at the sky and wonder about their places in the universe.

After considering the issues of intelligence and technology, how do we estimate the next factor in the Drake Equation? The fact that *we* have the smarts and tools to learn about the universe doesn't mean the outcome is inevitable and the factor is close to one. (That's flaky logic.) For every planet with a race like us, there could be millions of planets where life never evolved high intelligence, millions of planets where brains never led to technology, or millions of planets where technology isn't used for space exploration and communication. We therefore leave this factor in limbo while we look at the other important requirement for companionship: persistence.

TIMING IS EVERYTHING

THE LAST FACTOR IN THE DRAKE EQUATION relates to timing. Species emerge and then go extinct, so the persistence of civilization and technology control the likelihood that there's anyone (or any of their surviving machines) to talk to. The age of the Milky Way and the lifetime of the Sun are ten billion years, so we need to estimate the longevity of a technological civilization as a fraction of that vast span.

Consider an indisputable fact. Our capability to explore space is incredibly recent on a cosmic timescale: less than a century or 10^{-8} of the age of the galaxy. The pessimist might argue that we're already coming close to self-destruction by nuclear weapons. We could take the typical duration of our major civilizations—one thousand years—as a lower bound, in which case $P = 10^{-7}$. Imagine you have a Christmas tree for two weeks. You've bought lights, but each one flashes on for only a second during the whole fortnight. Even with a large number of lights, there is almost no chance two will be on at the same time. With this kind of communication, ETs would be very lonely.

The upper bound on persistence is pure speculation. If we or other civilizations can immunize ourselves from the forces of natural selection, then we may last far longer. Longevity shades into immortality, where the number of truly advanced civilizations in the galaxy steadily increases over time.

THE DOOMSDAY HYPOTHESIS

Immortality is cool, but there's a more sobering possibility called the Doomsday argument. In his Princeton office, J. Richard Gott III talks genially about the demise of mankind. He speaks with a syrupy Kentucky drawl and dresses smartly in pastel shades that make a splash in the drab academic landscape. Gott visited the Berlin Wall in 1969 as a student and, having visited other landmarks like Stonehenge, wondered how long it would endure. Assuming that he was visiting the Wall at a random time in its existence, Gott reasoned there was a 75 percent chance he was seeing it *after* the first one-quarter of its lifetime. It had been up for eight years when he saw it, so he predicted it would not be there after 1993 (1961, plus 3 times 8). In fact, it fell in 1989.

For humans, Gott switched to the norm in physics of estimating quantities with 95 percent confidence, but the logic is identical. It's recognizable as a form of the Copernican principle. If there's nothing special about your timing, you have a 95 percent probability of seeing something in the middle 95 percent of its existence. That means the thing has a 95 percent chance of lasting between one thirty-ninth and thirty-nine times its current age. For humans, this translates into a range of 5,100 to 7.8 million years. By this reasoning, we won't last forever.[18]

Philosophers have embraced, embellished, and occasionally panned the Doomsday argument. John Leslie, professor emeritus at the University of Guelph, asks us to consider two urns containing numbered balls. One urn has ten numbered balls and the other has one hundred million, but you don't know which is which. Suppose you draw a ball with the number seven on it. It's more likely that it would be drawn from the urn with balls numbered one to ten than from the urn with balls numbered from one to one hundred million. Now imagine the urns are possible human races and the balls are individuals ranked in birth order. If your rank is seventy billion among all the humans who have ever lived, it's more likely that the total number will be one hundred billion than many trillions. By this reasoning, we're nearer the end of humanity than the beginning.[19]

The Doomsday argument says that intelligent civilizations are short-lived and isolated in cosmic time. Regardless of the longevity of civilizations, we can draw two conclusions. If we are *not* alone, we're unlikely to be the first. There have been Sun-like stars with enough heavy elements to build planets for twelve billion years. That's three times longer than Earth has had life. Unless four billion years is an unusually short time to develop civilizations, ETs could have reached our state of development billions of years before us. That connects to the second conclusion: since we've only just attained the capability for space travel and communication, any civilization we encounter is likely to be far more advanced than we are. Arthur C. Clarke said, "Any sufficiently advanced civilization is indistinguishable from magic."

ROOM AT THE BOTTOM

In 1959, physicist Richard Feynman gave a talk about the future of technology called "There Is Plenty of Room at the Bottom." In it, he predicted nanotechnology, the ability to manipulate devices and construct things on the scale of atoms. About the same time, DNA codiscoverer James Watson said, "Life is digital information." What if we liberated ourselves from the architecture of one example of biology? How might that affect the prevalence and persistence of intelligent life?

At first glance, modern organisms seem like masterworks of information compression. Every human red-blood cell contains a genome with a compact disk's worth of genetic information, housed in a package one hundredth of a millimeter across. But if life could operate without the infrastructure of a cell, biochemical processes could be more highly concentrated and operate more efficiently. We consider cells with nuclei to be the most advanced and successful life-forms on Earth, but there are fascinating subcellular forms that have important capabilities (though biologists vigorously debate whether or not they're "alive"). They live in parasitic and symbiotic relationships with their hosts. Some cause diseases, others are beneficial. They lack cell walls, most of them don't metabolize, and some have no genetic code.

Consider Spiegelman's monster, for example. In 1997, Sol Spiegelman was working with strands of viral RNA in the lab. Even though no cells were present, the RNA fragments readily reproduced. As they did so, strands that started out 4,500 nucleotide bases long started slimming down, sloughing off parts of the genome that weren't needed. Smaller strands reproduced more rapidly, and the end point was a tiny genome fifty base pairs long that outreproduced everything else. Earlier experiments by Manfred Eigen had shown that RNA strands the size of Spiegelman's monster formed spontaneously from a broth of single nucleotides.[20]

In the conventional wisdom of biologists, this "survival of the smallest" is a problem for evolution because it prevents larger and more complex organisms from emerging. Small organisms have limited scope for genetic development, preventing interesting evolutionary development.

Perhaps conventional wisdom is wrong, or at least myopic because it's based on the "Just So Story" we tell of life on Earth. The speed and efficiency of small replicators might be the most common way that biology develops. Richard Dawkins generalizes the central paradigm of biology to a slogan that he considers perfect for a T-shirt: "Life results from the non-random survival of randomly-varying replicators." Complexity may arise from networks of small molecular units, and if this happens very quickly we can toss out the need for several-billion-year-old stars. Intelligence may arise in networks of small biological entities, as in the opening vignette, rather than in centralized organs such as brains. Such intelligence might be more adaptable and durable than our kind, which depends on survival and the regulation of fragile bodies.

Figure 127. The Doomsday Clock of the Bulletin of Atomic Scientists has been as close as three minutes to midnight (global catastrophe). The probability of companionship in the cosmos depends strongly on the persistence and durability of intelligent civilizations. If destruction after a few hundred years of technology was typical, we would be very lonely in the galaxy.

THE NUMBER OF COMPANIONS

Bringing all this together, what can we say about the output of Drake's equation? Our simplified version is $N = 10^{10} \times f_h \times f_i \times P$. If we accept moons beyond a traditional habitable zone for life but rule them out for intelligent life, then assume that most fertile terrestrial planets spawn life, and take the "pessimistic" value 10^{-7} for persistence, $N = 1000 f_i$. If *every* living planet develops a short-lived intelligent civilization, there will be roughly one thousand scattered across the galaxy (Fig. 127).

This sounds like a scenario for cosmic kinship, but it's not. The Milky Way would be sprinkled with civilizations, but they burn out so quickly that no two are active at the same time, and distances between them are thousands of light-years. And if less than one in a thousand planets develops a civilization, we're utterly alone.

An "optimistic" estimate of the durability of civilizations might key it to the average duration of species on Earth, about ten million years. In this case, P is 10^{-3}, and civilizations would have to be incredibly unlikely on fertile planets—less than one in ten million—for us to be alone. Even if as few as one in one thousand living planets evolves a civilization, the situation is transformed. There are ten thousand venues with technology and space travel.

Now we're in the realm of *Star Trek* and *Star Wars*. The average gap between civilizations is hundreds of light-years, but civilizations last millions of years, so this is a trivial distance. Almost anyone we meet is millions of years more advanced than us. The "long tail" of civilizations lasting billions of years or more could accumulate the wisdom of the galaxy and place it in a network of beacons so that newbies like us would just need to find the nearest probe to get up to speed. Civilizations that durable aren't limited to our paddling pool; knowledge could travel among the galaxies. Philip Morrison has said, "SETI is the archaeology of the future."

Speculation is fun, but without evidence it's like a dog chasing its tail. The debate over the existence of ETs might never be settled by observations, but it certainly can't be settled without them. We have spoken into the night, and we have listened. By attempting to communicate, we move beyond asking "Where are they?" to ask "Who are they?"

COMMUNICATING IN THE COSMOS

KENT CULLERS IMAGINES sitting at the control console of the Allen Telescope Array. The year is 2015. The facility was funded by Microsoft cofounder Paul Allen. It consists of 350 radio dishes scattered across a northern California valley. They work in parallel, and the powerful processors that collect the radio signals from space digest gigabytes of data each second. Radio waves easily penetrate the gas and dust of interstellar space, and the Allen Array can see across the galaxy. Apart from crushing raw processor power, all that's needed is patience. Cullers is a patient man; until his recent retirement, he worked on SETI for thirty years.

This peek into the near future is no stretch for Cullers. When he was young, his father described the planets so vividly that Cullers could imagine being there. In 1980, he was inspired by a report on the unfunded Project Cyclops, which could have detected a twin of our technological civilization around any of the nearest million stars. Cullers has the vision to think of how extraterrestrial intelligence may communicate in the broadest terms—he's been blind since birth.

SETI PIONEERS

SETI researchers have ridden a roller coaster of great expectations and dashed hopes for decades. At the turn of the twentieth century, the pioneers of newly developed radio technology realized it might be a good way to communicate with the stars. Brilliant and eccentric inventor Nikola Tesla had already pioneered alternating-current (AC) power generation (opposed by Thomas Edison, with his direct-current, or DC, system) and, while attempting to send power directly through the air, claimed his transmitter was picking up extraterrestrial signals. Twenty years later, Guglielmo Marconi also thought he had detected radio waves from space.[21] Both men probably heard distant lightning in the form of "whistlers"; the radio frequencies they worked at were too low to be transmitted through the Earth's atmosphere from an extraterrestrial source.

In 1960, the year before he developed the equation that bears his name,

Frank Drake pointed an eighty-five-foot radio dish based at the National Radio Astronomy Observatory in Green Bank at the nearby Sun-like stars Epsilon Eridani and Tau Ceti. He tuned the dial of his receiver to 1420 MHz, the frequency of a fundamental transition of hydrogen, the most abundant element, and listened for a few weeks. He heard only radio static. In a stroke of whimsy, Drake named his search Project Ozma after a character in the kids' books by Frank Baum that contained strange and exotic characters from far away.

In 1977, a search at the Ohio State University Radio Observatory turned up a single but powerful radio blip that looked artificial. The signal strength was recorded; later, when the astronomer on duty saw how strong the signal had been, he wrote, "Wow!" on the printout. The signal never repeated.

SETI has always had an uneasy relationship with the popular culture, where the belief in aliens is widespread. A whiff of disrepute dogs the subject, often unfairly. After the Green Bank meeting that launched the Drake Equation, dolphin researcher John Lilly warped out of orbit, experimenting with LSD and writing about conversations with ETs. In 1978, famously anti-intellectual senator William Proxmire handed SETI his "Golden Fleece" award as an egregious waste of taxpayer money. Astronomers responded by proposing the senator for membership in the Flat Earth Society. SETI's relationship with Congress remained problematic; then in 1993, Senator Richard Bryan from Nevada axed funding for SETI from NASA's appropriation, saying that after years of searching they "have yet to bag a single little green fellow." This grandstanding was pretty rich given that he also supported renaming Nevada highway 375, which passes near the notorious Area 51, as the "ET Highway."

Kent Cullers has heard all the standard complaints against SETI: that it assumes too much about the nature of intelligence, that it's keyed to specific notions of language or senses, that it's hopelessly anthropocentric (Fig. 128). The first artificial signal may be a numerical sequence, or a set of prime numbers, or binary pulses that define a Turing machine, the universal computer. Cullers accepts that the odds of success are long, and he knows they can't even be logically estimated. He's in it for the long haul, having already overcome tough personal odds. When he was in college, most physics texts hadn't been translated into Braille. "I had to ask a million questions and get my classmates to explain things to me," he says. He tries to rid himself of all assumptions and concentrate on the difficult problem of detecting signals in the presence of cosmic noise.

BEATING THE ODDS

Jill Tarter is also undaunted, after decades of the "Great Silence." She notes that SETI has improved its capabilities by fourteen orders of magnitude since

Project Ozma. Rather than dwelling on a history of failure, she thinks the search is only just beginning to get interesting.

Tarter is another person who is utterly committed to staying the course. She has had to deflect a number of blows associated with the gender imbalance in science; in high school, she was steered toward home economics, and she was the only woman in her year of the engineering physics program at Cornell. She fell into SETI almost by chance while a grad student at Berkeley, at a time when SETI was still generating snickers from other scientists. Ellie Arroway, female lead in the movie *Contact*, was partly based on Tarter, and she helped Jodie Foster prepare for the part. In fact, Ellie Arroway was an amalgam character, according to Ann Druyan, who was a producer on *Contact* and who cowrote the TV series *Cosmos* with her late husband, Carl Sagan. The character was inspired by Hypatia, "Ellie" is a reference to Eleanor Roosevelt, "Arroway" is an anglicized version of Voltaire's true name, and Jill Tarter and Carolyn Porco were the contemporary women scientists who shaped the role.

Figure 128. The message above was beamed out to the globular cluster M13 in 1974, using the Arecibo radio dish in Puerto Rico. To understand it, the alien recipient would need to recognize that the sequence of 1679 radio pulses is a number with prime-number divisors 73 and 23. They would then have to figure out that it should be turned into a grid (one arrangement is random, but 73 rows and 23 columns gives the pattern shown), with "on" pulses dark and "off" pulses white, designed for a visual sense. Even then, the meaning of the graphics is not obvious without the key, provided adjacent to the images. M13 is a globular cluster with very low heavy-element abundance, and so perhaps few planets, and far enough away that a return message could not come for fifty-four thousand years.

"I'm in the first generation of humans that can actually do a search to try and answer the ancient 'Are we alone?' question," says Tarter. "For millennia, humans had just been asking the priests and philosophers what they believed." Tarter has a personal chair at the SETI Institute, named to honor another SETI pioneer, Bernard Oliver, who edited the Project Cyclops report that first captured her imagination. The SETI Institute was founded in 1984. The research has corporate and philanthropic support; Paul Allen and Steven Spielberg have both contributed to current efforts, and Tarter won a TED Prize in 2009 for having an idea that would "change the world." Fund-raising continues for what may be a multigenerational exploration.

CURRENT SEARCHES

Despite a bumpy road with federal funding, SETI has spawned the largest and most distributed research community in any scientific field. The SETI League is a loose consortium of amateur radio astronomers who typically use cast-off satellite dishes to listen for artificial signals. Their ambitious goal is five thousand dishes around the world, giving continuous coverage of the sky.

Meanwhile, anyone with a PC and patience can participate in this quixotic quest. SETI@home was the pioneer in a new wave of projects to harness the idle power of PCs for scientific research. You can get a freeware program that downloads chunks of data from the Internet and analyzes them as a background task, communicating results to a central server. You can also help study climate change, or look for gravity waves, or conduct biomedical research, or help with the design of a particle accelerator. SETI was the first appealing application for distributed computing, and it's still the largest. SETI@home has more than six million participants in 255 countries (including sixty from tiny Vanuatu, so don't be the last person to join the project). Since 1999, three million years of processing time have been harnessed in this "supercomputer" that does 770 trillion calculations per second.

Radio SETI looks hopefully for a needle in a "cosmic haystack." Astronomers scan the quiet part of the radio spectrum for signals that are too narrow to be caused by any known astrophysical process. Pulses that repeat or have a recognizable pattern are candidates for being ET messages. The strategy combines shallow scans of the whole sky with deeper staring at particular stellar targets.

Figure 129. Built into a natural depression in karst limestone in Puerto Rico, the three-hundred-meter Arecibo dish is the world's largest telescope. It has often been used for SETI, and it appeared in the movie *Contact*. The sensitivity was calibrated on the Pioneer spacecraft, which still emits a feeble one-watt radio signal far beyond the edge of the Solar System at a distance of six billion miles. Arecibo could have detected a signal as weak as airport radar on planets around any of the eight hundred stars that Project Phoenix used it to search. In transmit mode, it can send a signal strong enough to be detected by an equivalent facility halfway to the center of the galaxy.

Project Phoenix ran from 1995 to 2004, targeting eight hundred stars out to a distance of 240 light-years. For each star, two billion narrow channels between 1000 and 3000 MHz were searched, with a total of eleven thousand hours of observing. The most powerful tool in the search was the three-hundred-meter Arecibo dish, the world's largest (Fig. 129). As Frank Drake likes to point out, if its surface of aluminum panels were not punctuated with a grid of holes to save weight, it could hold all the beer drunk in the United States in a year or about 357 million boxes of cornflakes. No ET signals were found.

The next frontier for SETI is the Allen Telescope Array. Forty-two six-meter dishes are under construction, with an eventual goal of 350 (Fig. 130). The project will survey a million stars out to one thousand light-years, scanning twelve billion channels from 500 to 11,000 MHz for each. From a combination of sensitivity, bandwidth, and number of targets, it will deliver another billionfold gain on the sum of all previous searches. Tarter is excited because a significant fraction of the cosmic haystack will be searched. But the Allen Telescope Array isn't fully funded, and Tarter admits it can be hard to fund-raise when people think you have your hand in the pocket of a Microsoft billionaire.

Figure 130. The Allen Telescope Array is planned to be a set of 350 six-meter radio dishes located in a lava-strewn valley north of San Francisco. Improvement in computers and signal processing means that large chunks of the radio spectrum can now be searched for signals that have no natural source. The array will provide a gain in speed on the sum of all previous SETI experiments by several orders of magnitude.

Inspired by Charles Townes, the inventor of the laser, SETI has also migrated to optical wavelengths. With our best pulsed lasers and a large telescope, we can generate a terawatt of power—that's 25 percent of the country's energy consumption! To avoid causing a widespread brownout, it's done for only one billionth of a second and beamed into space. If ETs were doing the same thing on a planet around a nearby star, we could detect the pulses for the tiny fractions of a second that they poke above the glaring light of the star itself. A laser one billion times more powerful could be seen across the galaxy.

THE MEDIUM IS THE MESSAGE

With no way to anticipate the thoughts or language of an alien intelligence, SETI researchers look for a signal that has no natural astrophysical explanation. But few of the things that go bump in the night—such as quasars and gamma-ray bursters—were ever predicted, so it might not be obvious what's natural and what's an ET. (After discovering metronomic signals from the first pulsars, Cambridge University researcher Jocelyn Bell ironically named them LGM-1, LGM-2, etc., after "little green men.") The current strategy is to look for radio pulses that are too narrow in frequency (one hertz) or optical pulses that are too rapid (one nanosecond) to create without technology.

Jill Tarter and Kent Cullers know it will be much easier to decide that a signal is artificial and nonrandom than it will be to decode its meaning. We can't talk to animals on Earth that share 95 percent of our DNA, so how likely is it that we'll understand aliens of unknown function and form?

Our own attempts to communicate can seem quaint. Take the Arecibo message, with its stick-man and radio-telescope silhouettes, like petroglyphs beamed into space (Fig. 128). Or the recordings attached to the Voyager spacecraft and launched in 1977, which are in an analog LP format that's already obsolete on Earth, although the cover explains how to play and enjoy the sounds (assuming the recipient has ears, an atmosphere, linguistic skills, and the inclination to create a retro technology). We've also been communicating inadvertently. A sphere of radio waves carrying our early TV transmissions has traveled at the speed of light and swept past ten thousand stars so far, with hundreds more reached each year. ETs either eagerly await our next sitcoms and reality shows or have crossed us off their list of promising civilizations.[22]

The timing issue looms large in SETI. If we look back at our progress, it's almost impossible to predict into the future (Fig. 131). To see how young our technology is, imagine a civilization that exceeded our level of development before the Earth formed and has been watching us intermittently since then. Let's squash the history of the Earth into twenty-four hours. Life forms after a couple of hours and stays small and uninteresting until an hour before midnight. Modern humans arrive on the scene one second before the stroke of midnight. All modern technology—nuclear power, space travel, and computers—is contained within one one-thousandth of a second, the duration of a single beat of a mosquito wing. Blink, and you'd miss it.

SETI has assumed a strong

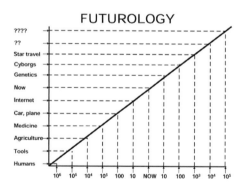

Figure 131. The difficulty of predicting the course of technology is illustrated here by looking backward in logarithmic intervals of time. We have improved our capabilities enormously since ten thousand years ago, but it would be exceptionally difficult to confidently predict our future this far ahead. The timing argument means that ETs will almost certainly be millions of years or more ahead of us in technological capability.

connection between intelligence, technology, and interstellar communication. However, cetaceans and cephalopods on the Earth show that intelligence need not be coupled to technology and that the use of technology for interstellar communication might be a specific cultural trait. SETI might sample only a tiny fraction of the intelligent species out there. What's striking is our need to communicate, a desire that almost elevates to a yearning.

WHY ARE WE SO LONELY?

THE UNIVERSE MAY BE BRIMFUL of microbial life. Since the "Great Silence" is not conclusive, there may be many intelligent creatures out there. We know from the timing argument that, if they do exist, they're likely to be much more advanced than us. So let's follow the chain of logic one last step. Our hand holds a bottle with a simple instruction written on it: DRINK ME.

OUR POSTBIOLOGICAL FUTURE

Hans Moravec has been dreaming of robots all his life. For the past thirty years, he's been making the dreams real in his lab, with a series of increasingly capable robots. In the 1970s, a shopping-cart robot, controlled by a computer the size of your living room, could painstakingly follow a white line. In the 1980s, TV cameras had enabled primitive stereo vision and a robot that could navigate a thirty-meter obstacle course in five hours. By 2000, Moravec built a robot that could navigate using photorealistic maps of its surroundings. In an analogy to animal intelligence, this is like the progression from a slug to a small fish. The exponential growth in processing power means we can make a robot with the functioning and intelligence of a small reptile (Fig. 132). Moravec projects reaching a human level of artificial intelligence in about twenty years.

In his book *Mind Children,* Moravec has considered the implications of the path we're on. He agrees with the more optimistic visions of science fiction— that humans will be liberated for creative and spiritual pursuits by the massive power of machines: "No longer limited by the slow pace of human learning and even slower biological evolution, intelligent machinery will conduct its affairs on an ever faster, ever smaller scale, until coarse physical nature has been converted to fine-grained purposeful thought."

There is, of course, a dystopian version of this future, put forward by Bill Joy, the cofounder and chief scientist at Sun Microsystems. Many of the useful machines of the future will be nanobots. They will monitor the environment, fix

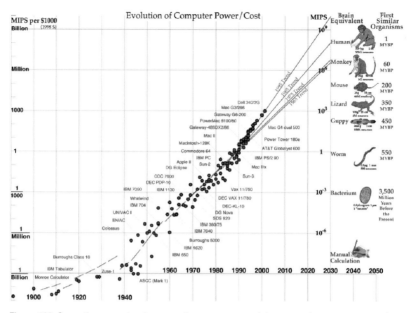

Figure 132. Computing power has been growing at an exponential rate over the past century, and recently it has been doubling every eighteen months. Harnessed into a robot, the best current computer would have the functioning and intelligence of a lizard, but the projection of current trends indicates human-level computational intelligence in only twenty to thirty years. We may soon be the limiting factor in this progression; when computers and robots program themselves, they'll improve more rapidly.

our illnesses from within, and go places where humans cannot. But if they have the power of self-replication—like miniature von Neumann probes—they could easily outcompete natural organisms, and their exponential growth would take over the world. It's called the "gray goo" problem. In our technological development, we're poised on a knife edge.

If we can make it through this stage, which may be the bottleneck that quenches many promising civilizations, we'll transcend the limits of natural selection. As anyone who has struggled with computers knows, software doesn't improve nearly as fast as hardware does. But if self-programming computers create improved versions of themselves, the progression will accelerate to what Ray Kurzweil calls the Singularity. He thinks the Singularity will extend immortality to the consciousness of individual organisms.

The nature of this transition is uncertain. It may lead to a purely computational future, where posthumans dispense with biology and become disembodied thought "collectives." It may lead to machine-human hybrids. We imagine increasingly bionic humans, with skeletons of advanced composite materials, while complex organs like brains and eyes remain biological. The transition might not be a one-way membrane. Posthumans can use their advanced capabilities to reinvent biology, evolving fluidly between constructed and natural forms. These are just our possible futures. Out in the vast cosmic "laboratory," the combined possibilities of biology, engineering, and computation are almost limitless.

SUPERCIVILIZATIONS

In the 1960s, the Russian physicist Nikolai Kardashev speculated about the technological capabilities of advanced civilizations.[23] A primitive, or K0, civilization isn't in control of the resources of its home planet. We are approaching K1 status, where we use all local resources, about 10^{12} watts. The energy we leak into space doesn't make us very noticeable at the distance of the stars. A K2 civilization takes advantage of all 10^{26} watts of their star, perhaps by building a hollow sphere around it (called a Dyson sphere, after the physicist Freeman Dyson). The Earth intercepts a minuscule portion of the Sun's energy; capturing all of it for a second would meet the world's energy needs for one million years. K3 civilizations are almost incomprehensible to us because they capture the entire resources of a galaxy.

Remembering that all of this could be a fantasy and we might be as advanced as life gets, how might these other civilizations travel and communicate? K1s would use technologies we're just beginning to develop, like nuclear propulsion and laser sails. The preferred mode of communication might be particle beams made of neutrinos or dark matter that travel at the speed of light and meet little resistance. Distinctions between travel and communication would be moot; teleportation could re-create conscious states or organisms at a remote location, carried by the fleet feet of electromagnetic waves. K2 engineers could harness stars and communicate by orchestrating stellar cataclysms like supernovae and gamma-ray bursts, visible in distant galaxies. If allowed by physics, K3 civilizations could master wormhole travel and might communicate by manipulating space-time. Our radio pulses seem feeble when we contemplate creatures that can send gravity waves, spawn baby universes, and travel the multiverse.

Curiouser and curiouser. With discussion of aliens that can harness galaxies, we seem to have ventured beyond the plausible. But as the Queen of Hearts reminded Alice, "Sometimes I've believed as many as six impossible things before breakfast." Let's look at what supercivilizations might get up to for entertainment.

REALITY AND SIMULATION

What would a posthuman intelligence do with its incredible capabilities? We've no idea because it would be to us as we are to a worm. The cutting edge of our technology manifests in video games, where we create artificial realities of increasing sophistication. How far can simulations be taken?

We might be living in a physical simulation built by a vastly superior race, an enhanced version of the scenario of the movie *The Truman Show*, in which the unwitting star of a TV show lived in an engineered reality under a painted dome. Physicists can figure out the energy needed to fabricate a simulation of a particular size. It turns out that only a K3 civilization could construct a detailed planet-scale simulation, with lower levels of construction accuracy for regions of the Solar System and beyond that we have barely explored. This scenario seems implausible because harnessing a galaxy's power is so fantastic.

It's harder to dismiss the idea that superintelligences would create a computer simulation of consciousness. Consider something called the simulation hypothesis: you exist in a virtual reality simulated in a computer built by some advanced civilization.

Nick Bostrom has argued that at least one of the following propositions must be true. The first is that almost all civilizations become extinct before advancing past our stage to the point where they could create such a simulation. The second is that almost no posthuman civilization chooses to run simulations of minds like ours. The third is that we're almost certainly living in a simulation. This is like the movie *The Matrix*, but without the hokey aspect that an advanced race would need to use our brains for battery power.

Bostrom proposes ideas like the simulation hypothesis with a straight face but a twinkle in his eye. His work mixes whimsy and logical rigor. The argument depends on substrate independence: the idea that consciousness can be realized equally in a silicon-based network in a computer and in a carbon-based biological network in a cranium. The requirements are far less than those for creating a physical simulation. The computing power needed to simulate a human mind is roughly 10^{17} operations per second. (For a comparison, the world's fastest supercomputer is a Cray XT5 Jaguar working at 1.8 petaflops, or 2×10^{15} operations per second, less than two orders of magnitude below this benchmark.) About 10^{35} operations per second would be needed to create the entire mental history of humanity, something Bostrom calls an "ancestor simulation." That's a mountain-sized computer using current technology, but using the quantum-computing techniques that are under development, it's a modest-sized machine.

Ancestor simulations would therefore be trivial for superintelligent civilizations to create, and they could make astronomical numbers of them. The simulation hypothesis is not ridiculous—it's based on a straightforward extrapolation of our own rather modest technological capabilities.

Returning to Bostrom's propositions, if the first is true, then intelligent life in the universe is rare, and we're poised on a precipice. Let's hope this is wrong. There's no way to evaluate the second proposition, since we can't know the recreational inclinations of a supercivilization. It seems to imply a lot of convergence in the motivations of advanced civilizations if essentially none of them chooses to run simulations of minds like ours, but that's just a hunch. If even a small fraction of them do create simulations, we're thrown directly into the implications of the third proposition.

If the first and second propositions are false, there *must* be many more simulated minds than nonsimulated minds running in organic brains. By the mediocrity principle, it's far more likely that you have a simulated mind than one of the very rare ones with biological neurons. *No!* Reading this, everything inside you rebels. *I'm real! I'm alive! I'm conscious of this moment.* Sorry, that proves nothing. Everyone will think their own lives are real and not simulated, but mostly they'll be wrong. There's no way we could figure out we live in a simulation unless the simulators wanted to let us in on the secret for their additional amusement.

Virtual reality is still primitive, but anyone who's watched a teenager lost in the finely rendered landscape of a role-playing video game knows where the technology is headed. The world that you think of as so "real" manifests as electrical signals in your brain, which is itself an electrical network. A highly advanced civilization could create such a richly realized world, and we wouldn't be able to tell the difference. When Alice had her adventure, what exactly did she experience? Was it a dream or a hallucination, and did she really wake up at the end?

The simulation hypothesis creates some intriguing issues of its own, such as the possibility of nested simulations. (A simulated civilization like ours might create its own simulations, making it unlikely that our own simulators are themselves real.) The regress isn't infinite because that would require infinite computing power at the basement level. It even points toward solipsism—what if some humans are simulated "shadow people"? And if this is true, how do you know they're not all simulated except you? Or me? If we dare, consciousness lets us face the nature of reality.

AT ONE WITH THE COSMOS

In a traditional Chinese tale, three philosophers come together to taste vinegar, which is a symbol for the spirit of life. First to drink is Confucius. "It is sour," he says. Next, Buddha drinks. He pronounces the vinegar bitter. Last to taste is Lao-tzu. He exclaims, "It is fresh!"

Sentience gives us the opportunity to forge a relationship with the universe we live in. Crudely speaking, there are three ways humans have chosen to understand their place in the universe. They aren't completely distinct, and many individuals choose to follow more than one path. In a sense, they're all vinegar.

The first is science. Science is one of the greatest achievements of the human intellect. We have an instruction manual for how the universe works and for the myriad ways that matter and radiation interact. We've projected our minds into the heart of an atom and through a universe of billions of galaxies. We told our story from the time of the hot, infant universe through the creation of atoms and our emergence from organic soup on one rocky orb lit by a sheltering star. The story is fantastic, and we believe it to be true.

But science is mute to meaning. We don't know why the universe exists or why some assemblages of atoms have the ability to question their existence. This book of science is powerful prose, but it's not literature, and it's not art. We'll never read it and say, "Of course, *that's* why!" More worryingly, science acts as a lens to magnify our best and worst features. We can use it to cure disease and feed the world or to kill one another and trash the biosphere. In that, at least, we have free will.

Then there's organized religion. Science and religion coexist uneasily, and in the United States the exchange is often acrimonious. Religious believers tend to view it as a moral failing not to posit a creator and a design for the universe. They think scientists are all materialists and reductionists. Many are, and a few have proposed that religion exists for evolutionary reasons and is a purely psychological construct.[24]

Astrobiology presents an interesting challenge for theistic religions. All evidence in front of us indicates that we're part of a continuous spectrum of biological activity on this planet, with no attribute separating us from other advanced creatures such as elephants and orcas. (Religious people may point to faith itself, but the supposition that animals don't have spiritual dimensions is not proven and therefore is itself an act of faith!) What we know is also consistent with terrestrial biology being one specific example of a more general set of biologies in the universe.

With no evidence for ETs, we could just dodge the issue. But the statistical arguments that sentience has arisen elsewhere cause adventurous theologians to consider the implications. Do intelligent aliens have souls? Did Christ die for their sins, too? What relationship do they have with the hypothetical Creator? Brother Guy Consolmagno, whom we met earlier on the Antarctic ice hunting for meteorites, has written a booklet called *Intelligent Life in the Universe? Catholic Belief and the Search for Extraterrestrial Intelligent Life*. In it, he considers these questions and finds no problem for Catholicism in ETs. "There's nothing in Holy Scripture that could confirm or contradict the existence of intelligent life elsewhere in the universe," he says.

The popular culture has gone beyond accommodation to meld ideas of science and religion. In Steven Spielberg's 1977 movie *Close Encounters of the Third Kind*, the alien ship acts like an austere Old Testament God; when Richard Dreyfuss is led inside near the end, he's in the crucifix position. Many movies since then have used aliens to tell stories of good and evil, damnation and redemption.[25] Spielberg did his take on the New Testament five years later with the iconic *E.T.* In that movie, a thinly veiled allegory of the story of Christ, the alien possesses a childlike purity and the power to heal. E.T. is misunderstood, persecuted, killed, and then resurrected. Aliens are a modern religion—they're vessels into which we pour our fears and longings.

The third path is spirituality. The word itself is wonderfully vague, broad enough to suggest metaphysics but without the baggage of conventional religions. (Two-thirds of scientists declare themselves to be strongly or somewhat "spiritual.") The umbrella of spirituality also encompasses nontheistic religions such as Buddhism, in which a central idea of interconnectedness is consistent, at least superficially, with modern science. Is this a true commonality or sloppy logic and wishful thinking?

COMPANIONSHIP OR LONELINESS?

Let's simplify the discussion by reducing the trinity to a duality. "A great truth is a truth whose opposite is also great truth," said the physicist Niels Bohr. What does astrobiology have to say about our potential isolation?

We may be alone. The contingencies of evolution might be so dominant that big brains are extremely unlikely and intelligence so rare that we're isolated in time and space. Or perhaps we're first. We would indeed be special. But only in the context of a religious tradition could we take solace in that fact. The hard facts of science would tell us that all of our hopes and dreams, all that we hold dear, are simply the products of circumstance, an accident.

We may not be alone. In this case, the possibilities multiply. Our biology might represent a general solution to the organization of matter, so that we share a kinship of something like DNA, and maybe function and form, with organisms across the cosmos. This can't be ruled out, but it's limited thinking: sure, they're out there, and they must be just like us. English astrophysicist Arthur Eddington said, "The universe is not only stranger than we imagine, it is stranger than we *can* imagine."

Other life-forms might be so advanced that we're inconsequential. Our religions would then be confections of ego, conceits as anthropocentric as the alien iconography of *Star Wars.* As a counterpoint to Brother Consolmagno's confidence that Christianity could expand to encompass aliens, consider the words of American patriot Thomas Paine from *The Age of Reason:* "To believe that God created a plurality of worlds, at least as numerous as what we call stars, renders the Christian faith at once little and ridiculous; and scatters it in the mind like feathers in the air."

Alien life might be so strange that communication is impossible or so strange that it's unrecognizable. Perhaps we're playthings of a superior race that scattered seed on the Earth four billion years ago. Biology elsewhere might use communal solutions, sharing both senses and perception, thus leaving us alone with our encased brains, our brooding thoughts. Consciousness—for want of a better word—may emerge on the scale of cells that fit on the head of a pin or over eons and on the scale of galaxies.

The great mystics of the world have spoken about the tension between being and nothingness, self and other. In the first sutra of the Siddha yogic tradition, consciousness is the cause of the universe just as surely as the universe is the cause of consciousness. The skeptic might call this solipsism as outrageous as the simulation hypothesis, but the scientific account of creation is equally fantastic. It says that all the trillions of habitable worlds in the universe, with their potential and actual life-forms, emanated from an iota of space-time, a dot of quantum possibility.

THE SEARCH FOR MEANING

In Albert Camus's novel *The Stranger,* the young Algerian man Meursault moves through his life in a dream. He lives utterly in the present, his actions spontaneous and not guided by any master plan. He kills a man for no apparent reason; eventually, he faces the guillotine. Camus's existential parable is sparsely written and uncomfortable to read. We have a powerful need to forge meaning in our lives. For surely that distinguishes us from bees or termites, with all their industriousness and genetically wired behavior? Yet science is consistent in ascribing no special reason to our existence. If we want to seek relief from what Camus called the "benign indifference of the universe," we must look within.

Astrobiology is young. We can't yet tell the full story of life on Earth, exploration of the Solar System is in its early phase, and we've not yet scratched the surface of the habitable planets beyond. In time, we may find out if we're special or just a footnote in the busy biological history of the outer part of the Orion arm of a spiral galaxy that we call the Milky Way.

Science is young. We've just recognized our place in the universe. We've every right to be amazed that we know as much as we do. Stephen Hawking puzzled in *A Brief History of Time,* "What is it that breathes fire into the equations and makes a universe for them to describe?" We still struggle with the question posed by Blaise Pascal 350 years ago: "Why is there something rather than nothing?"

Humans are young. Capable of individual acts of great beauty and kind-

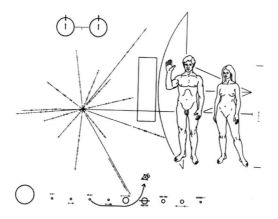

Figure 133. The plaque attached to the leg of the Pioneer 10 spacecraft has left the Solar System, although it will be tens of thousands of years before it reaches another star system. It was last heard from in 2003, when it was 8 billion miles from Earth. This message in a bottle tossed into the ocean of space is more a message to ourselves than a realistic attempt to communicate with alien intelligence.

ness, we are also collectively aggressive and shortsighted. Intelligence alone will not ensure our survival if technology outstrips wisdom. Scared of our mortality, we toss messages into the vast ocean of space and wonder if we're alone (Fig. 133). The best we can do is create meaning with each breath, each original thought, each act of compassion or love. There is bliss in our science and our art—we must treasure both.

NOTES

CHAPTER 1: THE UNFINISHED REVOLUTION

1. The inscription does not appear in Plato, or in the writings of his student Aristotle. Like many apocryphal sayings, it is apt if not literally accurate. The quote is first referenced more than a thousand years after Plato, but its spirit is contained in his commentary on the proper training of philosophers in book VII of *The Republic.*

2. There's no evidence that the Greeks tried to prove the existence of atoms, although if they had held the concept firmly enough they could have gotten close without any high technology. For example, if you take a drop of oil and let it spread out on the surface of a pond or small body of water, it will eventually reach monolayer thickness. The tiny sphere has dissembled into a large disk, and simple geometry can show that the thickness of the layer is thousands of times smaller than the thickness of a human hair.

3. Translation and proper attribution of quotes from the Greek philosophers is difficult. Democritus wrote over sixty works but only a couple of hundred fragments survive, many interpreted by Aristotle, who considered him a rival. Most Democritus quotes are culled from a fifth-century anthology of Stobaeus.

4. Greek philosophy included rationalists like Plato, who believed that the world could be understood by pure thought, and empiricists like Aristotle, who approached science through observation and classification. Modern science is based on observation, but it's remarkable, even to many scientists, that the universe is so well described by physical laws that are best expressed in mathematical form.

5. Earlier civilizations, like those of the Egyptians and the Babylonians, made careful observations of the patterns in nature. In fact, the Babylonians had a calendar accurate to five minutes per year, far superior to the Greek calendar that followed. Similarly, the Egyptians noted triplets of integers that could be used to make right-angled triangles; this allowed them to divide up the fertile land of the Nile delta and regulate agriculture. Their knowledge of astronomy had practical applications, as in the casting of horoscopes. But Babylonians didn't make models of the cosmos or search for explanations of celestial phenomena, and Egyptians didn't discover an abstract formalism—the Pythagorean theorem—that would let them calculate any right-angled triangle.

6. This process started seventy years earlier with John Dalton, a self-taught English scientist who never attended a university. Dalton experimented with the way gases combined and noticed they always combined in fixed proportions and that some gases could not be changed by chemical means. He deduced that substances

are made of fundamental and indivisible components called atoms, giving each element its particular chemical properties. Atoms were not truly accepted as a concept until the late nineteenth century.

7. Theories are intellectual frameworks for explaining a range of natural phenomena. The definition is more precise than the colloquial use of the word, which tends to mean a guess or a hunch, as in "evolution is just a theory." The best theories make extensive and unambiguous predictions, they are supported by a web of evidence, and they are subject to refinement or refutation by experiment or observation. Those that stand the test of time and testing get elevated to the status of "laws of nature."

8. Another classic trap is confusing correlation and causation. Bertrand Russell told the story of a chicken that came to associate the rising of the Sun with getting fed. Every day the Sun would come up, and the farmer would scatter seed. One day the farmer came out and throttled the chicken for the dinner table, a catastrophic failure of induction. In a bizarre true example, western anthropologists encountered a South Seas tribe in the early twentieth century that put lice on the heads of fevered children, thinking that doing so would cure them. Lice require a narrow temperature range, so they avoid the heads of sick kids.

9. Karl Popper formulated a classic view in the 1930s in which science progressed by falsifying theories. The first observation of a black swan falsifies the hypothesis "all swans are white." In practice, progress is rarely that simple, because observations designed to extend the reach of a theory are difficult or uncertain, and such observations rarely reject a theory decisively. In the gravity example, observation of Mercury's advancing perihelion didn't render Newton's theory useless but did show that there are physical regimes where Einstein's theory provides a more accurate description.

10. The Copernican Revolution is a complex example of how science advances. Scientists are supposed to choose the model that fits the data best, but in this case there was a strong aesthetic preference for a model that explained the retrograde motions of Mars and Jupiter at opposition without using the ad hoc device of the Ptolemaic model. The Copernican model flew in the face of common sense since we can't physically feel the Earth's motion. Supporters of the Copernican idea were forced to argue that the stars were so far away that parallax could not be detected. At the time Copernicus died and his book was published, there was no decisive evidence that supported his model.

11. Galileo didn't invent the telescope, although he allowed people to think he had. It was invented by Dutch optician Hans Lippershey in 1608. Ceramic and glass-making skills were very advanced in Holland. Once eyeglasses had been invented, it was only a matter of time before people experimented by combining lenses to magnify and demagnify. The telescope and the microscope were invented around the same time. (People began to wear eyeglasses then, too, as shown in contemporaneous paintings.) Galileo was the first to use the telescope for careful and systematic observations of the night sky.

12. Newton's mathematical formulation of gravity was masterful. About forty years ago, the Nobel Prize–winning astrophysicist S. Chandrasekhar revisited Newton's

analysis and found he could not improve on the *Principia* despite having 350 years of mathematical advances at his disposal. But Newton didn't solve all the mysteries of gravity. What was this strange force that acted with infinite range in the vacuum of space? In answer to this question, Newton gave the famous gnomic response, "I frame no hypothesis."

13. Thomas Wright used a utilitarian argument. "Such vast room in Nature, as Milton puts it, cannot be without its Use," he wrote in *An Original Theory or New Hypothesis of the Universe* (1750). He was also unabashed in celebrating the joys of being an astronomer: "This idea has something so cheerful about it, that I own I can never look upon the Stars without wondering why the whole World does not become Astronomers."

14. Herschel was the first person to map out the stars in three dimensions. His method was painstaking. First, he pointed identical telescopes at two different stars. Then he masked the aperture of one telescope until the two stars appeared to be of equal brightness. The ratio of the full aperture of one telescope to the partial aperture of the other gave the relative brightness. Next, he placed the stars in distance by assuming that all stars emitted equal amounts of light. Then the relative brightness of two stars gives their relative distance, by the inverse-square law. For example, a star that is four times fainter will be two times farther away. Doing this for thousands of stars was very time-consuming! The assumption that all stars emit equal amounts of light is not correct, but as a statistical method for estimating distance Herschel's method was effective.

15. Increasing the size of a telescope helps to see fainter objects by gathering more light, but it also allows smaller angles to be resolved or measured on celestial objects. Angular resolution scales inversely with the size of the aperture. The eye is twenty times too small to see the tiny angle of the parallax shift. Even larger telescopes must be figured exquisitely to realize the sharp images that their size can deliver in principle.

16. The stars were Cepheid variables, named after the constellation of their discovery. Henrietta Leavitt at the Harvard College Observatory discovered a relationship between the period of variation and the flux or luminosity of the star. The brightness variations occur at a particular stage in the star's evolution. Energy cannot escape from the star fast enough, so it builds up in the outer layer, causing it to heat up and expand. The energy then escapes, the star dims and cools, and the cycle repeats. Hubble took photographs of the Andromeda nebula for months and looked for these variable stars within the nebula. The period of variation told him the luminosity or true brightness, and by combining it with the apparent brightness he deduced the distance to the nebula.

17. Hubble recognized he had made a bold assumption when he inferred that Andromeda was hundreds of thousands of light-years away. He'd assumed that the physics of Cepheid variables was the same at remote locations as it was nearby in the Milky Way. If the universe works according to different laws at different locations, it's difficult to do cosmology. Hubble called this the principle of the "uniformity of nature."

18. The Doppler effect is familiar, but it's a flawed analogy for the Hubble expansion.

A Doppler shift is caused by the relative motion of a source of waves and the observer. If you're standing still and a police-car siren approaches, the waves get bunched up in the direction of motion. In the universe, the redshift is caused by the expansion of space itself. As light moves toward us from another galaxy, the intervening space is actually growing due to cosmic expansion, and light waves are stretched or redshifted as they travel.

19. Einstein's theory is superior to Newton's because it applies in a wider range of situations, and Newton's theory fails entirely in situations of very strong gravity. (In most parts of the universe, gravity is weak.) The deflection of light predicted by general relativity was observed during a 1919 eclipse expedition led by Sir Arthur Eddington, as light from a background star grazed the edge of the Sun. Einstein became a celebrity overnight and was the world's most famous scientist for most of his life. He worked hard to avoid nuclear proliferation and was offered the presidency of Israel twice, declining it both times.

20. The basic equation of general relativity relates the density of energy and matter to the curvature of space. Not only is light deflected by gravity, it also loses energy escaping gravity. This shift to longer wavelength or lower frequency is called a gravitational redshift. Black holes correspond to gravity so intense that all the energy is lost and no radiation escapes. The lowering of frequency means that clocks slow down as gravity gets stronger. Atomic clocks flown at high altitude actually tick slower than identical clocks at sea level.

21. The term "big bang" was coined by the English cosmologist Fred Hoyle, a proponent of the alternative steady-state theory. Hoyle claimed he never intended the moniker to be derogatory, but he thought it was outrageous to imagine that the material contents of a vast universe could be created instantaneously, from nothing. However, microwave background radiation permeates all space and does not originate from any type of astronomical object. It's perfectly explained by a model where the universe was much hotter and denser in the distant past, and it's very difficult to explain with any other idea. In ironic deference to the originator of the term and the grandiosity of the idea, astronomers do not capitalize big bang.

22. Understanding dark matter and dark energy is the biggest challenge in cosmology. Dark matter can be "explained away" if Newtonian gravity is wrong, but there's no evidence of that. Astronomers have ruled out everything from black holes to microscopic dust grains; the current best bet is that dark matter is made of exotic subatomic particles predicted by physics unification theories. Dark energy acts like antigravity. The vacuum of space can behave in this way, but the energy needed to cause the cosmic acceleration is an embarrassing factor of 10^{80} less than simple calculations predict. Both dark matter and dark energy expose the limitations and deficiencies of fundamental theories of physics.

23. Telescopes are characterized by their collecting area and their resolution, which means the sharpness of the images that can be made. The blurring of light by the Earth's atmosphere limits the depth of vision of a telescope, which is the reason the modest 2.2-meter Hubble Space Telescope competes successfully with much larger ground-based telescopes. Astronomers are perfecting a technique called adaptive optics, which uses actuators to rapidly adjust the shape of the mirror

and compensate for turbulence in the atmosphere. The images can be sharpened by a factor of thirty or more, which will make the upcoming generation of large telescopes more powerful than any in history.

24. The popular perception that associates blue with cold (cold lips, Arctic ice) and red with hot (sunburn, salsa, fire engines) is very deep-rooted. But spectral color is based on energy across the electromagnetic spectrum. A hot plate emits infrared radiation—waves of invisibly long wavelength—when it is switched off. As it warms up, its spectrum shifts to shorter wavelengths until it enters the visible spectrum, glowing dull red. When it is fully on, it is hot enough to glow orange or even yellow, and some metals can be heated enough to glow white or white-blue before they melt. White stars are hotter than yellow stars, which are in turn hotter that orange stars. The interpretation of color is also confused when an object is painted or has its color built into it. A blue object is not hotter than a red object; it just reflects blue light and absorbs red light, while the red object reflects red light and absorbs blue light.

25. The detection of gravity waves is a profound extension of our capabilities to view the universe. The entire history of astronomy has been based almost exclusively on the detection of light or other forms of electromagnetic radiation. This is one way to view the "stuff" of the universe that depends on the complex ways that radiation interacts with matter. The new Laser Interferometer Gravitational-wave Observatory (LIGO) hopes to detect ripples in space-time caused by cataclysms like supernovae, mergers of black holes, and the big bang itself. For the first time, we will be able to "see" with gravity.

26. This progression of nested motions must end, since the universe has nowhere to go! Microwave photons from the big bang define the true reference frame of the universe, since they permeate space and travel in all directions. With careful measurements, astronomers have shown that the microwaves are a little cooler (or redshifted) in one direction and a little hotter (or blueshifted) in the opposite direction. The Milky Way, the Andromeda galaxy, and the local "group" of galaxies move at 1,300,000 miles per hour with respect to the microwave background radiation. This sounds fast, but at only one five-hundredth of the speed of light it's a subtle effect.

27. The term "astrobiology" is only about ten years old. For decades before that, astronomers had used the term "exobiology," meaning "biology beyond Earth," which sparked wry criticism from biologists who pointed out that it was a subject without subject matter. Astrobiology is intended to convey the goal of understanding life in a cosmic context. In addition to direct searches for life beyond Earth, the subject has been propelled by our improving knowledge of the history of life on Earth.

CHAPTER 2: LIFE'S ORIGINS

1. The creation of light elements in the first few minutes after the big bang is called cosmic nucleosynthesis. In addition to helium, the main products were deu-

terium, or heavy hydrogen, at an abundance of one part in ten thousand, an isotope of helium called helium-3 at one part in one hundred thousand, and lithium at one part in one billion. The fact that cosmic measurements of each of these four species—which are found in very different places in the universe—match direct predictions of the big bang model is very powerful support for the idea that the universe had an early, hot phase.

2. Spectroscopy is an amazingly sensitive technique for measuring the chemical composition of remote objects. It's not difficult to detect elements at a level of less than one part in one trillion from a high-quality spectrum of a faint star. The scarcity of heavy elements—carbon at one part in a thousand, nickel at one part in a million, gold at one part in a billion—seems at odds with our everyday experience. We have to recall that the Earth is an unusual environment, a rocky cinder left over from star formation. Most of the universe is made of stars, and stars are mostly made of hydrogen and helium.

3. Cosmic abundance of the elements implies we've made measurements that truly represent the universe. In fact, the most detailed information is derived from the Sun and from meteorites, which represent primordial material in the Solar System. However, the spectra of hundreds of thousands of stars show similar abundance patterns, and familiar spectral features of carbon, nitrogen, oxygen, silicon, and magnesium are seen in the spectra of galaxies ten billion light-years away. Chemistry is universal.

4. Every atomic nucleus is held together by the strong nuclear force. This binding energy has a tiny amount of equivalent mass, according to Einstein's $E = mc^2$. So the mass of a helium nucleus is the mass of four nuclear particles plus an extra bit corresponding to the binding energy. In nuclear fusion in the Sun, the four particles are combined into a single helium nucleus. The total mass of the initial particles is more than the mass of the end result, so when hydrogen is turned into helium, 0.7 percent of the mass of a proton is turned into radiant energy, and since the conversion factor is the square of the speed of light, a lot of energy is released in each reaction. Sunlight!

5. "Shine On You Crazy Diamond" is on the album *Wish You Were Here*, and it contains at least a dozen references to stellar evolution and nucleosynthesis. The song is a double metaphor; its subject is also Syd Barrett, the eccentric and troubled founder of the group who died in 2006. Like a massive star, he shone brightly and burned out quickly due to drugs and mental illness, leaving the group three years after it was formed. White dwarfs are in fact not far removed from diamonds in composition and density.

6. The concept of time as an emergent property comes from the subject of thermodynamics, the behavior of ensembles of atoms. In the classic example, two types of gas in a box are separated by a partition. When the partition is removed, the gases mix; once they are mixed, they will never unmix. That's because there are a huge number of ways the atoms can be totally mixed and a far smaller number of states where they are mostly separated, so the situation moves toward the most probable state. Mixing or disorder is related to entropy. However, this simple physical system doesn't solve all the mysteries of the arrow of time; the real world con-

tains many systems far from equilibrium where order emerges spontaneously from chaos.

7. Watches and clocks are everyday objects with a long history. "Watch" refers to the practice of dividing up the time of keeping lookout on a ship, and it dates back four hundred years. The French philosopher Blaise Pascal was the first person recorded to wear a watch on his wrist. "Clock" is from the Middle German word for bell. The oldest clocks had no face or hands, since most people were illiterate and innumerate. They could be heard from miles away and so were useful in regulating village life.

8. The best atomic clocks are accurate to a nanosecond per day, or one part in one thousand trillion. This means an atomic clock takes three million years to lose or gain a second. In case you are thinking this is too esoteric, atomic clocks form the basis of the Global Positioning System, a technology that is increasingly entering our everyday lives. Not happy with being eclipsed by physicists, astronomers fought back. Pulsars are the collapsed corpses of massive stars, spinning at rates ranging from one to a few thousand times per second. They were discovered in 1967, and several thousand are known. The spin rate is stable (except for a gradual spin-down caused by the emission of gravity waves, as predicted by general relativity) and extremely well measured. The most accurate pulsar clock has a precision of one third of a nanosecond per day.

9. It's difficult to find rocks older than a 4.4-billion-year-old zircon because this was probably the time at which the crust cooled. The Moon is geologically inactive, so old rocks are more plentiful on it; the oldest samples brought back by Apollo astronauts are between 4.4 and 4.5 billion years old. But the best tracers of the age of the Earth are meteorites because they represent primitive material left over from the formation of the planets. More than seventy meteorites have been dated by radiometric techniques, giving an age of the Solar System (and therefore the Earth) of 4.54 billion years, with a margin of error of less than 1 percent.

10. In Michael Crichton's book *Jurassic Park,* scientists cloned dinosaurs using DNA extracted from the guts of mosquitoes that had fed on dinosaurs' blood and then been entombed in amber. While there is certainly amber that is 130 million years old, from the time when dinosaurs ruled the Earth, DNA is extremely fragile and is mostly destroyed by the polymerization process that forms amber. Cloning a dinosaur from the material found in amber would be like reconstructing a library from a few scattered book pages.

11. During the critical period from six hundred to five hundred million years ago, the only good evidence of the diversity of life comes from a handful of Lagerstätten, which arise from rare situations where an entire water ecosystem was entombed and preserved. For example, if a shelf of oxygen-free mud falls to the seafloor, decomposition will be suppressed long enough for casts of soft body parts to be created. The most famous of these formations is the Burgess Shale in British Columbia, from 505 million years ago.

12. Chemistry is mostly concerned with the exchange of electrons when atoms or molecules combine. The loss of electrons or the gain of oxygen is called oxidation, and the gain of electrons or the loss of hydrogen is called reduction. The original

Miller-Urey experiments made the assumption that the Earth's atmosphere was "reducing," or full of hydrogen-rich molecules that could act as electron donors. Under these conditions, it is energetically favorable for molecules to combine and gain in complexity.

13. The validity of all these experiments depends on there being not much free oxygen in the early Earth's atmosphere. While this is still disputed, the best reading of the available evidence is that oxygen did not begin to build up until photosynthesis began, some time after the first emergence of life. The fact that Miller-Urey experiments do not generate all of the amino acids present in life may not be a problem, since the earliest life may not have used the full set of twenty. In support of this, gene-sequencing methods have been used to identify amino acids that are most commonly found in the most ancient living things. They turn out to be the amino acids that are most readily produced in Miller-Urey experiments.

14. In his 1947 essay "What Is Life?" the English geneticist J. B. S. Haldane stated bluntly, "I am not going to answer this question." Biologists have tended either to dodge the question or to answer it very specifically in terms of the attributes of terrestrial biology. However, astrobiologists have to be careful to be not too Earth-centric, since the universe may include many mechanisms, biochemical or otherwise, to generate something that everyone would agree was alive. The Russian biochemist A. I. Oparin published a classic book on the origin of life in 1936, in which he used the example of an army of robots that builds an army of even better robots. The philosophical questions that swirl around the definition of life are profound.

15. Life that does not use light and photosynthesis might use chemical energy or geothermal energy. Going back to the source, chemical energy is based on the formation of those elements in previous generations of stars, and geothermal energy is based on either pressure caused by the mass of the Earth or on radioactive decay, which also requires heavy elements to have been created by stars. Sunlight is a by-product of fusion and so depends on the pressure created by a large mass. In a very general sense, the source of all energy to power life is the force of gravity.

16. An important subtlety in this argument is the idea of activation energy. Even though chemical reactions occur extremely quickly, in fractions of a second, things do not spontaneously combust or instantly oxidize. A little extra push is needed to make most chemical reactions occur, and this is called the activation energy. The net effect is to inhibit the tendency toward disorder and dispersal of energy—the Second Law of Thermodynamics. In the complex dance of a biochemical network, the Second Law can be thwarted long enough to allow organisms to live for many years.

17. The act of measurement decreases uncertainty and leads to new information. Think of the game of twenty questions. Each question has a yes or no answer, one bit of information. Each answer decreases uncertainty, because we now know that the object is bigger or smaller than a toaster, or that it is an animal as opposed to another type of living creature. Twenty questions encapsulate 2^{20} possibilities, or a million different things. So it's not surprising that people often win the game, even though that seems unlikely.

18. Information theory was developed to understand computers and signal processing. Making connections to the information content of biological systems is difficult (and often misleading) because there are many ways to define information. Living systems store information as "specified complexity" in local violation of the thermodynamic tendency for information to be lost or degraded in exchanges of matter and energy. Life also has the attribute of transmitting information, in its ability to replicate.

19. Variations in reproduction occur even *without* a change in the environment. The finite number of macromolecules within a cell leads to statistical variation. A single bacterial cell in a continuously stirred flask will divide in about an hour. Those two cells divide in another hour, but not at exactly the same time. After many cycles of division, the population is growing continuously, with no synchronization of division. Two organisms with the same DNA, in widely separated but identical environments, grow at different rates.

20. Aristotle wrote exhaustively about nature in *The History of Animals*, from 350 B.C.E. He benefited from the largesse of his patron Alexander the Great, who had his generals send Aristotle examples of all the plants and animals that were encountered during Alexander's conquests. Despite the quote, Aristotle was not an evolutionist in the modern sense.

21. The joint probability that the cards will randomly occur in sequence is 1 in $13 \times 12 \times 11 \times 10 \times 9 \times 8 \times 7 \times 6 \times 5 \times 4 \times 3 \times 2 \times 1$. But when the cards are removed from the "gene pool" one by one because they stick to one another, the odds go down dramatically to 1 in $13 + 12 + 11 + 10 + 9 + 8 + 7 + 6 + 5 + 4 + 3 + 2 + 1$. In the more realistic situation where the cards or atoms are drawn from a much larger set of each type in the environment, the odds of n atoms spontaneously joining together are 1 in 10^n, relative to 1 in $10n$ if they assemble sequentially. The huge difference between a geometric and a linear progression is why the odds of spontaneous assembly become tiny as n gets large.

22. Yet another obstacle to the smooth development of life is the fact that molecules can be left-handed or right-handed. The chemical reactions that generate life's ingredients produce equal numbers of each type. (This is also true of organic material found in meteorites.) In living organisms, however, only one handedness is used, and molecules of the wrong handedness cannot function. As life began, it would have been inefficient for there to have been two "versions," so it's likely that some form of competition favored one over the other. Biologists have some ideas about this but don't know how or when it happened.

23. Another plausible site for early life to develop is under our feet. All rocks, especially sedimentary ones, contain tiny pores rich in hydrocarbons, water, and minerals. Sterling Colgate and his collaborators have calculated that the top kilometer of the Earth's crust has enough pore space to build combinations of fifty monomers by chance interactions. Even if only one of these polymers (one out of 10^{30}) is capable of self-replication, in three hundred million years ten thousand autocatalytic polymer chains would form spontaneously. Life could develop inside rocks, immune from heavy bombardment and climate change. Even today, the subterranean biomass far exceeds the biomass in the oceans and on the land.

24. To address the criticisms and further explore the implications of the idea, Love-lock and a collaborator developed a model called Daisyworld. The Daisyworld planet has only two species: light daisies and dark daisies. Light daisies reflect light and so cool their environment; dark daisies absorb light and so have a warm-ing effect. With a realistic model of daisy growth, when the Sun gets brighter the populations of light and dark daisies adjust to maintain an optimum temperature for overall daisy growth. Daisyworld is only a thought experiment, but it demon-strates that species, operating only to ensure their own survival, combine to pro-duce a harmonious result for all. Self-regulation is an emergent property.

CHAPTER 3: EXTREME LIFE

1. Actually, early humans were fairly robust. Tens of thousands of years ago, with simple technology, they lived at the edge of polar regions and in the arid deserts of southern Africa and Asia. Other mammals live across the full range of Earth habi-tats, sometimes using hibernation to survive climatic extremes. Lower-order forms of life such as blue-green algae are, however, unrivaled in their ability to adapt and persist.

2. This is a tricky argument, since we cannot do the experiment. There is a counter-vailing view, presented most forcefully by Daniel Dennett, that convergent evolu-tion leads inevitably to similar solutions to the problem of adapting to a changing environment. Vision and flight have evolved independently at different times and in different orders of creatures. Despite this, it is indisputable that evolution on the Earth and in other cosmic settings is subject to random cosmic influences.

3. Ribosomal RNA is used as a proxy for the evolutionary relationships that are ex-pressed fundamentally in DNA sequences. Mutations occur in DNA, and it's DNA, not RNA, that's passed to subsequent generations. RNA sequences do experience mutations from transcription errors, but these are transient and not part of the evolutionary equation. Ribosomal RNA was the first and most successful tracer of phylogenetic trees, but researchers are trying to derive similar information from protein sequences as well.

4. Ernst Mayr, the evolutionary biologist who recently died at the age of one hun-dred, gave the classical definition of species: a group of actually or potentially in-terbreeding populations that are reproductively isolated from other such groups. Definitions of species often depend on appearance or morphology, and as a result there is no universally accepted definition. The answer to the question, "How dif-ferent is different enough?" is often "It depends." More problematic is the fact that the standard definition leaves out the large set of organisms that reproduce asex-ually. Recent evidence shows that humans and chimpanzees probably interbred millions of years after both separated from a common ape ancestor, further mud-dying the issue of defining when a species emerges.

5. The idea of a molecular clock is very powerful because in principle it allows the re-construction of the history of species all the way back to the last common ances-tor. But converting base-pair differences into linear time is not a trivial challenge. Molecular clocks can be sloppy or even completely unreliable. Accumulation of

genetic differences depends on population size and ecological factors as well as mutation rates. This has led to uncertainty by a factor of two in the branching time of rodents and primates, and it has sparked a controversy over the way in which species diverged before the Cambrian explosion, five hundred million years ago. For the history of life before the Cambrian explosion, when the fossil record is almost non-existent, there is no easy way to check the reliability of molecular clocks.

6. Lateral gene transfer undermines the phylogenetic tree based on DNA base pairs because bundles of genetic material are transmitted between quite different organisms. About 10 to 50 percent of the genes in the bacterial and archaeal lineages may have been transferred in the past. Bacterial genes are exchanged by cell-to-cell contact when DNA is copied from a plasmid or chromosome, and sometimes naked DNA is absorbed from the environment. We're subject to this; about forty of our genes are shared with bacteria.

7. *Nanoarchaeum equitans* has a genetic code of five hundred thousand letters, or nucleotide base pairs, compared with twelve million for common yeast and three billion for humans. This archaean microbe is a symbiont, which means it depends on the existence of another, larger microbe; the simplest parasites or symbionts probably couldn't have survived on the early Earth. The smallest viable genome, according to biophysicist Harold Morowitz, is a strain of cyanobacteria with a genetic code of 1.7 million letters.

8. Many extremophiles do not depend on the Sun's energy. All life-forms on Earth use one of two general mechanisms to support their metabolic processes: photosynthesis or chemical synthesis. Photosynthesis is based on sunlight, combining carbon dioxide and water to produce sugars and oxygen; chemical synthesis liberates hydrogen from compounds containing sulfur, manganese, iron, and other heavy elements. On the Earth, photosynthesis became the dominant mechanism because it's chemically efficient, and the supply of solar radiation is essentially infinite. Many nonphotosynthetic organisms consume carbohydrates made by plants and so create a food chain that depends on the Sun. But there have always been evolutionary niches that use chemical synthesis—in fact, on the early Earth it was the only mechanism. In a recent bizarre twist, researchers discovered photosynthetic bacteria near a deep-sea vent off the coast of Mexico, at a level far too deep for sunlight to penetrate. These microbes apparently collect and process the dull light from the hot vent itself. Life is endlessly inventive.

9. The primary survival strategy of a tardigrade is called cryptobiosis. This state of suspended animation is truly deathlike; the metabolism lowers to 0.01 percent of normal or is even undetectable, and the water content lowers to 1 percent. The creature forms something called a tun, a reduced and folded body. Tuns can survive vacuum, complete absence of water, and high doses of radiation. Cryptobiosis may have been invented very early in the history of life. Water bears may be able to teach us the tricks we will need to survive long voyages in space.

10. The deep sea remains almost completely unexplored. Three-quarters of the Earth's surface is covered by water, and 90 percent of that volume is a cold, dark environment at depths of more than one thousand meters. Of the special regions where black smokers exist, fewer than 1 percent of the sites have been visited by

submersibles. Although the ecosystem near an individual black smoker will rise and fall with the activity of the vent, in general the ocean floor is a far more stable environment than the surface of the Earth. Similar geochemistry could have existed since soon after the Earth formed, making this an excellent bet as the place life started.

11. Researchers at Lawrence Livermore Lab have used a twenty-foot-long gas gun to simulate the effects of a shock impact on comet material when it hits the Earth. The results were intriguing and somewhat counterintuitive. Even though the impact conditions created an almost instant rise of temperature to 500 °C and of pressure to four hundred thousand atmospheres, in different versions of the experiment 40 to 95 percent of the amino acids in the comet mix survived. Even more surprising, some of the collision energy created peptide chains from the initial amino acids. Violent impacts can make larger organic molecules! All of this occurred at the leading edge of the collision. Material at the trailing edge of the comet is treated more gently and would survive as a warm organic puddle.

12. Comets were the subject of an acrimonious debate in the 1970s, when two researchers claimed evidence of viruses in a comet nucleus. Without a lab sample to work with, such claims will always be suspect. The spectra of individual elements and simple molecules have unique and identifiable features, but as molecules get larger their spectra get more complex and more difficult to interpret. Life's macromolecules would be very difficult to identify with any reliability based on spectroscopy alone.

13. The bottleneck for sending rocks to another star is the time required for them to get ejected from our Solar System. It's as if the passenger has to wait ages for a train but then rapidly gets to his or her destination. Martian rocks take about thirty million years to be ejected from a solar system like ours. However, a solar system with a Jupiter-like planet near the orbit of Mars takes only a few million years to eject planetary debris. Given a range of solar-system architectures, it's likely that many of them can transmit material in less than the time we know that microbes can survive a space environment.

14. The argument from design nevertheless persists, most recently in the Intelligent Design movement, which poses a serious challenge to science instruction in American schools. The few scientists who argue for Intelligent Design are guilty of a misapplication of statistical arguments. But the persistence of the idea owes more to the surprise and bewilderment that nonscientists tend to feel when faced with the intricacies of nature and the desire of some religious denominations to impose their worldview in public schools.

15. This type of success does not imply that we are close to improving on nature with Life 2.0, at least for a while. With the artificial base pair inserted into the DNA ladder, the rate of typos increases during copying from one in ten million base pairs to one in a thousand. Researchers are confident they can improve this; the goal will be to know if our two base-pair system is the best possible solution or just the result of a fortunate accident.

16. Altering the genetic code is profound because nature has been using the same twenty amino acids since the primordial soup. It's also tricky because organisms

need fidelity in replication, so they evolved many ways of making sure that only those twenty amino acids get incorporated into proteins. Schultz and his group managed the feat by hijacking a codon that acts as a full stop in the syntax of biology and getting it to load the unnatural amino acid, so that transfer RNA would recognize it and use it to create a protein.

17. Synthetic biology's goals may sound grandiose compared to its achievements to date, but it's still a very young field, beginning with a 2004 conference at MIT, and a fifth held at Stanford in 2011. The idea of engineering a life-form is totally different from the standard approach in biology. As Knight puts it, biologists find complexity in the lab and delight in it, wanting to know how an organism works and writing papers about it. Engineers see the same complexity and want to get rid of it. Their goal is to make reliable component parts that work the same way every time. This leads to tension between the two disciplines—mutation is essential to biological evolution, but it's a headache for the bioengineer.

18. Venter kicked the Human Genome Project into high gear in 1999 by founding a company to commercialize genetic information, working in competition with the nonprofit research funded by the National Institute of Health. He was one of the five people whose genetic material was sequenced. Since 1995, over 1000 species have had their genomes sequenced; most are microbial.

19. The early history of the Game of Life mirrors the early development of computers. Conway and his friends explored its first patterns using graph paper, blackboards, and game boards with pieces. The game was so popular among early PC users that it's estimated to have chewed up more CPU cycles in the 1980s than any other type of computation. In principle, the game size is infinite. This is dealt with in practice by joining the edges of the computer screen (in software) to form a continuous, finite playing surface.

20. The amount of information in the world is increasing at a staggering rate. In 2008, the amount of new information was 500 exabytes, or 5000 billion billion bytes. This is 60 million times the information content of all the books in the Library of Congress and ten times the information content of all the words ever spoken. In the history of information, there have been four phases. For the first few billion years of life on Earth, the information content in the genome increased at a rate of only 0.01 bits per year. With humans and other advanced mammals, that rate increased to one bit per year. The invention of books and transmitted culture bumped up the rate vastly, to ten trillion bits per year. Computers have now boosted the rate by another factor of a million. In terms of the rate of processing that information, Moore's Law and its bandwidth equivalent currently stand at one billion bits per second but project to one thousand trillion bits per second, equal to the capacity of the human brain, by around 2020.

21. Robots are still hidden away in factories, but the most ubiquitous robot will eventually be your car. After decades of being mechanical and electrical conveyances, cars are getting packed with complex electronics. They will soon be entertainment and information centers, controllable by voice command. When roads are embedded with sensors to allow drive-by-wire, cars will have many of the capabilities of robots.

22. Robots are in their early, pioneering phase. Sony caused a stir in 2004 when it marketed childlike robot companions that could walk down stairs, dance, and do Tai Chi. In a head-to-head comparison of machines and organisms, the best robots have the complexity and capability of cockroaches. A group at the Swiss Federal Institute of Technology created a mechanical insect with touch sensors, learned behavior, and the ability to secrete pheromones. The robot bugs were accepted into cockroach "society" and were even able to modify the behavior of the real insects. This is a long way from the supercilious C3PO and the saturnine R2D2 of *Star Wars* fame, but a combination of Moore's Law and advances in nanotechnology will get us there more quickly than most people think.

CHAPTER 4: SHAPING EVOLUTION

1. It seems remarkable that a star can form, but all that's really required is a gas cloud localized enough to have an edge. Gas near the edge feels more gravity toward the center because there is more mass there. So it begins to move inward. As the cloud contracts, the density increases, and the inward tug increases. The result is called gravitational collapse, and in principle it can lead to star formation in as little as tens of thousands of years.

2. Meteorites condensed soon after the Solar System formed, so they contain pristine material from the epoch of formation. In the 1970s, researchers found meteorite inclusions containing xenon-129, which forms from the radioactive decay of iodine-129 in only seventeen million years. Xenon is inert and does not form minerals, so the meteorite material must have formed in the short interval between the time the iodine formed and the time it would have decayed away. Not only that, but iodine is created by the death of massive stars, so we have evidence that a nearby supernova seeded the solar nebula with radioactive material and probably led to the collapse of the gas cloud itself.

3. When different minerals condense at different temperatures, it is called a condensation sequence. As the solar nebula cooled from about 3000 °F (2000 °C), aluminum and titanium condensed first at 2400 °F (1300 °C) into metallic oxides. Iron and nickel formed mineral grains at 2100 °F (1100 °C). Next came silicates at 1900 °F (1000 °C), the basis of most terrestrial rocks. Carbon materials condensed at the much cooler temperature of 80 °F (27 °C), and hydrogen-rich molecules condensed into ices—water-ice, frozen methane, and frozen ammonia—from −280 °F (−173 °C) to −100 °F (−73 °C). Icy worlds are found only in the outer Solar System, at the orbit of Jupiter and beyond.

4. Since we think of comets as dirty snowballs, it's surprising that they're not the source of Earth's oceans. The key clue to the origin of water is the ratio of water containing deuterium, the isotope of hydrogen, to the normal form of water. In the oceans, "heavy" water is found at a rate of 150 parts per million. However, in the four comets that have come close enough for direct measurement, the ratio is twice this value. But the ocean ratio agrees with that of a common form of meteorite called carbonaceous chondrites.

5. The actual danger from meteorites is very small. There is some evidence that a Milanese monk was killed by one in 1650, and a dog may have been killed by a Martian meteorite in Nakhla, Egypt, in 1911. Apart from that, an Alabama woman had the closest call when, in 1954, a meteorite ricocheted through her house while she was asleep, grazing her hip and giving her third-degree burns. Interestingly, two houses in one small Connecticut town were hit within eleven years. In 1992, a thirty-pound meteorite smashed through the trunk of Michelle Knapp's Chevy Malibu, narrowly missing the gas tank. It fused with the car. She was offered $69,000 for the wreck. Her misfortune turned into fortune.

6. If you find yourself worried by space junk, stay inside during major meteor showers. The danger is also largest between midnight and dawn, when the Earth's spin combines with its orbital motion to give the largest chance of a projectile making it to the ground. Another good tip is to sleep standing up, advice ignored by Mrs. Hodges in Alabama (see above), since you will then present the smallest cross-sectional area to impactors, most of which arrive nearly vertically.

7. Darwin presented natural selection as a steady and gradual process, but in the early 1970s Niles Eldredge and Stephen Jay Gould developed the theory of punctuated equilibria, under which periods of rapid development of new species alternate with long periods of stasis. Catastrophic impacts cause more dramatic effects since they remove species for reasons that may not relate to their fitness to the environment. Impacts can act as a spur to evolution because they clear out ecological niches into which new species can radiate.

8. The Permian extinction rate is measured for marine invertebrates and extrapolated to all species. The severity of a mass extinction depends on whether it is defined in terms of the individual species or larger groupings of creatures. At the time of the Permian event, much life on Earth was still in the oceans, so it's possible that a dramatic geological or climatic change could have caused the extinction; no crater has been found. To support an impact hypothesis, it would have to be shown that the decline in diversity was essentially instantaneous, and this is limited by the time resolution of geological layering.

9. After some controversy in the scientific community, evidence for periodicity in the extinctions has not proved convincing. Since major impacts are random, not regular, the fact that the last one took place sixty-five million years ago doesn't mean we have thirty-five million years until the next one. If you take cards from ace to ten out of a pack, shuffle them repeatedly, and try to draw the ace, on average it will come up every ten draws. But it might take thirty draws or forty or it might take only one. The next impact can happen anytime.

10. Supernovae and hypernovae have been implicated in several mass extinctions, but a "smoking gun" has never been found. The star that emits the lethal radiation leaves only a pulsar or a black hole and a diffuse expanding nebula, which eventually dissipates. Astronomers can't project the expansion back accurately enough to pinpoint when the explosion occurred. Complex stellar motions in the Milky Way confound any attempt to reconstruct our immediate environment more than one hundred million years ago.

11. Geological evidence points to a relatively rapid rise in the oxygen abundance

about 2.3 billion years ago. In particular, "banded iron" formations (which are the source of 90 percent of the world's commercial iron) and reactive minerals like uraninite (UO_2) and pyrite (FeS_2) point to the earlier time when there was little free oxygen in the atmosphere or oceans. This leads to a puzzle, since the first oxygen-producing cyanobacteria in the fossil record date back 3.5 billion years. Various mechanisms have been proposed to explain why the oxygen was being removed from the atmosphere for the first billion years of photosynthetic life. The timing of the rise correlates with evidence for a severe "snowball Earth" episode.

12. The growth, coalescence, and breakup of continents has likely had important effects on the concentration of atmospheric carbon dioxide and oxygen, with corresponding impacts on the biosphere. When there is a single supercontinent, atmospheric carbon dioxide reduces because there is less volcanism and weathering, and as the supercontinent breaks up, greenhouse gases are released by increased volcanic activity. But after the breakup, the buildup of mountain ranges and the increase in rainfall and erosion remove carbon dioxide from the atmosphere. The creation of a supercontinent buries organic carbon that could otherwise soak up oxygen, thus allowing that gas to build up in the atmosphere.

13. Actual conditions during Snowball Earth—the mean temperature, and whether or not ice covered all of the oceans—are still uncertain. It's difficult to make reliable models for climate change when so many parameters are varying dramatically. However, most scientists are convinced that the Earth suffered profound climate change over a period of 250 million years. Taken as a whole, the geological evidence—worldwide glacial deposits, with carbonate layers above them from the melting phase, and banded iron formations pointing to a fluctuating level of oxygen—is convincing.

14. Interpreting transitions in the biosphere or events like mass extinctions is difficult because many things are going on at once and the geological record is imperfect. Major volcanic events line up with three mass extinctions in the past 250 million years, and the odds of that happening by chance are 0.01 percent. (To confuse matters further, one is the K-T event.) Impacts are attractive as an explanation, but there are extinctions without a "smoking gun" crater and huge craters with no matching event in the fossil record. In addition, there is evidence that pre-Cambrian evolution occurred at a snail's pace over hundreds of millions of years, so some bolides must fall without ruffling the biosphere.

15. Lyell and his colleagues subscribed to a view called uniformitarianism, which held that all geological and biological changes were due to steady, inexorable forces of nature. An old Earth was a natural implication of this view. The counter view, put forward by biblical literalists, was called catastrophism, which held that geological upheavals and extinctions were disasters inflicted by a wrathful God (like the Flood). Ironically, modern science recognizes that catastrophes do play an important role in evolution.

16. Darwin worked at the macro level, observing plants and animals. The rules for genetic variation were not appreciated until Gregor Mendel's work was unearthed in 1900, and the microscopic mechanism for coding genetic information was not revealed until the mid-twentieth century. Darwin's theory was incomplete, and it

had a shadow cast over it by physicist Lord Kelvin, who argued that the Sun was only a few hundred million years old. Darwin suspected that larger spans of time were required to enable the diversity of species.

17. It's important not to oversell the explanatory power of evolution for individual adaptations such as eyes and wings. The fossil record is too fragmentary to follow the trajectory of morphological changes. Fossils speak to form but not necessarily to function, so paleontologists need a lot of inference to talk about the lifestyle and environment of an organism. But the specific functional attributes of creatures as diverse as bats and sharks and snakes are magnificently adapted to the constraints and opportunities of their environments.

18. Natural selection is a game anyone can mimic. Take sheets of blue and red paper and cut them into several hundred small squares or cards. Shuffle a large stack of equal numbers of blue and red cards. Lay them out in a ten-by-ten grid. Now get a friend to play "predator." They eat cards from the grid by removing them, but since blue cards are slightly tastier or easier to catch than red cards, they remove five blue cards for every four red cards. Meanwhile you replace the missing card from the stack in your hand. Even though the selective advantage of being red is only 20 percent over that of being blue, as the game goes on the proportion of red cards in the grid steadily increases until being red becomes a winning strategy.

19. Convergence should not be confused with homology, where animals share a trait due to descent from a common ancestor. Chordates—which include vertebrates such as fish, birds, amphibians, and mammals—all have camera eyes because they share a common ancestor. The twin eyes evolved differently in different species, favoring night vision, black and white perception, color perception, positioning on either side of the head for prey, or side by side for predators, and so on. Many animals have five fingers, but the fingers evolved for digging in an armadillo, gripping in a salamander, and manipulating in a human.

20. Harold Morowitz has no idea what alien life might look like, but he suspects it will use the citric-acid cycle, which is fundamental to terrestrial metabolisms. From a database of 3.5 million organic molecules, Morowitz applied the six rules of the cycle and quickly homed in on a small subset that contained all the actual compounds used in the cycle. They are evidently not randomly selected. And macromolecules such as chlorophyll may also represent widespread or even universal solutions for biological energy capture.

21. Symbiosis is another mechanism that stands apart from Darwin's theory of natural selection. However, the sequence of cooperation, dependence, and then incorporation among organisms has been crucial in the evolution of life. The actual origins of symbiosis are difficult to extract from the genetic record, and there are many ideas on how it might have occurred. But we live with the consequences—the human body hosts hundreds of types of bacteria that are essential for healthy functioning, and viruses install genetic material in humans constantly, sometimes in beneficial ways.

22. As with earlier stages in evolution, the rapid diversification of animals did not require a corresponding increase in the complexity of the genome. A handful of hox genes control the shape and arrangement of body modules. Switch off a sin-

gle hox gene in a housefly, and legs grow where antennae should be; with incorrect function of another hox gene in an infant, flippers will grow instead of hands. These genes are responsible for the segmented bodies on some animals, for the bilateral symmetry in vertebrates, and for the fact that we have fingers and toes. Life on Earth isn't as strange as we might imagine because of this genetic parsimony.

CHAPTER 5: LIVING IN THE SOLAR SYSTEM

1. As writer and astronomer David Darling has pointed out, Ward and Brownlee were influenced in their thinking by their former colleague Guillermo Gonzalez, who coauthored *The Privileged Planet*, a book that explicitly connected the Rare Earth hypothesis to Intelligent Design. Evidence may eventually show that intelligent life is rare in the universe, and scientists are entitled to hold religious convictions, but the science of astrobiology doesn't support any particular theistic or spiritual framework.

2. Ward and Brownlee use a version of the Drake Equation, a set of multiplicative factors that combine to give the fraction of Earth-like planets in the galaxy that have all the special conditions they believe are needed for intelligent life to develop. Ten fractions multiplied together inevitably leads to a very small number of predicted Earths. But there may plausibly be pathways to long-lived complex life that do not require all of the factors to apply, in which case this calculation might be unduly pessimistic.

3. Planets on tight orbits around their parent stars will experience a phenomenon called tidal locking, which is responsible for the fact that we see the same face of the Moon all the time. Over time, the star acts as a brake on the rotation of the planet until it shares the rotation of the star. Such a planet would have one side basking in heat and red light and the other plunged into cold and permanent darkness. A sufficiently thick atmosphere might smooth out this variation and make the planet habitable.

4. Orson Welles, who was a brilliant twenty-two-year-old actor and impresario at the time, apologized publicly for the effect of his broadcast, but he was probably being disingenuous when he declared himself "stunned" by the reaction. He had deliberately reached for a hyperrealistic style of simulated news broadcast and knew that anyone tuning in late would probably be fooled. Welles was an iconoclast throughout his turbulent career—the creator of several brilliant movies but often at odds with the Hollywood establishment.

5. Lowell had been inspired by reports by Italian astronomers of *canali*, or channels. When this story was picked up by American newspapers, the word was translated as "canals," with the implication of intelligent engineering. In those early days of photography, telescopic observations were made with the naked eye. Features would shimmer and come and go due to circulation in the atmosphere, but in the still desert air the features sharpened, and it's easy to understand why Lowell was fooled into seeing linear markings. He knew that his telescope didn't have the res-

olution to see the canals themselves, so he presumed the dark marks were strips of irrigated vegetation next to the canals.

6. The gullies evaded detection by previous orbiters because they are very delicate features. The Viking orbiter had mapped the entire planet in the late 1970s with a resolution of about two hundred meters, twice the size of a football field. The camera on the MRO has a resolution of a meter, the size of a small boulder. This gain of a factor of two hundred made a vast difference in the ability to resolve narrow channels and see subtle features of erosion and sedimentation.

7. Without rocks to take into the lab and subject to radioactive-dating techniques, astronomers can measure the age of a planet or moon surface only by counting craters. The idea is simple, although there are subtleties to take into account to get reliable answers. In general, surfaces get more cratered with time. In the first half-billion years of the Solar System, large and small craters formed. Then only small craters formed, and more recently very few craters formed because the impact rate was low. On Mars, the large craters are all found in two regions, so these must be the oldest surfaces since they bear the imprint of the epoch of heavy bombardment. The northern half of Mars has few craters of any kind, so it's the youngest surface. Sure enough, that's the part of Mars with volcanoes, indicating that lava flows eradicated craters, and few have formed since.

8. Even though this example seems like a slam dunk, some scientists have argued that it doesn't yet prove water beyond a reasonable doubt. Geologists have noted pale deposits with a similar shape on the Moon, which is unarguably dry. By far the best explanation for the Mars gully changes is layering by a slurry of water and dirt and gravel, but it might take spectroscopic evidence or a runoff "caught in the act" for us to be completely certain.

9. A spectrometer aboard Opportunity found that the outcrop had iron-rich minerals that are found on Earth only through the action of water. Also, there was a lot of sulfur, suggesting that 40 percent of the rocks were made of magnesium sulfate (or Epsom salts). Bromine and chlorine are also present, and their abundance relative to sulfur in the outcrop changes in a way that suggests a sequence in which the water evaporates slowly and the minerals become saltier and saltier as the level falls. The pattern of sediments suggests a lake.

10. The "salty sea" is another good example of how hard it is to prove a hypothesis with limited evidence. Steve Squyres and the Mars rover team claimed that the geology on the Meridiani Plain was suggestive of sandstone that had been steeped in acidic, salty water that later evaporated. But two other teams proposed that the formations could be explained by a meteorite impact or a massive volcanic eruption laying down a bed of ash. Since the formations are 3.5 billion years old, deciding between these ideas may be very difficult.

11. The steady loss of Mars's early atmosphere would have taken about one billion years, after which Mars was cold and dry. However, some examples of glaciation, sedimentation, and surface water date to times after the first billion years. This is a challenge for the Martian models because carbon dioxide cannot contribute to global warming once it has escaped into space. Perhaps local conditions are conducive to surface water, as in the gully formations. Impact heating could lead to

local, temporary oases. Liquid water has probably been in the Mars crust throughout its history.

12. The evidence for the origin and fate of Martian water comes, as on Earth, from the ratio of water that contains the hydrogen isotope deuterium to normal water. As molecules bounce around in an atmosphere, the lighter ones move faster and tend to escape, while the heavier isotope molecules get left behind. The current ratio of heavy to light water in the Martian atmosphere is three times higher than the value in ancient Martian meteorites, which indicates that much of Mars's water has been lost in space.

13. One of the controversial aspects of the ALH 84001 research was the argument that, while each of the lines of evidence could have an alternative, nonbiological explanation, the combined implication of the arguments made the claim more secure. This is not persuasive because each of the pieces of evidence must stand on its own merits; if they all fall away, we're left with nothing. The bar should be set high for all claims of life. Carl Sagan once observed in the *Cosmos* television series, "Extraordinary claims require extraordinary evidence."

14. Methane was detected in the Martian atmosphere by Mars Express in 2004, but it has become a case in point of how hard it is to be sure of biomarkers. Methane on Mars would be zapped by UV radiation in a few hundred years, so it must be constantly replenished. On Earth, much methane comes from biological sources (such as ruminating cows). However, the concentration on Mars is tiny—one methane molecule for every one hundred million carbon dioxide molecules; at that level, it could have several types of nonbiological origin.

15. Mars sample return leads to the concerns of forward contamination—terrestrial microbes surviving the journey to Mars—and back contamination, Earth being infected by Martian microbes. The second issue sounds more ominous, but safety procedures have been in place since the Apollo astronauts brought back Moon rocks. The first issue is a bigger worry for scientists; once contaminated, a pristine world may be changed forever. A 1967 UN treaty addresses planetary contamination, and NASA has a senior official with this portfolio, someone with the impressive title of planetary protection officer.

16. While some may find the prospect chilling, the industrial approach to terraforming is cost-effective and involves a modest extrapolation of current technology. Self-replicating machines or factories may be only a few decades away. It's a cheap solution because only the prototype has to be shipped—all of the other costs are local. Machines like this could even take on the tough job of terraforming Venus, after the planet had been prepped by impacting it with asteroids to eject the thick atmosphere and by installing a sunshade. No, really.

17. Global warming has turned into a political hot potato in the United States, but it's a scientific issue. We must distinguish the questions of whether or not the Earth is warming (it is), whether or not human activity is primarily to blame (not decided, but increasingly likely), what the effect of the heating will be (climate-change models are still not reliable, but most of the outcomes are bad), and what we want to do about it. This last is a societal issue, but choosing not to spend money answering the first three questions would be myopic and a gross disservice to future generations.

18. The core of the Viking was a sophisticated science package to look for biological activity in the Martian soil, but it was Sagan who argued for a camera at the last minute, in case there were "Martian polar bears." He was joking, but the camera provided important geological insights, and the evocative images took root in the public imagination. He also published a famous paper with Ed Salpeter speculating on the possibility of large buoyant organisms circulating in the temperate zones of Jupiter's atmosphere. David Grinspoon's speculation about Venusian cloud life has an echo of this earlier idea.

19. Galileo was one of the most successful missions in NASA's history, despite being cursed by a primary data antenna that stubbornly refused to unfold. NASA engineers figured out a workaround. Launched in 1989, it carried out a fourteen-year exploration of Jupiter and its moons before being interred in the atmosphere of the giant planet. Along the way, Galileo made the first two flybys of asteroids, and it also photographed Comet Shoemaker-Levy's breakup and death plunge into Jupiter in 1994. In its extended Europa mission, Galileo got within two hundred kilometers, taking pictures that could see objects as small as six meters across.

20. This sounds outrageous, but magnetic sensing is seen across 3.5 billion years of evolution of life on Earth, up the tree of life from bacteria to higher vertebrates. Orientation and navigation by microorganisms along magnetic-field lines offer an advantage over random motion, so there is selective advantage in having the magnetic sense. Rather than a quirk or fluke, magnetic sensing may be an ancestral trait of all animals, and use of this tool could plausibly be widespread (and even more sophisticated) in life-forms beyond Earth.

21. Evidence of water based on modeling moons and planets is very indirect and should be taken with a pinch of salt. (In fact, dissolved salts may help keep water liquid at temperatures below freezing.) However, these models are now quite sophisticated, and they usually indicate a subsurface zone of water, heated from the bottom by volcanic vents in the larger moons. The crust of ice or rock provides a perfect insulating layer and protection from cosmic rays.

22. Enceladus was considered too tiny to be geologically active, but it has warm spots, and it is spewing out plumes of salty ice crystals hundreds of miles high. Io is also tiny, but it has prodigious volcanism caused by a gravity squeeze from Jupiter. It resurfaces itself with several inches of sulfuric lava every year and might have "aquifers" of liquid sulfur dioxide. David Grinspoon has mused on the potential of Io to support life. He notes that sulfur has a complex chemistry with other available elements, it can store energy, and under certain conditions it can form polymers. Once more, we're limited by our imaginations and the experience of our own carbon chemistry.

CHAPTER 6: DISTANT WORLDS

1. Three years before the Mayor and Queloz discovery, Alex Wolszczan used high-precision radio-timing techniques to find three terrestrial planets orbiting a pulsar. This early work has been unfairly neglected because the planets probably formed from debris left over after the death of the massive central star, and since

pulsars are shrunken neutron corpses that emit no light the planets are unlikely to be habitable. But the existence of pulsar planets reminds us that planets can form in the strangest places.

2. Marcy and Butler used some open time on the Lick Observatory three-meter telescope to quickly confirm the discovery of Mayor and Queloz, who were based at the Geneva Observatory. However, the Swiss team published their work in *Nature*, which imposes a gag rule on authors so that the journal can have the scoop. For six weeks, Marcy and Butler were media stars, appearing on TV and the front pages of major papers, while the Swiss were held silent. Even though the Americans were careful to give proper credit, this created some hard feelings.

3. Jupiter exerts the same gravity force on the Sun that the Sun exerts on Jupiter. The giant planet doesn't simply orbit the Sun; rather, both bodies orbit a common center of mass. Jupiter is one thousand times less massive than the Sun, so the center of mass is one thousand times closer to the Sun than to Jupiter. (You can think of balancing a meter rule with the Sun at one end and Jupiter at the other—the balance point and the center of rotation will be one millimeter from the Sun's end.) Orbital velocity is inversely proportional to mass. Instead of trying to measure Jupiter's orbital velocity of thirteen kilometers per second, astronomers must measure the thousand times smaller orbital velocity of the Sun. This is called a reflex motion.

4. One crucial advance used at Lick Observatory was a cell of iodine gas inserted in front of the telescope, which imprinted a series of narrow spectral lines on the star spectrum, acting as a wavelength reference. Also, the optics of the spectrograph produced features as sharp and well dispersed as any in history. Marcy and Butler then used software to measure spectral features to within one-thousandth of a pixel on the CCD. This last method was needed to improve from a precision of kilometers per second to meters per second.

5. Planet orbits are scattered at random orientations or inclinations in the universe. If a planet happens to be orbiting directly in the line of sight, an observer sees the full effect of the motion. If the planet happens to be orbiting transverse to the line of sight, there is no Doppler effect because no part of the motion is toward and away from the observer. In an individual case, analysis of the Doppler curve can give only a lower limit to the mass. (To be precise, it measures $M \times \sin i$, where i is the inclination angle.) With a large sample, a Doppler method underestimates the range of planet velocities, and therefore the planet masses, by a factor of two.

6. Planets on very tight orbits imprint Doppler variations on timescales of days to weeks. That had led to a claim that some of the signals detected might actually be due to stellar pulsations, where the star itself is wobbling like a water balloon. For existing exoplanets, the claims have been mostly rebutted, but as the precision of the Doppler technique reaches one meter per second or better, there may be a planet detection "floor" imposed by complex gas motions in the stars being observed. This will set a practical limit to the lowest-mass planet that the Doppler method can discover.

7. Amateur astronomers play an increasingly important role in advancing the subject. Many are engineers with a high level of technical skill, and a few thousand

dollars buys a CCD camera that's close to research grade. In addition to locating exoplanets, amateurs have generated important data on variable stars and near-Earth objects. Paul Comba of Prescott, Arizona, has discovered more than fifty asteroids, and Robert Evans, a pastor in New South Wales, Australia, has discovered more than forty supernovae in galaxies beyond the Milky Way, more than any other astronomer, amateur or professional.

8. Lensing is an extremely subtle effect. Even a galaxy deflects light by no more than an arc second, which is the angle between the two sides of a quarter seen at a distance of a kilometer. A star would deflect distant light by the angle between the two sides of a human hair seen at that same distance, and the deflection for a planet would be even less. Luckily, the temporary amplification is stronger—at least 30 percent and up to a factor of ten. Since lensing doesn't depend on the brightness of the planet or its distance from a star, it can be used to detect planets thousands of light-years away.

9. The first microlensing survey was developed for a very different purpose: to understand the nature of the dark matter that holds the Milky Way (and other galaxies) together. Several dozen events were seen, and all the lenses were low-mass stars, such as white dwarfs. The survey found that about 20 percent of the extended halo of the Milky Way is in the form of dim stars, meaning that 80 percent is in the form of exotic and mysterious subatomic particles, of a kind not yet observed in any physics lab. The nature of this dark matter is one of the biggest enigmas in astrophysics.

10. Paczynski was given little hope for recovery but opted for brain surgery, a procedure about which doctors do not give any guarantees. The delicate operation was a success and Paczynski continued to beat the odds—teaching, lecturing, and combing the skies for things that go bump in the night. Unfortunately, the tumor recurred and he died in 2007 while this book was first in production.

11. Computational situations involving the gravity of many objects are called n-body problems. Techniques are similar whether the calculation involves particles in a forming solar system, stars in a globular cluster, or galaxies in the expanding universe. If there are n objects in a simulation, the number of forces that must be calculated is $\frac{1}{2} \times n \times (n-1)$, which goes up by n^2 when n is large. In practice, various tricks are used to ensure that the number of calculations scales no faster than n, such as the fact that the gravity force of the more distant objects can be ignored, since gravity diminishes as the square of the distance.

12. Spitzer works at infrared wavelengths, so it is very sensitive to cool dust and gas. While Spitzer does not directly detect planets, it has made detailed observations of the disks from which stars and planets form. In several cases, there are gaps in the disks that point to places where giant planets have swept up material. The stars are young, so the planets must be at least that young. Spitzer has also found indications of planets around brown dwarfs, or failed stars, affirming that planets form in all types of stellar environment.

13. Astronomers used the Spitzer Space Telescope to detect infrared radiation from two additional Jupiters on much smaller orbits, but the telescope did not have the resolution to image the planets, so a subtraction technique was used. Both plan-

ets were previously discovered by optical transits. First, they measured the infrared radiation from both star and planet. Then, when the planet dipped behind the star, they measured the radiation from just the star. The difference between the two was the radiation from the planet.

14. The angular resolution of a telescope, also called the diffraction limit, sets the scale of the finest feature it can see. In arc seconds, the angular resolution is given by 250,000 × (wavelength/diameter). A twice bigger telescope makes twice sharper images, and a telescope makes twice sharper images with blue light than with red light. The 2.2-meter Hubble has a resolution of 0.05 arc seconds, but the ten-meter Keck telescope has its theoretical resolution of 0.01 arc seconds degraded to 0.5 arc seconds or worse by turbulent motions in the Earth's atmosphere. Recently, Keck has used interferometry to recover its ideal resolution.

15. Angel has proposed to send one million or more gossamer-thin light deflectors into Earth orbit using space guns and then shepherding them into a loose formation at a gravitational balance point between the Earth and the Moon. The robotically controlled array would deflect just enough sunlight to mitigate the effect of the next fifty years of anticipated global warming. He doesn't flinch in estimating the cost—a trillion dollars—which he points out is a small fraction of the industrialized world's GNP over the next twenty years.

16. Radio astronomers have used interferometry for decades, in part because radio waves are hundreds of thousands of times longer than light waves. Interferometry is also being developed on the two ten-meter Keck telescopes and the four eight-meter telescopes of the VLT. Keck and the VLT have longer baselines and therefore higher resolution than the LBT—sixty and two hundred meters, respectively. Both facilities have also built a set of small "outrigger" telescopes that combine light from different angles to get smoother, rounder images.

17. Planet hunting is one of the most exciting uses of SIM, but the mission will transform other aspects of astronomy by measuring stellar distances using the parallax method with an accuracy hundreds of times better than current measurements. SIM will measure the masses of all types of stars from supergiants to brown dwarfs, map out the structure and mass of the Milky Way in great detail, detect binary black holes, and probe the central regions of distant quasars. That's not bad for a telescope with a one-foot aperture.

18. The first reliable measurement of the speed of light was made in 1676, soon after Galileo's death, by the Danish astronomer Olaf Romer. He carefully observed the innermost moon of Jupiter, which zips around the giant planet in two days and is eclipsed once each orbit. When the Earth was farthest from Jupiter, the eclipses were about eleven minutes late and when the Earth was closest to Jupiter, six months later, they were about eleven minutes early. This means light covers the radius of the Earth's orbit in about eleven minutes.

19. Particle energy comes from Einstein's famous equation $E = mc^2$. Since c is such a large number, a tiny amount of mass contains a vast amount of energy. To see the dramatic difference between chemical energy and mass energy, consider a hamburger. If you eat a quarter-pound hamburger, the 250 calories you gain are equal to a million joules, or enough energy to keep a one-hundred-watt lightbulb

lit for three hours. If you could tap the full mass-energy of the hamburger, it would be 10^{16} joules, or enough to power a small town for a year. For comparison, sending a Shuttle-sized payload to a nearby star on a one-way subrelativistic journey would cost 10^{20} joules, ten times the U.S. annual energy consumption.

20. As any object approaches the speed of light, its mass increases. This bizarre effect of special relativity is easy to observe in subatomic particles. The denominator of the mass term involves the square root of $1-v^2/c^2$, which becomes very small as v gets close to c. Energy put into accelerating the object goes increasingly into raising the mass instead. As a result, the speed of light is an absolute barrier—it would require an infinite amount of energy to make any object, including subatomic particles, move at the speed of light. Still, NASA hosted a workshop on faster-than-light travel in 1994.

21. A survey of a thousand households in 1995 found that 60 percent of them would go to space for a vacation, and one in five would spend a year's salary to do so. At the Virgin Galactic price tag of two hundred thousand dollars, that's six million U.S. households participating. If the cost came down to fifty thousand dollars, the number is twenty-five million. The entertainment analogy isn't far-fetched. If these households bought only one ticket every thirty years at the current projected price, the annual revenue would be ten billion dollars per year. This matches the box-office receipts of Hollywood, which sells about a billion ten-dollars in tickets each year.

22. By going into hibernation, interstellar travelers will experience dislocation in time as well as in space, since they will be out of synch with the human culture they left. In the Inquisitor War science-fiction novels by Ian Watson, the last thought a person has going into suspended animation stays with them the entire voyage—either a torture or a blessing, depending on the thought.

CHAPTER 7: ARE WE ALONE?

1. The classic Fermi question is one he asked his students: how many piano tuners are there in Chicago? It seems impossible to do much better than a guess, but by estimating the number of people in Chicago and the fraction of families that have pianos, and by assuming that the average piano needs tuning once per year and that a tuner can work on three pianos per day, you can get an estimate of about one hundred, which is not far off the number listed in Chicago's yellow pages. The story of Fermi's "Where are they?" conversation was first told by Eric Jones in the pages of *Physics Today* in 1985.

2. Explaining the absence of something opens up the realm of possibility enormously compared to prosaic science, which has to account for particular observations. Stephen Webb wrote a book detailing fifty possible solutions to the Fermi paradox, and there are dozens more in the technical and popular literature. Webb divided solutions into three broad types: those that say ETs are here and we've made contact; those that say they exist but communication and interstellar travel are difficult or not pursued; and those that give reasons why ETs do not exist.

3. Supernatural beliefs are pervasive. The strongest are tied to religion—in a 2005 Fox News poll, 85 percent of Americans believed in heaven and 82 percent believed in miracles. A 1996 Gallup poll found that half of adults believe UFOs are evidence of alien contact, 5 percent have seen one, and 71 percent think that the government knows more than it is letting on. By contrast, and to the dismay of scientists and educators nationwide, the belief in (and understanding of) the bedrock theories of evolution and the big bang hovers down around 30 percent.

4. Some of the belief subcultures are undeniably fascinating. Any Freudian psychiatrist would be intrigued by the stories of probe insertion and experimentation from people claiming to be alien abductees. Belief is resilient; there are still web sites and books about crop circles as alien messages, fifteen years after the two English blokes who dreamed up the hoax owned up to it. *Science News* convincingly debunked crop circles by creating a large number of replicas of a predetermined pattern in the middle of the night. Thirty-five years after convicted felon and confessed fraudster Erich Von Däniken sold millions of books with purported evidence for ancient astronauts in the historical and archeological record, the same hackneyed ideas are being sold to a new generation of believers.

5. This idea was first proposed by John von Neumann. He used the separate notions of a constructor, which can manipulate matter in its environment to make copies of itself, and the program, which contains instructions for replication. There's an obvious parallel with the mechanisms of life itself.

6. One chilling possibility that takes its cue from a science-fiction short story is called the "berserker" scenario. Imagine a fleet of malignant von Neumann probes that swoops in on fledgling technological races and obliterates them. We would then be alone and the airwaves silent for a chilling reason. No sane civilization would actually create berserkers, but they might result from a software mutation in more benign probes.

7. The special conditions required to make carbon were first noted by astrophysicist Fred Hoyle fifty years ago. Making carbon is tricky for a star because it first must combine two helium nuclei to make beryllium, which decays incredibly quickly with a half-life of 10^{-16} seconds. The alternative is fusion by a rare triple collision of three helium nuclei. The only reason this works is the existence of a particular resonance state of the carbon nucleus that allows the three helium nuclei to "stick." Another special resonance state allows some carbon to be turned into oxygen, but not so much that it's all used up. These examples of fine-tuning led Hoyle to speculate in an article titled "The Universe: Past and Present Reflections" that "a superintellect was monkeying with the laws of physics!"

8. The anthropic principle was proposed in its modern form by Brandon Carter, who originated the equally intriguing Doomsday argument. He presented it at a meeting celebrating the five-hundredth anniversary of the birth of Copernicus, and it is provocative in part because it appears to subvert the Copernican idea that we have no special place in the universe. In fact, Carter never said that life or humans are central or pivotal, but he did make the milder statement that our situation is "inevitably privileged to some extent."

9. Without an underlying physical theory, the mere hypothesis of many universes

does not account for the peculiar properties of this one. To believe this is to fall prey to what's called the inverse gambler's fallacy. The gambler's fallacy is the belief that a particular situation, like throwing a double six in craps, is more likely if it hasn't happened for a while and less likely if it has just happened. Vegas makes a lot of money on that one. The inverse gambler's fallacy is using the fact that a double six was just thrown to infer a long run of throws. We live in a special—double six—universe, but the odds of it happening are always one in thirty-six, even if it happens only once; the supposition of multiple universes does not explain it.

10. M-theory got its name from Ed Witten, a brilliant Princeton physicist and winner of the Fields Medal, the National Medal of Science, and a MacArthur Fellowship. He gave no particular guidance to what the "M" stands for; suggestions have included Mystery, Magic, Membrane, and Missing. M-theory is incomplete, and its implications are still being uncovered, but it appears to be the most promising theory for understanding the underlying nature of matter and space. It has already led to new insights into black holes and gravity.

11. The Drake Equation has been recast a number of times over the years, most usefully by Jonathan Lunine, in a form that mirrors the gum analogy. Lunine's version starts with the number of stars in the galaxy and multiplies it in turn by the fraction of stars with heavy elements similar to the Sun's; the fraction of stars of mass suitable for longevity and large habitable zones; the fraction of such stars with terrestrial planets; the fraction of those planets that are continuously habitable; the fraction of those where life does arise; the fraction of those where complex or eukaryotic life develops; the fraction of biospheres where intelligence emerges; the fractional lifetime of a technological civilization compared to the age of the galaxy; and the fraction of those civilizations that choose to communicate.

12. A colonization argument based on von Neumann probes applies even if the galaxy is sparsely populated. Once a civilization decides to explore or colonize with self-replicating probes, it will traverse the galaxy in a small fraction of the galaxy's age, regardless of whether the probe speed is 10 percent or 0.1 percent of the speed of light and for any reasonable model of diffusion of the probes. Even if civilizations are rare, this diffusion is likely to have first happened long ago, though the civilization that spawned the probes might now be extinct.

13. Self-selection rears its ugly head here: we cannot use our existence to argue for the inevitability of our existence. If lab experiments reached the point where they could simulate prebiotic conditions on Earth and other likely terrestrial environments and show whether or not life emerges in a statistical sense, that would be an independent handle on f_l. This factor may be measurable within ten to fifteen years by the direct inspection of the atmospheric composition of remote terrestrial planets.

14. Insect colonies display amazing variety and richness of behavior, which should make us cautious about defining the capabilities and limitations of other forms of intelligence. Some researchers have argued that simple communication, computation, and intentionality in plants reflect a kind of intelligence, and others observe that microbial communities have analogs of cooperation, division of labor,

communication, and sociality. Observers would see a complex and coordinated social life.

15. This discussion glosses over the role of consciousness. While it's easy to imagine selective evolutionary pressure that can lead to large brains capable of planning and strategy, the survival value of self-reflective awareness or consciousness is not as clear. Do increasing brain complexity and its feedback in behavior lead inexorably (and inevitably) to Descartes's internal "theater of the mind" that frames an external world and allows us to contemplate our place in the universe? Nobody knows the answer.

16. Careful experimental design is crucial because it's very easy for human experimenters to give visual cues to their animal subjects. But intelligence has turned up in some surprising places. The corvid bird family—which includes crows, ravens, and jays—has long been known for resourceful behaviors. For example, crows living in cities use cars to crack nuts for them, waiting with pedestrians for the red light that will let them retrieve the food safely. A New Caledonian crow named Betty fashioned an unfamiliar object, a stiff piece of wire, into a hook to snag a small bucket of food. This level of spontaneous problem-solving rivals or exceeds the strategies used by apes. Both types of animal live in complex social communities.

17. Bronze was more malleable and less brittle than naturally occurring stone or iron; it led to new tools, new weapons, and the plow, enabling humans to harness animal power for the first time. Bronze is an alloy of copper and tin, both of which are found naturally in almost pure form, but it's stronger than each. This makes sense in terms of modern atomic theory, where atoms of tin act as "grit" within the lattices of copper atoms, preventing them from sliding over one another and stiffening the material dramatically.

18. The assumption that there's nothing special about the timing of observation is crucial. Gott tested his idea with an article in *The New Yorker* in which he predicted the longevity of forty-four Broadway shows, with excellent accuracy so far. You could apply the argument to estimate the longevity of the marriage of a random couple you met at a wedding, but not to the new bride and groom, since you're observing them at the beginning, by definition. The Copernican principle is used to say there's nothing special about our current status in terms of the overall history of life in the universe.

19. Many people get uneasy hearing this argument for the first time, thinking there must be something wrong with it. Brandon Carter, Nick Bostrom, and many other philosophers have weighed in on the validity of the reasoning, and the obvious critiques have been rebutted. In any exponentially growing category that ends suddenly, you are more likely to be near the end than the beginning. But if humans endure by transitioning to a postbiological future, then the reference class and the logic of the argument change.

20. Spiegelman was working in the artificial conditions of the lab, but the naturally occurring potato spindle tuber viroid is not much larger; it's a loop of DNA made of 359 bases. For comparison, a simple bacterium might have ten million bases, and the human genome consists of five billion bases. Prions are about the size of

small viruses and are immune to boiling, the effects of acid, and UV radiation. Made of protein, they're the only self-reproducing biological entities that have no DNA or RNA. However, they are not viable on their own so they cannot be the basis for life.

21. Dreams of communication with ETs go back even farther. German mathematician Karl Gauss proposed clearing huge areas of Siberian forest into a triangle with adjoining squares, to show Martians we know the Pythagorean theorem. Austrian astronomer Joseph von Littrow wanted to fill a circular trench in the Sahara Desert with kerosene and ignite it. Other astronomers suggested constructing mirrors and flashing signals in Morse code. Speculation about life on Mars spurred these ideas, none of which was carried out.

22. It's easy to poke fun at the SETI messages, but their anthropocentric flavor shouldn't obscure a major strength of the approach—pattern-recognition techniques are very powerful in their ability to discriminate noise and information, even if the meaning of that information is not clear. Artificial signals of intelligent intent will stand out from all known natural processes. Even the goal of communication might not be out of reach. In 1960, Dutch mathematician Hans Freudenthal proposed "Lingua Cosmica," a signaling strategy based on the hopefully universal attributes of logic and mathematics.

23. From Oparin's work on the origins of life to Safronov's insights into planet formation, the Russians have been influential through the history of astrobiology. Konstantin Tsiolkovsky was a deaf, self-educated rural schoolteacher, yet he made brilliant designs of rockets and orbiting space stations years before the Wright brothers' first flight. He was a proponent of the Russian philosophy of Cosmism, which held that space travel was part of a utopian vision for mankind. Iosif Shklovskii hosted the first international SETI conference in 1971, and it was the annotation and expansion of his 1962 book by Carl Sagan that opened the eyes of many U.S. scientists to the potential of astrobiology. Kardashev was Shklovskii's star student.

24. The English evolutionary biologist Richard Dawkins, in particular, has become a lightning rod for the science-religion wars with his 2006 book, *The God Delusion*. Regardless of your personal views, it's indisputable that nonreligious, agnostic, and atheist people are a seriously underrepresented minority in the United States. A 1999 Gallup poll asked voters if they would pick as president a well-qualified member of their party if they had various attributes. Women, blacks, Catholics, Jews, Baptists, and Mormons came in over 90 percent approval, 59 percent would vote for a gay candidate, and atheists languished in last place at 49 percent.

25. The precursor of biblical allegory in science-fiction film was the 1951 classic *The Day the Earth Stood Still*. An alien who chooses the name Carpenter (Christ's profession) comes to Earth with great powers but a message of peace. He is misunderstood, apparently killed, and resurrected. He ascends to the heavens at the end of the film.

GLOSSARY

Accretion: In the history of the Solar System, the early process by which gas and dust steadily clumped by gravity and grew into moons and planets.

Adaptive optics: Use of flexible mirrors to compensate for the blurring effects of the Earth's atmosphere; required to image exoplanets.

Adenosine triphosphate (ATP): Integral molecule that controls energy for all cellular processes in life on Earth.

Aerobic organisms: Organisms that require oxygen to survive.

Age of the Earth: Based on the decay rate of multiple radioactive isotopes, measured to be 4.54 billion years, with an accuracy of 1 percent.

Age of the universe: Detailed observations of the expansion rate and early state of the universe lead to a calculated age of 13.7 billion years since the big bang.

ALH 84001: Martian meteorite containing evidence for microbial life that ultimately proved inconclusive.

Allen Array: When complete, a set of 350 antennas designed to conduct a sensitive search for radio signals from distant civilizations.

Amino acids: The building blocks of proteins, as specified by the genetic code. Life on Earth uses twenty out of a much larger possible set.

Anaerobic organisms: Organisms that do not need oxygen to survive.

Anthropic principle: The idea that certain characteristics of the physical universe are carefully tuned to allow the existence of carbon-based life-forms.

Archaea: the most ancient of the three major branches of life on the Earth; the others are eukarya and bacteria.

Arecibo: A three-hundred-meter-diameter radio dish in Puerto Rico, sometimes used to send SETI signals.

Artificial life: The study of life, its processes, and its evolution through computer models, robotics, and synthetic biology.

Asteroid Belt: A set of large rocky bodies found on circular orbits between Mars and Jupiter.

Astrobiology: The study of life in the universe, including the history and limits of life on Earth.

Bacteria: The smallest type of living organisms.

Big bang: The tremendous release of energy in the beginning of the universe from which all matter in the expanding universe derived.

BioBricks: Standard biological components, created in the laboratory.

Biochemistry: The chemistry of life.

Biological landscape: The idea that terrestrial biology is one example of a wide array of potential biologies and is not necessarily an optimal solution.

Biomarker: Indirect tracer of extraterrestrial life, usually anticipated to be the spectral signature of gas in a planet's atmosphere that indicates metabolic processes.

Cambrian: A major period of geological time, when life proliferated in the oceans of the Earth, running from 542 to 488 million years ago.

Carbon-based life: Life that requires carbon for its critical functions by using it in long, information-storing molecules.

Carbon cycle: The cycling of carbon dioxide between the atmosphere and the Earth's crust.

Cassini: Highly successful mission to Saturn and its moons, including deployment of the Huygens lander to the surface of Titan.

Catalysis: Speeding up a chemical reaction by introducing an agent that is not changed by the reaction.

Cell: The smallest unit of life processes; highly organized chemical factories.

Cellular automata: Models often used in theoretical biology where a grid of cells grows according to simple rules that relate to the state of the initial cell.

Cephalopod: Invertebrates that can have well-developed senses and large brains, such as the octopus.

Chemical reactions: Processes where elements and compounds combine and separate. Chemical reactions affect electrons but not atomic nuclei.

Chemistry: The study of the composition, structure, properties, and reactions of atoms and molecules.

Civilization types: Defining civilizations by whether they harness their resources from a star, planet, or galaxy. Originally defined by Nikolai Kardashev.

Comet: Small Solar System body made of rock and ices, occupying a spherical cloud and spending most of its time far beyond the orbit of Neptune.

Complexity: An important, but varied, concept in biology. It can refer to sophistication of genes, metabolic pathways, brain architecture, or functions of the organism.

Contingent evolution: Evolution that is subject to random and unpredictable influences.

Convergent evolution: Similarities of organisms that arise when they evolve in similar environments.

Copernican Revolution: Profound change in thought in the sixteenth century, when the Earth was understood not to be the center of the universe.

Cosmic abundance: The average abundance in the universe of the stable elements.

Cryptobiosis: A state where all metabolic functions slow down enormously, or cease.

Cyanobacteria: The photosynthetic bacteria that produced the oxygen in the Earth's atmosphere.

Cyborg: Cybernetic organism, or a self-regulating integration of artificial and natural systems.

Dark energy: Enigmatic component of the universe causing cosmic acceleration.

Dark matter: Enigmatic component of the universe that permeates galaxies and the space between them. Dark matter outweighs normal matter by a factor of five or six.

Deoxyribonucleic acid (DNA): This long molecule in the shape of a double helix is the key to life. DNA carries the genetic code in a four-letter chemical alphabet.

Design: The idea that features of the universe and living organisms have an intelligent cause, rather than being the result of undirected processes.

Digital information: Any form of information that comes in the form of (or can be converted into) discrete levels suitable for manipulation by a computer.

DNA bases: The bases adenine (A), cytosine (C), guanine (G), and thymine (T) connect the two DNA strands. A pairs with T and C pairs with G in the genetic code.

Domains: The major classifications of life are eukarya, bacteria, and archaea.

Doomsday argument: A statistical argument that we are a substantial way through the entire span of the human species.

Doppler effect: Change in wavelength of any wave due to motion of the source of waves. This has been the tool for discovering the vast majority of exoplanets.

Drake Equation: Astrobiology pioneer Frank Drake formulated this way of calculating the number of intelligent communicable civilizations in the Milky Way.

Dyson Sphere: A theorized shell constructed around a star so an advanced civilization could use all the energy from the star. Named after physicist Freeman Dyson.

Earth-crossing asteroids: Asteroids with orbits that can cross Earth's orbit of the Sun, the most likely kind to cause mass extinctions.

Eccentricity: The amount by which the orbit of a planet or moon deviates from a circle.

Electromagnetic spectrum: The full range of electromagnetic radiation from long to short wavelengths. Only the narrow range of visible light can be detected by the eye.

Enceladus: A small moon of Saturn, only five hundred kilometers in diameter, with subsurface water that occasionally erupts as geysers.

Encephalization quotient (EQ): A determination of animal intelligence from the ratio of brain size to body mass.

Endospore: A cell that allows certain organisms to be dormant for a certain amount of time.

Endosymbiosis: From the Greek for "living together," the situation where an organism lives within the cells or body of another organism.

Entropy: The measure of disorder in a physical system, which tends to increase with time. Entropy is also related to the number of possible states of a system.

Enzyme: A protein that catalyzes, or accelerates, a chemical reaction.

Eons: The largest divisions of time in the Earth's history; the four eons are the Hadean, the Archaean, the Proterozoic, and the Phanerozoic.

Eras: Eons are divided into eras; for example, the Phanerozoic eon is divided into the Paleozoic, Mesozoic, and Cenozoic eras.

ETI: Extraterrestrial intelligence, the hypothetical existence of sentience beyond Earth.

Eukarya: One of the domains of life, which includes all plants and animals.

Eukaryotic cell: A kind of cell with a nucleus separated by a membrane from the rest of the cell, the most complex cell type.

Europa: Sizeable moon of Jupiter that is covered with a fractured icy crust overlying a water ocean, perhaps the most likely place to find life beyond Earth.

Evolution: In biology, the change in inherited traits of a population from generation to generation.

Evolutionary adaptation: An advantageous characteristic of an organism that lets it better survive and reproduce in its environment.

Exoplanet: Also called an extra-solar planet; any planet orbiting a star beyond the Solar System.

Expanding universe: The idea, first supported with evidence by Edwin Hubble, that the universe is getting larger as all galaxies move away from each other.

Extremophile: Organisms that live in what are considered hostile environments to humans, such as extremely saline or hot environments.

Fermi paradox: Enrico Fermi's idea that if extraterrestrial intelligence did exist, we should know of its existence. Also called Fermi's question: "Where are they?"

Fine-tuning: The fact that many of the constants of nature have values within a narrow range suited to the existence of carbon-based life.

Fossil: The remains of a living organism that have been turned to stone over a long period of time.

Fossil record: The story of the Earth's geological history as told through fossils.

Gaia hypothesis: The proposal that the living and nonliving parts of the Earth operate as a complex interacting system, though not actually as a single organism.

Gas giant planets: In the Solar System, Jupiter, Saturn, Uranus, and Neptune. In general, a planet made primarily of gaseous hydrogen and helium.

Gene: The minimum amount of genetic material that expresses a characteristic of a living organism. A gene is a sequence of several hundred bases along the DNA molecule.

General relativity: A theory of gravity proposed by Albert Einstein and confirmed soon after, stating that space is warped by mass, including the universe itself.

Genetic code: The way DNA base pairs are interpreted to provide instructions for an organism's genes.

Genetic drift: The influence of chance on the survival of alleles, or variants of a gene.

Genetic engineering: The science of restructuring an organism's genome.

Genome: The full sequence of DNA base pairs in an organism.

Genotype: The exact genetic makeup of an organism, or its specific set of genes.

Geocentric cosmology: The description of a spherical universe with the Earth stationary at the center, associated most strongly with Aristotle.

Geological processes: The four main geological processes are tectonics, erosion, volcanism, and impact cratering.

Geological timescale: The division of the history of the Earth; divided into eons, then eras, then periods, and further into epochs and ages.

Geology: The study of the history, origin, and structure of the Earth.

Global warming: The rise in the Earth's temperature generally accepted to be caused at least in part by human generation of greenhouse gases.

Greenhouse effect: A situation where the introduction of greenhouse gases to the atmosphere causes solar radiation to be trapped and the atmosphere to heat up.

Greenhouse gases: Gases that trap heat and radiation in the Earth's atmosphere; the most important examples are carbon dioxide and methane.

Habitable zone: The area surrounding any star in which an orbiting planet or the moon of that planet could have liquid water on its surface.

Hadean: The first eon of the Earth's history, before about 3.8 billion years ago.

Heavy bombardment: A time in the Earth's history when impacts from debris left over from the creation of the Solar System were common, ending about 3.8 billion years ago.

Heavy elements: All elements except hydrogen and helium; also referred to as "metals" by astronomers.

Heliocentric cosmology: The description of the universe with the Sun at the center and the planets in orbit around the Sun, proposed by Copernicus and verified by Galileo.

Heredity: Characteristics of organisms that are passed on through generations.

Hubble Space Telescope: The premier observing facility in astronomy. HST is NASA's flagship mission and an important contributor to the study of exoplanets.

Ices: Materials that solidify at low temperatures, such as water and methane.

Impactor: An object from space that strikes a planet or moon.

Impacts: Randomly occurring collisions of the Earth with space debris.

Impact sterilization: The sterilization of a planet due to an impact from a cosmic body.

Inflation: Early phase of extremely rapid expansion that caused the universe to become smooth and nearly flat.

Information: A quantity that can be measured and transmitted. In biology, information is stored in the genetic code.

Intelligence: The capacity for abstract thought, coupled with the mastery of tools or technology. Intelligence is found only in animals with large and complex brains.

Interferometry: Combining radio or optical telescopes to achieve the angular resolution equivalent to a single huge telescope.

Interstellar clouds: Regions of dust and gas between star systems.

Interstellar communication: The use of pulsed and coded radio waves or light waves to send messages to other stars.

Interstellar travel: Travel between stars, requiring thousands of years with current population technologies.

Kingdom: The second largest classification of living organisms.

K-T event: The mass extinction between the Cretaceous and Tertiary periods about sixty-five million years ago, commonly attributed to an impact from space.

Kuiper Belt: A zone with large chunks of space debris just beyond the orbit of Neptune.

Lander: A spacecraft designed to land on the surfaces of planets or moons.

Last common ancestor: The root point of a diverged set of species. The last universal ancestor is the hypothetical single-celled organism that gave rise to all life on Earth.

Lateral gene transfer: Transfer of genetic material by a means other than reproduction.

Law of conservation of energy: The fact that the total amount of energy in a system is constant, even though energy may change forms.

Law of gravitation: Newton's expression governing the gravity force between any two objects superseded in situations of intense gravity by Einstein's general relativity.

Laws of thermodynamics: Rules governing the behavior of heat. As energy changes forms, the proportion of energy in the disordered form of heat increases.

Lensing: A gravitational phenomenon of general relativity, where an intervening object focuses and brightens the light from a more distant object.

Life: Challenging to define in any way that has meaning beyond the Earth, but probably requiring the localized use of energy and the storage of information in molecular forms, evolution, and adaptation to the environment.

Light travel time: The time for light to travel a certain distance. Light travel times are hours in the Solar System, years to nearby stars, and millions of years to the nearest galaxies.

Lipid: A common molecule in cells that is integral to the structure of cell membranes.

Many worlds: The idea that the Earth is just one among many worlds in space, including potential applications of geology, chemistry, and biology beyond the Earth.

Mars Exploration Rovers: The twin rovers Spirit and Opportunity have explored Mars since 2003, providing much of the evidence that Mars has hosted water.

Mass-energy conversion: The conversion of mass into a much larger amount of energy, according to Einstein's equation $E = mc^2$.

Mass extinction: At least five times in the history of life on Earth a sizeable percentage of species has been extinguished in a geologically short time, potentially by an impact.

Mathematics: The study of numbers and their properties, and the symbols and operations that can apply to numbers.

Matter-antimatter annihilation: Refers to the enormous release of energy as particles come into contact with their corresponding antiparticles.

Metabolism: The chemical reactions that govern the functioning of living things.

Metallicity: The proportion of all the elements heavier than hydrogen and helium in an astronomical object.

Meteorite: A stony or sometimes metallic object landing on the Earth from space that represents relatively pristine material from the formation of the Solar System.

Microbe or microorganism: A living creature too small to be seen by the naked eye.

Microlensing: Temporary brightening of a star's light when an unseen planet passes in front of it, gravitationally focusing its light.

Microscope: A device for magnifying invisibly small objects, which was essential for the development of the science of biology.

Milankovitch cycles: The collective effects of the Earth's motion on climate.

Milky Way: Our galaxy, a large system of several hundred billion stars, including the Sun.

Mitochondria: The structures in eukaryotic cells that help make energy through the construction of ATP from oxygen.

Molecular fossil: Highly indirect evidence of life, in the form of isotopic imbalances in rock that indicate the presence of an ancient metabolism at work.

Multicelled organism: Beginning about 1.2 billion years ago, the lineage of life that led to plants and animals.

Multiverse: A speculative theory suggesting that conditions at the time of the big bang led to a suite of universes, each with different physical properties.

Mutation: Change to the genetic material of an organism, caused by copying errors or by external agents such as radiation and chemicals.

Nanobot: Miniaturized probe for remote sensing or, potentially, interstellar travel.

NASA: National Aeronautics and Space Administration, the U.S. governmental agency that launches most planetary missions and telescopes in space.

Natural selection: A mechanism of Darwin's theory of evolution, where individuals well adapted to the environment reproduce more than those less well adapted.

Observable universe: The region of space within which light has had time to reach us since the big bang.

Orbiters: Spacecraft designed to orbit planets or moons for scientific observation.

Organic chemistry: The chemistry of molecules containing carbon.

Origin of life: A historical event on Earth, probably occurring about four billion years ago. The origin of life is subject to investigation but the details may never be known.

Panspermia: The hypothesis that life on Earth and elsewhere may have been seeded by material that travels between habitable bodies.

Parallax: The subtle angular shift when an astronomical object is observed from different points of the Earth's orbit of the Sun. Leads to the most reliable measure of distance.

Periodic table: A way of organizing all the elements according to the number of outer electrons. Elements in a particular column share many chemical properties.

Periods: The third largest divisions of time in the Earth's geological history.

Phenotype: The actual physical characteristics of an organism, governed by genes, the environment, and random variation.

Photosynthesis: Probably the most important biochemical pathway of life on Earth. The use of sunlight to produce sugar, and then ATP, the fuel for all living things.

Phylogenetic tree: A diagram showing evolutionary relationships among species relative to a common ancestor, usually measured through deviations in RNA or DNA.

Phylum: The level of classification of organisms below a kingdom.

Physical universe: Space containing all matter and energy; may be substantially larger than the observable universe.

Physics: The study of the forces of nature and the laws that govern the way matter and radiation interact.

Pioneer 10: Launched in 1972, the first human probe to leave the Solar System, carrying a message on a plaque.

Pixel: The very small electronic components that make up digital detectors and images.

Plate tectonics: A theory of geology where continental-size plates of the crust and upper mantle move over the semiliquid rock layer below.

Polycyclic aromatic hydrocarbon (PAH): Small carbon-ring molecules seen in comets and interstellar space, possibly a basis for life.

Polymerase chain reaction (PCR): A technique for isolating and then exponentially amplifying DNA using enzymes.

Principle of mediocrity: Equivalent to the Copernican principle, the idea that the Earth and its biology are not special in the universe.

Prokaryotic cell: A cell without a defined nucleus.

Proterozoic: The eon representing the time before large complex life-forms developed on the Earth, from 2.5 billion to 542 million years ago.

Protoplanetary disk: A disk of gas where planet formation will eventually occur.

Quorum sensing: The ability of bacteria within a colony to communicate and coordinate behavior.

Radioactive dating: The technique for determining the age of a material by measuring the amount of a radioactive isotope and its decay product.

Radioactive decay: The random process where a heavy atomic nucleus spontaneously decays.

Rare Earth: The idea that the conditions that led to complex life on Earth are so rare that we may be the only planet in the galaxy with complex life.

Reflex motion: The wobble or periodic motion of a star due to the influence of a much smaller orbiting planet or companion.

Replicator: A molecule that can store information and reproduce.

Ribozyme: From ribonucleic acid enzyme, an RNA molecule that catalyzes a chemical reaction.

RNA World: A hypothesis that on the early Earth a phase of RNA-based life preceded the current DNA-based life.

Sample return mission: A space mission with the intent of bringing back physical samples of another world to Earth.

Scientific method: Systematic observation of nature, with the goal of finding patterns and providing physical explanations for phenomena.

SETI: The Search for Extraterrestrial Intelligence.

Simulation hypothesis: An argument framed by philosophers that we might possibly be the simulated computational creations of an advanced civilization.

Singularity: A term coined by inventor Ray Kurzweil, referring to a time in the future when humans will attain a postbiological state.

Snowball Earth: The term referring to long ice ages that engulfed our planet several times between 750 and 580 million years ago, and possibly 2.2 billion years ago.

Solar composition: Any gas or material composed of 76 percent hydrogen and 22 percent helium, with much smaller trace amounts of all other elements.

Solar cycle: The Sun's dynamical engine operates on an eleven-year cycle and affects the Earth's weather.

Solar luminosity: 4×10^{26} watts, the luminosity of the Sun.

Solar sail: A hypothesized method of space travel that uses a large, reflective sheet that can propel a spacecraft by pressure from sunlight.

Species: A basic unit of biological classification; for multicelled organisms a population whose individuals can breed and produce fertile offspring.

Spectroscopy: The technique of dispersing light into an array of wavelengths in order to see the narrow atomic or molecular features that are indicative of temperature and chemical composition.

Stellar fusion: The process that leads to all elements heavier than helium in the universe as atoms collide in the high-temperature cores of stars.

Stellar recycling: The mixing and subsequent ejection of heavy elements created in star cores later in their lives when the stars begin to lose mass.

Stromatolite: Fossilized bacterial colonies that first developed about three billion years ago.

Supernova: The violent death of a massive star produces many of the heavy elements on which life depends, and can affect life on a planet sufficiently nearby.

Symbiotic relationship: The benefit of a parasite and host organism sharing resources.

Telescope: Device for gathering and focusing light or other electromagnetic radiation. The largest optical telescopes are currently ten or eleven meters in diameter.

Terraforming: Altering a planet to resemble the Earth in physical properties, or at least to be habitable.

Terrestrial Planet Finder (TPF): NASA's ambitious future mission to discover and characterize Earth-like exoplanets.

Terrestrial planets: In the Solar System, Mercury, Venus, Earth, or Mars. In general, a small planet made primarily of rocky material.

Theory of evolution: A hypothesis proposed by Charles Darwin suggesting that uneven reproductive success selects certain species over others in response to the environment.

Thermodynamics: The study of heat (or disordered) energy, how it moves and changes.

Tidal heating: Heating caused when a small body is in a tight elliptical orbit around a larger body.

Titan: Large moon of Saturn with a thick nitrogen atmosphere and shallow seas made of liquid ethane and methane.

Trace fossil: Indirect evidence of a living organism rather than the fossilized organism.

Transit: Situation where an exoplanet periodically passes in front of its parent star, dimming it slightly.

Tree of life: In biology, a metaphor for the steady diversification of life from a common ancestor.

Tunguska: A massive explosion in a remote part of Russia in 1908, probably caused by a meteor or piece of a comet.

21-cm emission line: A low-energy spectral feature that results from a change in the spin state of cold atomic hydrogen, used in searches for extraterrestrial intelligence.

UFOs: Unidentified flying objects, usually purported to be visitations by aliens, but there is no compelling evidence to support this assertion.

Vesicle: A basic container or compartment in biology, a stage in the evolution to a cell.

Von Neumann machines: Computers or robots that are able to reproduce themselves, proposed by the mathematician John von Neumann.

Voyager: Twin spacecraft that explored and then exited the Solar System carrying gold records with sounds and images of Earth.

Zircon: Very stable crystal, a sample of which is the oldest rock found on Earth.

Zoo hypothesis: A solution to the Fermi paradox that suggests that alien civilizations know we are here but have chosen not to contact us.

READING LIST

CHAPTER 1: THE UNFINISHED REVOLUTION

Bennett, J., Shostak, S., and Jakosky, B. (2007), *Life in the Universe*, San Francisco, Pearson Education. Second edition of a clearly written, introductory textbook that covers all aspects of the search for life beyond Earth. Richly illustrated.

Chela-Flores, J. (2007), *The New Science of Astrobiology*, New York, Springer. Up to date and reasonably priced introductory paperback, with broad coverage from history of life on Earth to solar system exploration and SETI.

Derry, G. N. (2002), *What Science Is and How It Works*, Princeton, Princeton University Press. Spirited and entertaining look at how we learn about the natural world; part textbook, part manifesto.

Dick, S. J. (2001), *Life on Other Worlds: The 20th Century Extraterrestrial Life Debate*, Cambridge, Cambridge University Press. Good historical background to astrobiology and the emergence of a new scientific field.

Dick, S. J., and Strick, J. E. (2004), *The Living Universe: NASA and the Development of Astrobiology*, New Brunswick, N. J., Rutgers University Press. A modern history of astrobiology, with emphasis on NASA's research programs.

Ferris, T. J. (2003), *Coming of Age in the Milky Way*, New York, Harper Perennial. Updated and re-released, this classic is a beautifully written history of astronomy, with a broad sweep from the ancient Greeks to modern cosmology.

Gingerich, O. (2004), *The Book Nobody Read: Chasing the Revolutions of Nicolaus Copernicus*, New York, Walker and Company. Detective story by a noted historian who tracks down copies of the book that changed the way we look at the universe.

Heilbron, J. L., editor (2005), *The Oxford Guide to the History of Physics and Astronomy*, Oxford, Oxford University Press. Short scholarly articles, alphabetically arranged, on scientists, concepts, and discoveries from the Renaissance to the present.

Hoskin, M., editor (1996), *The Cambridge Illustrated History of Astronomy*, Cambridge, Cambridge University Press. Covers the development of astronomy as a science, from ancient observatories to the Copernican Revolution.

Lunine, J. I. (2005), *Astrobiology: A Multidisciplinary Approach*, San Francisco, Pearson Education. High level and challenging, but worth the effort. This textbook covers the subject from biology and astronomy to the most recent discoveries.

Sagan, C., and Druyan, A. (1997), *The Demon-Haunted World: Science as a Candle in the Dark*, New York, Ballantine Books. Eloquent exposition of the power of the scientific method, and a pointed debunking of pseudoscientific beliefs.

Sagan, C., and Druyan, A. (2006), *The Varieties of Scientific Experience*, New York, Penguin. Based on Sagan's Gifford Lectures on Natural Theology, this book argues for the primacy of the scientific method and science's ability to inspire.

Singh, S. (2004), *Big Bang: The Origin of the Universe*, New York, HarperCollins. Sweeping history of cosmology, with emphasis on the evidence for the universal expansion and birth in a hot, dense state.

CHAPTER 2: LIFE'S ORIGINS

Adams, F. (2002), *Origins of Existence: How Life Emerged in the Universe*, New York, Free Press. Easy-to-read treatment of cosmic evolution as a context for the development of life.

Chown, M. (2001), *The Magic Furnace: The Search for the Origin of Atoms*, Oxford, Oxford University Press. Engaging account of the creation of elements within stars, and stories of the scientists who unraveled this mystery.

Dalrymple, D. G. (2004), *Ancient Earth, Ancient Skies: The Age of the Earth and Its Cosmic Surroundings*, Stanford, Stanford University Press. Readable account of the methods used to date the Earth, stars, and universe.

Dyson, F. (1999), *Origins of Life*, Cambridge, Cambridge University Press. Highly influential slim volume by a preeminent physicist; updated from the original edition, strong on the thermodynamic basis for life. Fairly high level.

Fry, I. (2000), *The Emergence of Life on Earth: A Scientific and Historical Overview*, London, Free Association Books. Origin-of-life research, from Pasteur to the Mars meteorite. Extensive bibliography and balanced presentation of ideas.

Hazen, R. M. (2005), *Genesis: The Scientific Quest for Life's Origins*, New York, John Henry Press. Very clearly written account of recent theories on the formation of life by a well-known popularizer of science.

Lovelock, J. E. (1995), *The Ages of Gaia: A Biography of Our Living Earth*, New York, W.W. Norton. This earlier book is better and more scientifically detailed than the 2002 paperback that followed on the same subject.

Margulis, L., and Sagan, D. (2000), *What Is Life?*, Berkeley, University of California Press. Masterful account of the early evolution of life on Earth, symbiosis, and the diversity of the microbial world.

Mason, S. F. (1991), *Chemical Evolution: Origin of the Elements, Molecules and Living Systems*, Oxford, Clarendon Press. Technical and encyclopedic survey of the chemical basis for biology.

Schopf, J. W. (2002), *Life's Origin*, Berkeley, University of California Press. Clearly written articles by major researchers in the study of the origin of life, written at a moderately technical level.

CHAPTER 3: EXTREME LIFE

Avise, J. C. (2004), *The Hope, Hype and Reality of Genetic Engineering*, Oxford, Oxford University Press. Overview of the techniques scientists have learned for manipulating genetic material. Assumes no biology background.

Darling, D. (2001), *Life Everywhere: The Maverick Science of Astrobiology*, New York, Basic

Books. Excellent general book on astrobiology, with anecdotes of many leading researchers, and coverage of extreme forms of life and the Rare Earth idea.

Gross, M. (2001), *Life on the Edge: Amazing Creatures Thriving in Extreme Environments*, New York, Perseus Books. Popular account of the creatures that inhabit environmental niches unsuitable for humans.

Howland, J. L. (2000), *The Surprising Archaea: Discovering Another Domain of Life*, Oxford, Oxford University Press. The story of the recent recognition of a whole new family of microbes at the base of the tree of life.

Kauffman, S. A. (1996), *At Home in the Universe: The Search for Laws of Self-Organization and Complexity*, Oxford, Oxford University Press. Personal account of complexity theory and its relation to biological processes.

Levy, S. (1993), *Artificial Life: A Report from the Frontier Where Computers Meet Biology*, New York, Vintage. Journalistic account of the emergence of this new field and vignettes of the often-idiosyncratic people who work on artificial life.

Reiss, M. J., and Straughan, R. (2001), *Improving Nature?*, Cambridge, Cambridge University Press. Nontechnical overview of genetic engineering, with a particular focus on the moral and ethical issues raised by the research.

Ward, M. (2000), *Virtual Organisms: The Startling World of Artificial Life*, New York, St. Martin's Press. Picks up where the Levy book ends, covering models of early life on Earth and robots.

Watson, J. D., and Berry, A. (2003), *DNA: The Secret of Life*, New York, Knopf. The codiscoverer of DNA cowrote this highly readable account of modern genetics, and it has a personal "behind the scenes" flavor.

Wharton, D. A. (2002), *Life at the Limits: Organisms in Extreme Environments*, New York, Cambridge University Press. Personal story of the study of extremophiles by an Antarctic researcher; requires some biology background.

CHAPTER 4: SHAPING EVOLUTION

Carroll, S. B. (2005), *Endless Forms Most Beautiful: The New Science of Evo Devo and the Making of the Animal Kingdom*, New York, W.W. Norton. Excellent account of the ways in which gene expression affect evolution.

Dawkins, R. (2004), *The Ancestor's Tale: A Pilgrimage to the Dawn of Evolution*, New York, Houghton Mifflin. Comprehensive and masterful unwinding of forty family trees, looking back over four billion years, by evolution's crispest expositor.

Gould, S. J. (2002), *The Structure of Evolutionary Theory*, New York, Belknap Press. Not for the faint of heart, this summary of Gould's life work weighs in at nearly 1,500 pages, but it is a gold mine of evolutionary ideas.

Jakosky, B. (1998), *The Search for Life on Other Planets*, Cambridge, Cambridge University Press. Introductory-level textbook, especially good on the early geological and biological history of the Earth.

Knoll, A. H. (2003), *Life on a Young Planet: The First Three Billion Years of Evolution on Earth*, Princeton, Princeton University Press. A top paleontologist gives an overview of early evolution on Earth.

Lunine, J. I. (1999), *Earth: Evolution of a Habitable World*, Cambridge, Cambridge University Press. Clearly written and comprehensive introductory textbook on the geology and biology of the Earth. Black and white illustrations.

Morris, S. C. (2004), *Life's Solution: Inevitable Humans in a Lonely Universe*, Cambridge, Cambridge University Press. In a treatment that runs somewhat counter to both Dawkins and Gould, the Burgess Shale researcher stresses evolutionary convergence.

Scott, E. C., and Eldridge, N. (2005), *Evolution vs. Creationism: An Introduction*, Berkeley, University of California Press. The dialectic of the debate with intelligent design and creationism is a good way to see the strength of the evolution idea.

Tyson, N. D., and Goldsmith, D. (2004), *Origins: Fourteen Billion Years of Cosmic Evolution*, New York, W.W. Norton. Companion to a two-part NOVA series, this book manages to combine cosmic and biological evolution in one treatment.

Ward, P. D. (2000), *Rivers in Time: The Search for Clues to Earth's Mass Extinctions*, New York, Columbia Press. How biologists and paleontologists deduced that external intruders have affected the course of evolution. Great detective story.

Ward, P. D. (2006), *Out of Thin Air: Dinosaurs, Birds, and the Earth's Ancient Atmosphere*, New York, John Henry Press. The interplay between oxygen levels in the atmosphere and the rise and fall of animal species.

Zimmer, C. (2002), *Evolution: The Triumph of an Idea*, New York, Harper Perennial. In this companion text to a 2001 PBS series broadcast, a science journalist gives an excellent overview of a compelling theory. Richly illustrated.

CHAPTER 5: LIVING IN THE SOLAR SYSTEM

Beatty, J. K., Petersen, C. C., and Chaikin, A., editors (1998), *The New Solar System*, Cambridge, Cambridge University Press. The 4th edition is a semi-technical overview of the physical properties of the planets, with many excellent illustrations.

Bond, P. (2007), *Distant Worlds: Milestones in Planetary Exploration*, New York, Springer. Recent book by a consultant to ESA on the fifty-year history of planetary exploration, with a comprehensive list of missions.

Crosswell, K. (2003), *Magnificent Mars*, New York, Free Press. A book Mars might write if it could tell its own story. Excellent survey of recent discoveries on Mars and the requisite colorful imagery.

Greeley, R., and Batson, K. (2001), *The Compact NASA Atlas of the Solar System*, Cambridge, Cambridge University Press. Cheaper version of the coffee table book from 1997, it has stunning photos from every NASA planetary mission.

Greenberg, R. (2005), *Europa, the Ocean Moon: Search for an Alien Biosphere*, New York, Springer. Story of the Galileo probe's mission to Jupiter's watery moon and the prospects of life there.

Grinspoon, D. H. (1997), *Venus Revealed: A New Look Below the Clouds of Our Mysterious Twin Planet*, New York, Addison Wesley. Grinspoon presents the geology and atmospheric conditions of Venus, and relates its evolution to that of Earth.

Harland, D. M. (2007), *Cassini at Saturn: Huygens Results*, New York, Springer Praxis.

Semi-technical account of the main scientific results of the Huygens lander and its exploration of Saturn's moon Titan.

Lang, K. R. (2003), *Cambridge Guide to the Solar System*, Cambridge, Cambridge University Press. Popular level treatment of the planets, great browsing material, and lavishly illustrated on almost every page.

Morton, O. (2002), *Mapping Mars: Science, Imagination, and the Birth of a World*, New York, St. Martin's Press. A young science journalist conveys the beauty of the red planet, and the urgency of our quest to understand it.

Sobel, D. (2005), *The Planets*, New York, Viking Adult. Beautiful series of essays on the planets, each richly evocative and full of literary and cultural allusions. A great counterpoint to the scientific exploration of the Solar System.

Squyres, S. (2005), *Roving Mars: Spirit, Opportunity, and the Exploration of the Red Planet*, New York, Hyperion. Riveting account of the ups and downs of developing a space mission, from the leader of the team that operates the Mars rovers.

Ward, P. D., and Brownlee, D. (2000), *Rare Earth: Why Complex Life is Uncommon in the Universe*, New York, Springer-Verlag. Influential and controversial book that concludes that the particular conditions needed for intelligent life occur very rarely.

CHAPTER 6: DISTANT WORLDS

Boss, A. (2000), *Looking for Earths: The Race to Find New Solar Systems*, New York, Wiley and Sons. Leading planetary theorist gives a nice history of the false hopes and final success in the detection of extra-solar planets.

Casoli, F., and Encrenaz, T. (2007), *The New Worlds: Extrasolar Planets*, New York, Springer Praxis. The most recent book on the discovery and properties of planets beyond the Solar System, written at a popular level.

Furniss, T. (2003), *A History of Space Exploration: And Its Future*, New York, The Lyons Press. Good basic survey of space history, from Goddard to NASA's Space Station. Well illustrated.

Gilmour, I., and Sephton, M. A. (2003), *An Introduction to Astrobiology*, Cambridge, Cambridge University Press. This introductory undergraduate textbook for an Open University course is particularly good on methods for discovering exoplanets.

Gilster, P. (2004), *Centauri Dreams: Imagining and Planning Interstellar Travel*, New York, Springer-Verlag. A look at NASA engineers who are working on propulsion systems that could take a payload to nearby stars within a generation.

Lemonick, M. D. (1998), *Other Worlds: The Search for Life in the Universe*, New York, Simon & Schuster. Excellent journalistic narrative of the discovery of extrasolar planets and the search for distant Earths.

Lewis, J. S. (1998), *Worlds Without End: The Exploration of Planets Known and Unknown*, New York, Perseus Books. Excellent introduction to the comparative study of planets, from a geological perspective.

Mayor, M., Frei, P.-Y., and Roukema, B. (2003), *New Worlds in the Cosmos: The Discovery of Exoplanets*, Cambridge, Cambridge University Press. The insider's account of the discovery of planets beyond the Solar System.

van Pelt, M. (2005), *Space Tourism: Adventures in Space Orbit and Beyond*, New York, Springer-Verlag. That the future of space travel will likely be propelled by tourism and recreation is explored in the this well-written discussion of the commercial sector.

Villard, R., and Cook, L. R. (2005), *Infinite Worlds: An Illustrated Voyage to Planets Beyond Our Sun*, Berkeley, University of California Press. Brilliant visualizations of the new exoplanets, none of which can yet be resolved by existing technology.

Zubrin, R. (2000), *Entering Space: Creating a Spacefaring Civilization*, New York, Penguin Putnam. This wide-ranging look at the future of space travel is written by a visionary engineer who helped define NASA's current plan for Mars exploration.

CHAPTER 7: ARE WE ALONE?

Bostrom, N. (2002), *Anthropic Bias: Observation Selection Effects in Science and Philosophy*, London, Taylor and Francis. High level but rewarding exposition of the observation selection effects that make anthropic reasoning dangerous.

Cohen, J., and Stewart, I. (2002), *What Does a Martian Look Like?: The Science of Extraterrestrial Life*, New York, Wiley and Sons. Clever use of science fiction, good and bad, to illuminate what we might expect from life beyond Earth.

Ekers, R. D., Cullers, D. K., and Billingham, J. (2002), *SETI 2020: A Roadmap for the Search for Extraterrestrial Intelligence*, Mountain View, SETI Press. Fascinating and comprehensive compilation of strategies in SETI, written by the experts.

Gott, J. R. III (2002), *Time Travel in Einstein's Universe: The Physical Possibilities of Travel Through Time*, New York, Houghton Mifflin. Written by a prominent theorist, this is a glimpse into the possibilities open to an advanced civilization.

Greene, B. (2005), *The Fabric of the Cosmos: Space, Time, and Texture of Reality*, New York, Vintage. Eloquent writing about the esoteric topics in cosmology and physics. Along with his earlier book, a great introduction to the theory of everything.

Grinspoon, D. (2004), *Lonely Planets: The Natural Philosophy of Alien Life*, New York, HarperCollins. Highly entertaining exploration of planetary science, the new discoveries in astrobiology, and the cultural backdrop to SETI and UFOs.

Kurzweil, R. (2005), *The Singularity Is Near: When Humans Transcend Biology*, New York, Viking Adult. Renowned inventor and optimistic futurist Kurzweil makes the case that humans will develop technology of almost unrecognizable capability.

Moravec, H. P. (2000), *Robot: Mere Machine to Transcendent Mind*, Oxford, Oxford University Press. Reads like science fiction, but this vision of the future of robots is written by someone with decades of experience in crafting cutting-edge robots.

Penrose, R. (2005), *The Road to Reality: A Complete Guide to the Laws of the Universe*, New York, Knopf. Monumental and often challenging, this book does a great job at conveying how mathematics underlies physical theories of the universe.

Rees, M. J. (1998), *Before the Beginning: Our Universe and Others*, New York, Perseus Books. Clear exposition of the multiverse idea from the Astronomer Royal for England, and clear overview of modern cosmology.

Shostak, S., and Barnett, A. (2003), *Cosmic Company: The Search for Life in the Universe*,

Cambridge, Cambridge University Press. A senior scientist at the SETI Institute lays out the case for trying to make contact. Lively and humorous writing.

Webb, S. (2002), *Where Is Everybody? Fifty Solutions to the Fermi Paradox and the Problem of Extraterrestrial Life*, New York, Copernicus Books. As the title says, this is a wide-ranging survey of many possible explanations for the absence of ETs.

MEDIA RESOURCES

WEB SITES

American Natural History Museum, http://www.amnh.org/science. This famous New York City museum does scientific research, and good overviews of paleontology and astronomy are available in the web areas of those two divisions.

Anthropic Principle, http://www.nickbostrom.com. The personal site of Oxford don Nick Bostrom is a great place to learn about the anthropic principle, the Doomsday argument, the simulation hypothesis, and other intriguing philosophical ideas.

Astrobiology Magazine, http://www.astrobio.net. An outstanding resource, supported by NASA, with thousands of articles archived since 2000, interviews, images, and an effective search function. Supports RSS feeds and XML delivery to handhelds.

Astrobiology Web, http://www.astrobiology.com. Commercial site with many articles and a large image gallery. Good place to sign up for NASA video feeds and get live NASA TV on the desktop.

The Extrasolar Planets Encyclopedia, http://www.obspm.fr/planets. Hosted by the Paris Observatory, the place to go for the current census of extrasolar planets, details on each system, and tutorials on planet detection.

Mars Exploration, http://mars.jpl.nasa.gov. Hosted by NASA's Jet Propulsion Lab, this site has details of all current and planned Mars missions, with special sections for children and educators.

NASA Astrobiology, http://astrobiology.arc.nasa.gov. From NASA's Ames Research Center, this is a central source for news on research and educational resources. Has a link to NASA's astrobiology road map.

NASA Astrobiology Institute, http://nai.arc.nasa.gov. Home of the "virtual" institute made up of sixteen US universities and research centers and six international partners. Great jumping-off point to explore all the projects being carried out around the country.

Planet Quest, http://planetquest.jpl.nasa.gov. Place to learn about current and future searches for terrestrial planets, hosted by NASA's Jet Propulsion Lab. Some great multimedia content and animations can be found here.

The Planets, http://www.nineplanets.org. Maintained by Bill Arnett, the best site for a multimedia tour of the Solar System, all planets and major moons. Accurate and up-to-date, with extensive links.

SETI at Home, http://setiathome.ssl.berkeley.edu. Place to download free software to analyze chunks of SETI radio search data and participate in the largest distributed computing project on Earth. Cool visual interface.

SETI Institute, http://www.seti.org. Nonprofit institution for research and education on the search for intelligent life in the universe, home to more than one hundred scientists. Previous and current searches are described.

Space, http://www.space.com. The best of the commercial sites for information about space and astronomy. Good news stories and discussion areas. The site has RSS feeds and a night-sky-viewing applet.

Talk.Origins Archive, http://www.talkorigins.org. Best single place to read about the scientific basis for the theory of evolution, this Usenet group has an extensive archive of articles and FAQs. Very high quality content.

Tree of Life, http://tolweb.org/tree. Learn how any organism on Earth is related to any other organism in the phylogenetic tree. This international collaboration of more than four thousand web pages is a rich mine of information.

CD-ROMS AND DVDS

Accelerating Media, http://www.acceleratingmedia.com. Excellent single source for DVDs or videotapes of more than two hundred conferences on space and planetary science, with titles on extremophiles, Mars exploration, the future of space travel, etc.

Cosmic Voyage, http://www.cosmicvoyage.org. IMAX movie, released in 2002 on DVD, with Morgan Freeman narrating. Evocative story line runs from the big bang through the evolution of life on Earth.

DNA Interactive, http://www.dnai.org. Produced by Cold Spring Harbor Lab, this interactive DVD has two hundred video clips, many animations and interviews with eleven Nobel Laureates on the mechanism of life revealed by DNA.

The Elegant Universe, http://www.pbs.org/wgbh/nova/elegant. Three-part TV series hosted by Brian Greene, on one DVD, covering cosmology and fundamental physics in an entertaining way. Companion web site has transcript and interactives.

Evolution, http://www.pbs.org/wgbh/evolution. Eight episodes on four DVDs of this PBS special, first broadcast in 2001. The companion web site has excellent materials and video clips; a companion book by Carl Zimmer is available.

Life Beyond Earth, http://www.shoppbs.org. Based on a two-hour PBS special hosted by Timothy Ferris, this single DVD is now a couple of years old but still noteworthy for the great visuals and Ferris's engaging storytelling style.

Origins, http://www.shoppbs.org. Four-part NOVA series, first broadcast in 2004 and hosted by Hayden Planetarium director Neil Tyson. Includes interviews with astrobiologists and planet-hunters. Two DVDs, companion book available.

The Planets, http://www.cosmicvoyage.org. Eight episodes on four DVDs, shown on A&E network in 1999, containing the "greatest" hits from twenty years of NASA Solar System exploration.

Powers of Ten, http://www.powersof10.com. Interactive CD-ROM, based on the famous Eames movie, showing scales in the universe. Added layer shows scales of time. Many outstanding features.

What We Don't Know, http://www.channel4.com/science. Recent TV series from UK's channel 4, now on DVD. Hosted by the Astronomer Royal, Sir Martin Rees, includes SETI, post-biological evolution, and the simulation hypothesis.

ILLUSTRATION CREDITS

CHAPTER 1: THE UNFINISHED REVOLUTION

Figure 1: Aristarchus and the relative size of the Earth, Moon, and Sun, courtesy Chris Impey. Figure 2: Playing card sequences, copyright Chris Impey. Figure 3: Woodcut by Camille Flammarion, 1870, this 1970 version courtesy Roberta Weir. Figure 4: Michelangelo Cactani, 1855, from *La Materia Della Devina Commedia* di Dante Alighieri. Figure 5: Robert Fludd, 1621, from *Utriusque Cosmi Maioris Scilicet et Minoris Metaphysica*. Figure 6: Thomas Digges, 1576, from *A Perfect Description of the Celestial Orbs*. Figure 7: Galileo Galilei, 1610, from *Sidereus Nuncius*. Figure 8: Isaac Newton, from *The History of the World*, by H. F. Helmolt, 1902, New York: Dodd, Mead, and Company. Figure 9: Orbits by Isaac Newton, 1687, from *Principia Mathematica*. Figure 10: Thomas Wright, 1750, from *An Original Theory or New Hypothesis of the Universe*. Figure 11: Portrait of William Herschel, 1845, courtesy University of Texas Libraries. Figure 12: William Parsons's Leviathan telescope, courtesy the Armagh Observatory. Figure 13: Stellar parallax, courtesy NASA. Figure 14: Photograph of Edwin Hubble, courtesy the Huntington Library, San Marion, California. Figure 15: Hubble cosmic expansion in 2-D, courtesy Richard Pogge, Ohio State University. Figure 16: Microwave sky, courtesy NASA/WMAP Science Team. Figure 17: Contents of the universe, courtesy NASA/WMAP Science Team. Figure 18: Hubble Ultra Deep Field, courtesy NASA/ESA, Steve Beckwith, and the HUDF Team, Space Telescope Science Institute. Figure 19: HST servicing mission, courtesy NASA Glenn Research Center. Figure 20: The electromagnetic spectrum, courtesy NASA. Figure 21: Astrobiology road map, courtesy NASA/CMEX.

CHAPTER 2: LIFE'S ORIGINS

Figure 22: Cosmic element abundance, courtesy NASA Remote Sensing Tutorial. Figure 23: Life cycle of massive stars, courtesy NASA. Figure 24: Curve of binding energy per nucleon, courtesy Lawrence Livermore National Laboratory. Figure 25: Parent and daughter radioactive decay, courtesy Pamela J. W. Gore, Georgia Perimeter College. Figure 26: Age of rocks on Earth, courtesy Steven Robinson, editor, History web site. Figure 27: The oldest known object on Earth, a tiny chip of zircon crystal, courtesy National Science Foundation. Figure 28: Praying mantis in amber, courtesy Keith Luzzi, Torra Adventures. Figure 29: Trilobite engraving by von Emanuel Kayser, in *Geologische Formationskunde*, 1902. Figure 30: Cyanobacteria, courtesy Jack Holt, Susquehanna University. Figure 31: Cyanobacteria, courtesy Frans Havven, Life Science Trace Gas Facility, Radboud University, the Netherlands. Figure 32: Miller-Urey apparatus, courtesy NASA/Goddard Space Flight Center. Figure 33: Flowchart of origin of life, courtesy NASA/Goddard Remote Sensing Tutorial. Figure 34: Energy chain of metabolism, courtesy National Institutes of Health. Figure 35: Chemical evolution to cells, courtesy Jack Szostak, Howard Hughes Medical Institute. Figure 36: Fatty acid vesicles, courtesy Jack Szostak, Howard Hughes Medical Institute. Figure 37: Cell growth and division, courtesy Jack Szostak, Howard Hughes Medical Institute. Figure 38: Chemical to Darwinian evolution, courtesy Jack Szostak, Howard Hughes Medical Institute.

CHAPTER 3: EXTREME LIFE

Figure 39: Tree of life, from *General Morphology of Organisms*, Ernst Haeckel, 1866. Figure 40: The phylogenetic tree of life, courtesy Carl Woese, University of Illinois. Figure 41: Unrooted bacterial phylogenetic tree, courtesy Norman Pace, University of Colorado. Figure 42: Overlap of genetic material, courtesy Nick Woolf, University of Arizona. Figure 43: Lateral gene transfer and the tree of life, courtesy W. Ford Doolittle, University of Dalhousie, and *Science* magazine. Figure 44: *Nanoarchaeum equitans*, courtesy Karl Stetter, University of Regensburg. Figure 45: *Deinococcus radiodurans*, a microbe able to withstand extreme physical conditions, courtesy Michael Daly, Uniformed Services, University of the Health Services. Figure 46: Tardigrade, courtesy Susie Balser, Illinois Wesleyan University. Figure 47: Grand Prismatic Spring, courtesy the National Park Service, U.S. Department of the Interior. Figure 48: The extremophile called Strain 121, courtesy Derek Lovley, University of Massachusetts. Figure 49: Hydrothermal chimney, courtesy Visions 2005 and the School of Oceanography, University of Washington. Figure 50: Extremophiles, courtesy Maryland Astrobiology Consortium, NASA, and Space Telescope Science Institute. Figure 51: ALH 84001, courtesy NASA Johnson Space Center. Figure 52: Bioengineering, courtesy Institute for Biomedical Technology, University of Twente. Figure 53: *H. influenzae* genome, courtesy Institute for Genomic Research. Figure 54: Cellular automaton, courtesy Kevin McDermott, using the open source automata generator, http://kidojo.com/cellauto. Figure 55: The Tierra digital environment, courtesy Thomas Ray, Anti-Gravity Workshop. Figure 56: Gecko Team bioengineering logo, courtesy Kellar Autumn, Lewis and Clark College, Portland, Oregon.

CHAPTER 4: SHAPING EVOLUTION

Figure 57: Inside the solar nebula, courtesy William Hartmann, Planetary Sciences Institute. Figure 58: Condensation temperature sequence in early Solar System, courtesy NASA. Figure 59: Early Earth cratering rate, courtesy John Valley, University of Wisconsin. Figure 60: Earth impact size and frequency, courtesy Chris Chapman and David Morrison, NASA Ames Research Center, kind permission of Springer Science and Business Media. Figure 61: Tunguska impact and explosion, 1908, attributed to Russian mineralogist Leonid Kulik. Figure 62: The 1833 Leonid meteor storm, from *Bible Readings for the Home Circle*, 1899. Figure 63: Main Asteroid Belt and Trojan asteroids, courtesy National Maritime Museum. Figure 64: History of the growth and extinction of animal species, courtesy Arthur Buikema, Virginia Tech. Figure 65: The Milankovitch cycles, courtesy Geology Department, University of South Carolina. Figure 66: Typical radiation flux on a terrestrial planet, courtesy John Scalo, University of Texas. Figure 67: History of the Earth's atmosphere, courtesy NASA/Goddard Remote Sensing Tutorial. Figure 68: Global carbon cycle, courtesy NASA and the Regents of the University of Michigan. Figure 69: Pangea, courtesy NASA. Figure 70: Chambered nautilus, courtesy Laura Walls, Project Muse. Figure 71: Cambrian fauna, courtesy Simon Conway Morris, University of Cambridge, and Masanori Gukuhari, Studio R. Figure 72: Genetic drift, courtesy Ronald Mumme, Allegheny University. Figure 73: Complexity pyramid, courtesy Albert-László Barabasi, Dana Farber Cancer Institute, Harvard University, and *Science* magazine. Figure 74: History of animal diversity, courtesy Sinauer Associates and W. H. Freeman. Figure 75: Animal brain-to-body-mass ratios, courtesy Academic Press and Elsevier. Figure 76: The Earth's climate history, courtesy Robert Rohda and the Global Warming project.

CHAPTER 5: LIVING IN THE SOLAR SYSTEM

Figure 77: Planet habitability flow chart, courtesy NASA/Goddard Remote Sensing Tutorial. Figure 78: Planetary habitable zones, courtesy James Kasting, Penn State University. Figure 79: Mars canals, from *Mars and Its Canals*, by P. Lowell, 1906. Figure 80: Mars composite by the Viking Orbiter, courtesy NASA Ames Research Center. Figure 81: Mars Sojourner rover, courtesy NASA/JPL/Cornell. Figure 82: Martian sand dunes, courtesy NASA/JPL/Cornell. Figure 83: View of Mars from Viking, courtesy NASA Goddard Space Flight Center. Figure 84: Mars Global Surveyor image, courtesy NASA Jet Propulsion Laboratory. Figure 85: A change in a Mars gully over a four-year period that indicates running water, courtesy NASA/JPL and Malin Space Science Systems. Figure 86: The Martian surface from Opportunity rover, courtesy NASA/JPL/Cornell. Figure 87: Icy crater on Mars as seen by Mars Express, courtesy European Space Agency. Figure 88: Detail of the Martian meteorite ALH 84001, courtesy NASA Johnson Space Center. Figure 89: Comparison of NASA missions to Mars, courtesy NASA Jet Propulsion Laboratory. Figure 90: The "face" on Mars as seen by Mars Global Surveyor, courtesy NASA Jet Propulsion Laboratory. Figure 91: Volcanoes on Venus seen by Magellan, courtesy NASA Jet Propulsion Laboratory. Figure 92: Channels on Titan seen by Huygens, courtesy European Space Agency. Figure 93: Cassini flyby of Titan that found a hydrocarbon lake called Kraken Mare, courtesy NASA. Figure 94: The icy surface of Europa as seen by the Galileo Orbiter, courtesy NASA Jet Propulsion Laboratory. Figure 95: Planned Europa hydrobot, courtesy NASA. Figure 96: Ice rafts on Europa, courtesy NASA Jet Propulsion Laboratory.

CHAPTER 6: DISTANT WORLDS

Figure 97: Doppler effect, courtesy NASA/Goddard Remote Sensing Tutorial. Figure 98: Doppler curve of the first extra-solar planet discovered, around the Sun-like star 51 Pegasi, courtesy Michel Mayor and Didier Queloz, Geneva Observatory, and Geoff Marcy and Paul Butler, Lick Observatory. **Figure 99**: Mass distribution of extra-solar planets, courtesy Jean Schneider, Paris Observatory. **Figure 100**: Masses and distances from stars for extra-solar planets, courtesy Jean Schneider, Paris Observatory. **Figure 101**: Orbit eccentricity and average distance from stars for extra-solar planets, courtesy Jean Schneider, Paris Observatory. **Figure 102**: First detection of an extra-solar planet by the eclipse technique, courtesy David Charbonneau, Harvard-Smithsonian Center for Astrophysics. **Figure 103**: Incidence of extra-solar planets as a function of heavy elements in the parent star, courtesy Debra Fischer, San Francisco State University. **Figure 104**: Signature of extra-solar planet using the microlensing technique, courtesy Lawrence Livermore National Laboratory. **Figure 105**: Water content of rocky planets from computer simulations, courtesy Sean Raymond, University of Colorado. **Figure 106**: Artist's impression of a planetary system in which two terrestrial planets straddle a gas-giant planet, courtesy NASA/JPL. **Figure 107**: Biomarkers in the Earth's atmosphere as seen from space, courtesy NASA. **Figure 108**: Theoretical spectrum of an Earth-like planet, courtesy Jimmy Paillet, Ecole Normale Superieure de Lyons. **Figure 109**: Mass vs. discovery year for exoplanets, courtesy Jean Schneider, Paris Observatory. **Figure 110**: Proposed Terrestrial Planet Finder coronagraph mission, courtesy NASA/JPL. **Figure 111**: Proposed Terrestrial Planet Finder interferometry mission, courtesy NASA/JPL. **Figure 112**: Robert Goddard's first flight of a liquid propellant rocket, courtesy NASA Goddard Space Flight Center. **Figure 113**: Design for Project Daedalus, courtesy the British Interplanetary Society. **Figure 114**: Average cost of Hollywood movies versus cost of typical space missions, copyright Chris Impey. **Figure 115**: Imagined space colony, courtesy NASA.

CHAPTER 7: ARE WE ALONE?

Figure 116: UFO image, from 1997 study by the Central Intelligence Agency, courtesy U.S. government. **Figure 117**: Western hemisphere distribution of UFO sightings, courtesy Larry Hatch, UFO Database, http://www.larryhatch.net. **Figure 118**: Fifty years of UFO sightings, courtesy Larry Hatch, UFO Database. **Figure 119**: UFO, in an altered photograph widely available from many sources on the Internet. **Figure 120**: Fine-tuning and extra dimensions, after an image from Jacob Naelyn. **Figure 121**: The Drake Equation, courtesy Frank Drake, SETI Institute and University of California, Santa Cruz. **Figure 122**: Schematic representation of the Drake Equation, courtesy http://www.aerospaceweb.org and Frank Drake, University of California. **Figure 123**: Age and heavy element abundance of stars in the neighborhood of the Sun, courtesy S. Feltzing, Lund Observatory. **Figure 124**: Growth in brain size of our human ancestors, courtesy William Tietjen, Bellarmine University. **Figure 125**: Changes in brain mass relative to body size of dolphin and orca ancestors, courtesy Lori Marino, Emory University, and John Wiley and Sons. **Figure 126**: Giant octopus, courtesy U.S. government. **Figure 127**: The Doomsday Clock, after a concept from the Bulletin of Atomic Scientists, courtesy www.aperfectworld.org. **Figure 128**: Binary-coded radio message sent toward the globular cluster M13, courtesy Frank Drake, University of California, and NAIC. **Figure 129**:

The Arecibo 305-meter radio dish, courtesy NAIC and the National Science Foundation. Figure 130: The Allen Telescope Array, currently under construction, courtesy SETI Institute, and Seth Shostak. Figure 131: The difficulty of predicting the future shown in logarithmic intervals of human progress, copyright Chris Impey. Figure 132: Evolution of computer power and cost and its projection into the future, courtesy Hans Moravec, Carnegie Mellon University. Figure 133: Plaque sent on the Pioneer 10 spacecraft, courtesy NASA.

INDEX

Page numbers in *italics* refer to illustrations. Page numbers greater than 311 refer to notes.